FRONTIERS IN GEOCHEMISTRY

Frontiers in Geochemistry

Contribution of Geochemistry to the Study of the Earth

EDITED BY

Russell S. Harmon

Department of Marine, Earth and Atmospheric Sciences, North Carolina State University

and

Andrew Parker

Department of Soil Science, School of Human and Environmental Sciences, University of Reading

A John Wiley & Sons, Ltd., Publication

Registered office
John Wiley & Sons Ltd, The Atrium, Southern Gate, Chichester,
West Sussex, PO19 8SQ, UK

Editorial offices: 9600 Garsington Road, Oxford, OX4 2DQ, UK
                       The Atrium, Southern Gate, Chichester, West Sussex, PO19 8SQ, UK
                       111 River Street, Hoboken, NJ 07030-5774, USA

For details of our global editorial offices, for customer services and for information about how to apply for permission to reuse the copyright material in this book please see our website at www.wiley.com/wiley-blackwell

Library of Congress Cataloguing-in-Publication Data

Frontiers in geochemistry : contribution of geochemistry to the study of the earth / edited by Russell Harmon and Andrew Parker.
     p. cm.
  Includes index.
   ISBN 978-1-4051-9338-2 (hardback) – ISBN 978-1-4051-9337-5 (paperback)
  1. Geochemistry–Congresses.   I. Harmon, R. S. (Russell S.)   II. Parker, A. (Andrew), 1941-
  QE514.F75 2011
  551.9–dc22
                                 2010046377

A catalogue record for this book is available from the British Library.

This book is published in the following electronic formats: ePDF 9781444329964; Wiley Online Library 9781444329957; ePub 9781444329971

Set in 9/11.5 pt Trump Mediaeval by Toppan Best-set Premedia Limited
Printed and bound in Malaysia by Vivar Printing Sdn Bhd

1   2011

Contents

Contributors

PHILIP BENNETT *Department of Geological Sciences, The University of Texas at Austin, Austin, TX 78712, USA*

DAVID R. COLE *School of Earth Sciences, The Ohio State University, Columbus, OH 43210, USA*

BRUNO DHUIME *Department of Earth Sciences, University of Bristol, Wills Memorial Building, Queens Road, Bristol BS8 1RJ, UK and Department of Earth Sciences, University of St Andrews, North Street, St Andrews, Fife, KY16 9AL, UK*

HUIZHEN GAO *Sandia National Laboratories, P.O. Box 5800, Albuquerque, New Mexico 87185, USA*

SIGURDUR R. GISLASON *Institute of Earth Sciences, University of Iceland, Askja, Sturlugata 7, 101 Reykjavik, Iceland*

KARSTEN M. HAASE *GeoZentrum Nordbayern, Universität Erlangen-Nürnberg, Schlossgarten 5, D-91054 Erlangen, Germany*

CHRIS J. HAWKESWORTH *Department of Earth Sciences, University of St Andrews, North Street, St Andrews, Fife, KY16 9AL, UK*

JOCHEN HOEFS *Geowissenschaftliches Zentrum,UniversitätGöttingen,Goldschmidtstraße 1, D-37120 Göttingen, Germany*

HEINRICH D. HOLLAND *Department of Earth and Environmental Sciences, University of Pennsylvania, Philadelphia, Pennsylvania, 19104, USA*

MORTEN JARTUN *Geological Survey of Norway, NO-7491 Trondheim, Norway*

BALZ S. KAMBER *Department of Earth Sciences, Laurentian University, 935 Ramsey Lake Road, Sudbury, ON P3E 2C6, Canada*

ANTHONY I. S. KEMP *School of Earth and Environmental Sciences, James Cook University, Townsville, QLD 4811, Australia*

YOUSIF K. KHARAKA *Water Resources Discipline, U.S. Geological Survey, 345 Middlefield Road, Menlo Park, CSA 94025, USA*

ERIC H. OELKERS *LMTG, UMR CNRS 5563, Université Paul-Sabatier, Observatoire Midi-Pyrénées, 14 avenue Edouard Belin – 31400 Toulouse, France*

CHRISTOPHER OMELON *Department of Geological Sciences, The University of Texas at Austin, Austin, TX 78712, USA*

ROLF TORE OTTESEN *Geological Survey of Norway, Trondheim NO-7491, Norway*

TOMAS PACES *Czech Geological Survey, Klarov 3, 118 21 Prague 1, Czech Republic*

HENRY P. SCHWARCZ *School of Geography and Earth Sciences, McMaster University, Hamilton, Ontario, L8S 4K1, Canada*

CRAIG D. STOREY *School of Earth and Environmental Sciences, University of Portsmouth, Portsmouth PO1 3QL, UK*

STUART ROSS TAYLOR *Department of Geology, Australian National University, Canberra 0200, Australia*

YIFENG WANG *Sandia National Laboratories, Mail Stop 0779, P.O. Box 5800, Albuquerque, New Mexico 87185, USA*

HUIFANG XU *Department of Geology and Geophysics, University of Wisconsin, Madison, Wisconsin 53706, USA*

Editors' Preface

This book is a contribution to the International Year of Planet Earth, arising from the Major Geosciences Program on *Contribution of Geochemistry to the Study of the Planet*, sponsored and conducted by the International Association of GeoChemistry (IAGC) during the 33rd International Geological Congress, held in Oslo, Norway from 6–14 August 2008. This symposium was dedicated to the internationally-renowned geochemist Wallace Broecker.

Since the era of modern geochemical analysis began in the 1960s, geochemistry has played an increasingly important role in the study of planet Earth. Today highly sophisticated analytical techniques are utilized to determine the elemental, organic and isotopic compositions of the Earth's cosmological sphere, its atmosphere and surficial skin, and shallow and deep interiors across a wide range spatial scales. We originally chose the topics to cover the whole range of geochemistry, both pure and applied, in an attempt to synthesize a coherent geochemical view of the Earth and its history. The first session of the program on historical perspectives comprised a review of selective areas of geochemistry and its applications and contributions to the study of the Earth. The second session focused on the present and future, and considered current and future developments in geochemistry.

The Introduction, by Ross Taylor, summarizes the importance of geochemistry to the study of the Earth generally, and sets the scene for the detailed accounts that follow.

The first section of the book, Historical Perspectives, contains six chapters that consider aspects of geochemical processes which led to the development of the solid Earth as it is today. Kamber examines the geochemical evolution of the mantle and lower crust through time. Hawkesworth, Kemp, Dhuime and Storey discuss the character and evolution of the continental crust, with a focus on using the radiogenic and stable isotope composition of zircon as a monitor of crustal generation processes. Haase reviews the development of the oceanic crust and the particular set of geochemical processes operating in this domain. Holland covers the evolution of the atmosphere, Gislason and Oelkers describe the crucial topic of the weathering of primary rocks and the carbon cycle, and Paces gives an account of the evolution of groundwater, which is of course critical in many surficial geochemical processes.

The second section of the book, Frontiers in Geochemistry, contains six chapters that show the rapidly-evolving analytical tools and approaches currently used by geochemists, which may be used to solve emerging environmental and other societal problems. Kharaka and Cole continue in the allied field of carbon sequestration, with Wang, Gao and Xu adding the significance of nanostructures. A description by Bennet and Obelon follows of the microbial processes which led to the evolution of life, and continue to control many environmental scenarios. Archaeological and anthropological applications are covered by

Schwartz, and finally Jartun and Otteson discuss the relatively new field of urban geochemistry, which of course has highly significant environmental consequences in the human sphere.

The contributors have provided not only a concise, comprehensive, and up-to-date account of the Earth's geochemical evolution, but have signposted the critical areas where further

research should lead, from the basic science, environmental and economic standpoints.

Russell S. Harmon
Raleigh, North Carolina, USA

Andrew Parker
Reading, Berkshire, UK

Introduction

STUART ROSS TAYLOR
Australian National University
October 2009

Geochemistry has now become so well-established in the study of geological problems, complete with societies, journals, books, university departments and professorships, that it is often forgotten how recently it developed, primarily as the result of the development of sophisticated analytical equipment. After the great scientific advances in understanding the Earth in the first half of the 19th century, geology was moribund during the period from about 1860 to about 1940 because it lacked the techniques to solve its important problems ... [and] geologists ... were inevitably doomed to working on trivia until new tools were forged' (Menard 1971).

In the meantime, the concept of 'multiple working hypotheses' became fashionable to deal with the many intractable problems and 'geologists in the 20th century became accustomed to carrying on interminable controversies about problems that they were unable to solve' (Brush 1996). Such debates often reached levels reminiscent of medieval religious disputes, classic examples and worthy of historical study, being the question of continental drift, the origin of granites and whether tektites originated from the Moon or the Earth. Many bizarre explanations appeared, a consequence of 'the inherent difficulties of the science [that] rendered it peculiarly susceptible to the interpretations of ancient miracle-mongers and their modern successors' (Gillispie 1951).

So the subject had to wait for the development of specialized techniques based on physics and chemistry, from optical spectrographs to mass spectrometers, in order to resolve its disputes.

Fortunately, the advent of sophisticated analytical techniques has helped to answer many of the questions posed by the field observations and so has enabled the many complex problems discussed in this book to be studied.

Chemical analyses of rock, minerals and meteorites have a long history, stretching back to the 18th century, but among the first attempts to assemble geochemical data in a coherent fashion was that of Clarke (1908) at the United States Geological Survey.

However the real beginnings of modern geochemistry began in the third and fourth decades of the 20th century through the insights of Victor Moritz Goldschmidt, developed only after he had worked and published extensively on crystallographic and geological problems. A good background in geology as well as in physics and chemistry remains as a *sine qua non* for geochemists.

Goldschmidt realised that first steps in understanding the distribution of the chemical elements in rocks and minerals required a knowledge both of crystal structures of minerals and of the sizes of ionic species, both little understood at the time. He published a comprehensive table of ionic radii in 1926, one year before that of Linus Pauling (Mason 1992). Perhaps as good example of his geochemical foresight as any can be found in a 1926 paper in which he drew attention to the separate behaviour of divalent europium from the other trivalent rare earth elements, on account of its much larger ionic radius. Europium has indeed turned out to be among the most useful of any member of the Periodic Table, important in astrophysics, meteoritics and in understanding of the geochemical evolution both of the Moon and of the continental crust of the Earth.

In the succeeding years, despite appalling political difficulties during the 1930s and 1940s (including narrowly escaping deportation to a Nazi death camp), Goldschmidt established geochemistry as a scientific discipline, utilizing the tools of X-ray diffraction, X-ray spectrography and atomic emission spectrography in Gottingen and Oslo, as elegantly described in the biography written in 1992 by one of his former students at Oslo, Brian Mason.

The subject, although much delayed by the disruptions of World War II, rapidly became established in the 1950s, as analytical instrumentation, particularly that of mass spectrometers, became reliable, and eventually, with the arrival of computers, largely routine. So the subject arose and has prospered from scientific and technical advances. Nevertheless, some cautions should be heeded. The sheer mass of data now routinely accessible may overwhelm the observer. Goldschmidt, as one observer reported to me, always spent much time in selecting samples for analysis; 'Six samples are enough for a scientist' as folklore has it.

Likewise, the impressive ability now to analyse minerals at a scale of microns raises problems of perspective. Ancient wisdom reminds us that one swallow does not make a summer and of the tendency to make mountains out of molehills: one zircon grain does not make a continent. Analysis on the scale of microns, impressive though it may be, must always be rooted in the realities of geology.

But the advances in analytical techniques and the amount of chemical and isotopic data now available enable us to address such broad geochemical questions as the location and behaviour of the chemical elements and their isotopes, the evolution of the oceans, the crust and that of the Earth itself, that are among the wide variety of subjects discussed in this book.

Although the topics addressed here are exclusively terrestrial, it should be recalled that the laws of physics and chemistry and the abundances of the chemical elements, on which geochemistry is based, apply with equal emphasis on the other rocky planets, although nature has a surprising ability to produce unexpected and unpredicted results with these constraints. The Earth is not the norm among planets, either in the solar system, or likely elsewhere.

A further cautionary tale may be noted as technology has advanced, with the ability to utilize increasingly esoteric isotopic systems to study not only geochronology but also geological phenomena (something that seems to have begun with the ^{87}Rb–^{87}Sr system). There has been a tendency to hail each system, as the technology to exploit it has developed, as the panacea. Their subsequent history, however, whether that of the Rb–Sr, Sm–Nd, Lu–Hf, Re–Os or W–Hf systems, has usually revealed unanticipated problems; nature is subtle, but paradoxes arise from faulty human understanding, not from chemistry and physics.

Following the spectacular advances pioneered by Goldschmidt, much progress in the mid-20th century resulted from applying his insights; Harrison Brown, Hans Suess, V. I. Vernadsky, Harold Urey, Frtz Houtermans, Bill Wager and Louis Ahrens among many others, may be mentioned. Geochemistry, that has flourished mostly among geologists rather than chemists, is now firmly established as a scientific discipline. But its future course is as impossible to predict as it was in 1930 or 1950, reminding us of the wisdom from folklore that it is difficult to make predictions, especially about the future.

REFERENCES

Brush, SG. (1996) *Transmuted Past*. Cambridge University Press, p. 55.

Clarke, FW. (1908) *The Data of Geochemistry*. US Geological Survey Bulletin 330.

Gillispie, CC. (1951) *Genesis and Geology*, Harvard University Press, p. 127.

Mason, B. (1992) *Victor Moritz Goldschmidt: Father of Modern Geochemistry*. The Geochemical Society Special Publication No. 4. San Antonio, Texas.

Menard, WH. (1971) *Science and Growth*. Harvard University Press, p. 144.

Part 1
Contribution of Geochemistry to the Study of the Earth

1 Geochemistry and Secular Geochemical Evolution of the Earth's Mantle and Lower Crust

BALZ S. KAMBER

Laurentian University, Sudbury, Ontario, Canada

The incompatible elements U and Th are related to Pb via radioactive decay. Extraction, modification and storage of continental crust have, over time, left an isotopic record in the continental crust itself and in the depleted portion of the mantle. Ancient lower crustal xenoliths require that crust has matured by upward transport of radioactive heat-producing elements; hundreds of millions of years after formation.

Recycling of continental material has contributed in at least three ways to the generation of enriched mantle-melt sources. First, this has occurred by delamination of lower crustal segments back into the mantle. Second, sediment has been recycled back into the mantle in subduction zones, and third, since the oxygenation of the atmosphere, seawater U, weathered from the continents, has been incorporated into hydrated oceanic crust with which it has ultimately been recycled back into the mantle.

The joint treatment of the lower continental crust and the mantle in terms of their geochemistry and their isotopic evolution may seem, at first, a less than obvious choice. They are, however, related in the sense that the evidence for their evolution is largely of indirect nature, either inferred from rare xenoliths or via products of partial melting. Any joint treatment of these two geochemical reservoirs also inherently carries with it the assumption that they have, at least in part, mutually influenced each other's temporal evolution. Before attempting to condense into an opening book chapter the relevant aspects of the exhaustive body of knowledge about the geochemistry of the mantle and the much sparser information regarding the lower crust, it is necessary to remind ourselves of the evidence for their mutually related evolutions.

INTRODUCTION

The view that the Earth has suffered some form of early global, planetary-scale depletion event is deeply rooted in classic geochemical texts, including those focusing on plumbotectonics, i.e. the reconstruction of planetary differentiation from a Pb-isotope perspective (e.g. Stacey and Kramers 1975). Most early attempts at modelling the isotopic evolution of the mantle postulated one or

Frontiers in Geochemistry: Contribution of Geochemistry to the Study of the Earth, First edition. Edited by Russell S. Harmon and Andrew Parker. © 2011 Blackwell Publishing Ltd. Published 2011 by Blackwell Publishing Ltd.

two pervasive differentiation steps, resulting, for example, in the increase of the U/Pb ratio of the silicate portion of the Earth (the bulk silicate Earth). The notion of an early depletion event was further cemented with the observation that Archaean komatiites and high-Mg basalts, in terms of their trace-element chemistry, appeared to resemble modern ocean-island picrites, yet their radiogenic isotope character was much more depleted (e.g. Campbell and Griffiths 1993). This finding seemed to suggest an early depletion event that imparted the long-term isotopic effect with superimposed much more recent (relative to 2.7 Ga) re-enrichment of the mantle, which explains the trace-element systematics but which had not yet translated into long-term isotopic evidence. More recently, the observation has been made that the bulk silicate Earth has a ^{142}Nd/^{144}Nd ratio different from the most common chondritic meteorites (Boyet and Carlson 2005). This has added new momentum to the idea of a very early silicate differentiation event that must have occurred within less than 1 half-life (103 Ma) of the short-lived parent of ^{142}Nd.

While the evidence for such an event appears as strong as ever, the critical question for this present treatment is whether that event was the principal cause for establishing the chemistry of the depleted mantle as it is sampled at most ocean ridges via the normal mid-ocean-ridge basalt (N-MORB). Namely, if the early depletion event imparted such a fundamental geochemical signal, which over time also was manifest as a long-lived radiogenic isotope signature, then the subsequent extraction, maturing and recycling of continental crust would only have played a secondary role in modifying the chemistry of the depleted mantle. Hence, the chemistry and radiogenic isotope composition of the depleted mantle would largely tell us about the early planetary depletion event and not about the history of extraction and recycling of continental crust.

In order to address this question, it is necessary to consider elemental systematics and the radiogenic isotope evolution of those elements that are most strongly enriched in continental crust, for their extraction will be most strongly reflected by the residual depleted mantle. While average continental-crustal absolute abundances are difficult to estimate on account of the sparse occurrence of *bona fide* lower crustal rocks, there is nonetheless wide agreement regarding the relative enrichment of elements. The elements most strongly enriched in continental crust are largely those that behave most incompatibly during mantle melting, plus an assortment of elements that are particularly soluble in hydrous fluids, and were therefore preferentially moved into the melt-source regions of the magmas that eventually differentiated to give rise to continental crust. The best studied of these is Pb (e.g. Miller et al. 1994) but other fluid mobile elements, such as B (e.g. Ryan and Langmuir 1993), W (e.g. Kamber et al. 2005; König et al. 2008), Li (e.g. Chan ct al. 1999) and As (Mohan et al. 2008) have also been documented. The extended trace-element diagram for average upper-continental crustal rocks, in which elements are arranged in order of incompatibility during mantle-decompression melting, illustrates not only the extraordinary enrichment of the most incompatible elements but also the strong deviations of the fluid-mobile elements from an otherwise predicable, smoothly decaying trend. Regardless of the particular significance of the elements that deviate from this trend, it is intuitively appreciable that the geoscientist interested in that aspect of mantle depletion potentially caused by the extraction of continental crust is best served by working with the elements that plot toward the left side of the abscissa of Fig. 1.1. From an isotopic point of view, it is therefore not surprising that the extent of variability in the U/Pb and Th/Pb isotope systems in crustal and mantle rocks is of the order of several tens of percent and has formed the very basis of the mantle-rock nomenclature.

Indeed, one of the strongest pieces of evidence for the mutual chemical interaction between mantle and crust is found in the Pb-isotope composition and U-Th-Pb systematics of the source of N-MORB basalts. The present-day Pb-isotope composition of N-MORB firmly shows that, on the billion-year timescale, the time-averaged Th/U ratio of the depleted mantle source was

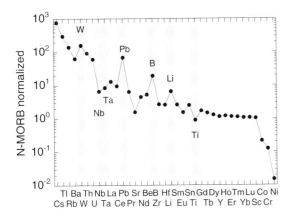

Fig. 1.1 Extended trace-element diagram in which elements are arranged, from left to right, in order of decreasing incompatibility during anhydrous mantle decompression melting. Shown is an average upper crustal river sediment composite, normalized to N-MORB (modified after Kamber et al. 2005). Boron value is an estimate, using the B/Be ratio 11 for arc rocks from Mohan et al. (2008). Grey bars highlight elements discussed in text.

ca. 3.6. This can be inferred from the $^{208}Pb/^{206}Pb$ ratio (Kramers and Tolstikhin 1997), which represents the decay products of the long-lived ^{232}Th and ^{238}U, respectively. Rather surprisingly, then, the measured elemental Th/U ratio of N-MORB is much lower, somewhere between 2.4 and 2.6. This observation is often termed the second terrestrial Pb-isotope paradox (e.g. Kramers and Tolstikhin 1997) or the kappa (as in $^{232}Th/^{238}U$) conundrum (e.g. Elliott et al 1999). This discrepancy is not an artefact of preferential U over Th partitioning into the N-MORB parental melt because the intermediate decay product systematics of the U and Th chains support a low Th/U ratio of the source rocks, i.e. the depleted mantle itself (Galer and O'Nions 1985). The solution to this paradox is now widely believed (e.g. McCulloch 1993; Elliott et al. 1999; Collerson and Kamber 1999) to be the preferential recycling of continental U under an oxidized atmosphere since the great oxygenation event at ca. 2.3 Ga (Bekker et al. 2004). This observation alone provides very robust evidence that the depleted mantle has not remained chemically inert and unchanged since an early depletion event.

The high-field-strength elements Th, U, Nb, and Ta offer further insight into the interaction between the depleted mantle and continental crust. These elements are all very incompatible and have very similar bulk partition coefficients during mantle-decompression melting. This is reflected in their close grouping in the extended trace-element diagram (Fig. 1.1). Yet the chemistry of upper continental crust shows a very distinctive deficit in Nb (and to a lesser extent Ta) relative to Th and U. This finding is very widely attributed to the preferential sequestering of Nb and Ta into a Ti-phase (e.g. rutile) in subducting slabs (e.g. Hofmann 1988). Extraction of continental crust, to the extent of its present mass of ca. 2.09×10^{25} g, has severely depleted the entire mantle in Th and U. It is estimated that between 30–50% of terrestrial Th and U are harboured by continental crust. By contrast, enrichment in the equally incompatible Nb is much lower, and, hence the mantle is proportionally less depleted in this element by a factor of at least three. It should come as no surprise then that the modern N-MORB Nb/Th ratio of ca. 18 is much higher than that of chondrites of ca. 8. If this greater-than-100% difference in a ratio that can be analysed to within 2–5% precision was caused by the early depletion event, it follows that ancient melting products of the depleted mantle should also have a ratio of ca. 18; but this is not in fact the case. For example, it is found that regardless of locality, high-Mg basalts and komatiites of the widespread 2.7 Ga mantle melting event have a Nb/Th of only 12 (e.g. Sylvester et al. 1997; Collerson and Kamber 1999), much lower than modern depleted mantle melts and much closer to the chondritic value. This observation shows that, at least for the very incompatible elements, the mantle has become more depleted as a function of how much continental crust was extracted. For these elements, the early depletion event played a less important role and, therefore, they are the tools with which to most effectively reconstruct the depletion history of the mantle.

TEMPORAL EVOLUTION OF THE DEPLETED MANTLE RESERVOIR

There are two principal methods to reconstruct the depletion history of the N-MORB mantle source. The first is to search for well-preserved N-MORB-like rocks of as large an age range as possible and to study their chemical and radiogenic isotope systematics. The second is to use forward modelling to approximate the isotopic contrast displayed by modern N-MORB and average continental crust. Examples of both approaches are reviewed here.

The reconstruction approach has the obvious advantage that each temporal observation from ancient N-MORB samples provides a time capsule for the evolution from the primitive to the present-day depleted mantle. In practice, it turns out that finding well-preserved N-MORB comparable basalts is difficult. The densest array of observations is, surprisingly, from the Archean eon. Many well-preserved greenstone belts exist, ranging in age from 3.7 to 2.6 Ga, and while some are clearly ensialic in origin (e.g. Blenkinsop et al. 1993), a sufficient number of uncontaminated mafic to ultra-mafic volcanic rocks are preserved. The situation for the Proterozoic is much less satisfactory. Apart from two ophiolites (Zimmer et al. 1995; Peltonen et al. 1996), the majority of other Proterozoic greenstones either formed in an arc or back-arc environment (e.g. Leybourne et al. 1997), were variably contaminated during magmatic ascent through pre-existing continental crust, or are not sufficiently well-preserved. For the Phanerozoic, the number of ophiolites and accreted ocean-floor assemblages is adequate. It must be stressed here that N-MORB of any age is particularly sensitive to continental contamination in those elemental systematics of most interest to this discussion, the systematics of those elements for which there is the most divergence between the mantle and continental crust.

In terms of suitable element systematics for reconstruction, any pair of elemental neighbours with sharply deviating behaviour on Fig. 1.1 are candidates. Namely, for elements with near-identical bulk partition coefficients during mantle melting, a suitably large-degree melt (such as the parental melt of N-MORB) will truthfully reflect the relative concentrations in the source. Subsequent fractional crystallization (up to ca. 6% MgO) will also not greatly affect the ratio of the elements of interest. Theoretically at least, it should be possible to track mantle depletion by study of the following ratios: Th/W, Nb/Th, Ta/U, Be/B, Pr/Pb, and Zr/Li. Note that, in all these examples, the element more enriched in continental crust is the denominator and hence all ratios are expected to have increased in the depleted mantle with increasing extraction of continental crust.

In reality, a number of factors conspire to render most of these ratios less than useful for the intended purpose. Insufficient data are available for Th/W, Be/B and Zr/Li. Post-emplacement elemental mobility may affect Pr/Pb and Be/B, and the redox-sensitivity of U has affected the mantle Ta/U ratio. At present, then, the only viable ratio is Nb/Th, which was used earlier to illustrate the fact that the N-MORB source mantle has become depleted by extraction of continental crust. Jochum et al. (1991) first proposed that the reconstruction of this ratio in the depleted mantle should be a reliable monitor of the mass of continental crust that had been extracted from the mantle through time, but their limited dataset and, by modern standards, insufficient analytical precision prevented these authors from drawing a conclusion. Collerson and Kamber (1999) applied a three-fold filter to the by then much improved literature database for Nb/Th in greenstones. They eliminated most rocks that had less than 6% MgO, excluded rocks with negative slopes in CI-normalized rare earth element (REE) patterns (to screen against ocean island basalts; OIB) and rejected rocks that had lower radiogenic $^{143}Nd/^{144}Nd$ ratios than widely accepted depleted-mantle evolution curves, (such as dePaolo and Wasserburg 1976) to avoid contaminated samples.

The Nb/Th curve for the depleted mantle, depicted on Fig. 1.2(a), was converted into the continental crust mass-versus age, curve (shown on Fig. 1.2(b)), that uses a primitive mantle Nb/Th starting value lower than in chondrites to

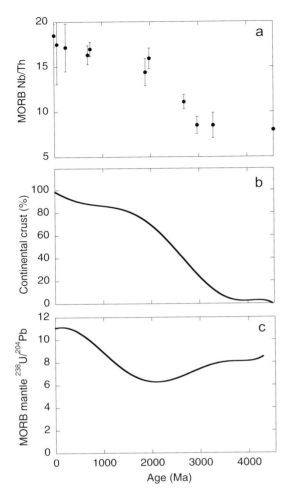

Fig. 1.2 Temporal evolution trends for (a) Nb/Th ratio in the depleted mantle; (b) continental crust mass estimated from Nb/Th ratio and Pb-isotope systematics; (c) modelled U/Pb ratio evolution in the depleted mantle. Modified from Kamber et al. (2003) and Kramers and Tolstikhin (1997).

allow for sequestration of ca. 15% Nb into the core (following Kamber et al. 2003) because Nb can become siderophile under very reducing conditions prevailing during metal removal into the core (Wade and Wood 2001). The curve suggests a sigmoidal evolution for Nb/Th in the depleted mantle, starting with relatively low ratios until

3.5 Ga, then increasing strongly between 3.0 and 2.0 Ga, and a slow increase ever since.

The second approach to track mantle depletion is to study the time-integrated effect of continental extraction and recycling on depleted mantle isotope systematics. Most readers are probably familiar with the long-lived $^{147}Sm/^{143}Nd$ system. Owing to the slightly higher incompatibility of Nd, continental crust has a lower Sm/Nd ratio than its mantle source and as a result, over time, will develop a lower $^{143}Nd/^{144}Nd$ ratio. The contrary situation is, of course, true for the depleted portion of the mantle. However, because neither Sm nor Nd are nearly as concentrated in continental crust as Th, and because Sm/Nd fractionation is much more modest than Nb/Th, it turns out that the present-day mantle $^{143}Nd/^{144}Nd$ ratio is not very sensitive to the extraction history and recycling rate of continental crust. Nägler and Kramers (1998) explained, in detail, that Nd-isotope systematics cannot easily discriminate between models with linear net growth of the continents or that producing the sigmoidal curve shown in Fig. 1.2(b). However, Nd-isotope systematics do argue against very early formation of voluminous continents and subsequent recycling (to lower the average continental age to ca. 2 Ga).

The only isotopic system that is truly sensitive to the mantle depletion history is U/Pb, because the mantle is so depleted in both these elements. Kramers and Tolstikhin (1997) explored the effects of a variety of mantle-depletion scenarios on the difference in predicted Pb-isotope compositions of the depleted mantle and average continental sediment. While their preferred solution for a continental-crust volume-versus-age curve is not unique, they identified a few key parameters. First, the strongest control over the position of the modelled Pb-isotope composition of the depleted mantle is exerted by the continental crust-extraction versus recycling balance, which must satisfy an average continental age of ca. 2 Ga. Second, the timing of preferential U-recycling is important. This is tied to the age of the great oxygenation event, because under an atmosphere devoid of free O, U remained immobile. Once free O accumulated in the atmosphere, U but not Th was transferred into the ocean, and from there into hydrated oceanic lithosphere and sediment. A proportion of this U became recycled into the mantle. The timing of the onset of this process is critical for modelling of the mantle Pb isotope curves and this marker has since been confirmed to have occurred between 2.4 and 2.2 Ga (Bekker et al. 2004). Finally, the timing of Pb loss to space (volatility) and to the core is also important. The preferred solution of Kramers and Tolstikhin (1997) is sensitive to relatively late Pb loss to the core, but this is not supported by W-isotope systematics (e.g. Yin et al. 2002). The information available at present supports the conclusion of Kramers and Tolstikhin (1997) that the U/Pb ratio of the depleted mantle was dynamic (Fig. 1.2c). The most important outcome concerns the significant difference in the position of the depleted mantle Pb-isotope evolution modelled with a dynamic U/Pb compared to that of a static U/Pb, such as could have been set by a single early depletion event. Figure 1.3 illustrates that the differences are greatest for the late Archaean and Palaeoproterozoic as well as for the modern mantle. We will return to this important point when discussing the Pb-isotope systematics of OIB.

Remembering that because this type of forward model only uses modern isotope-compositions as input parameters, it can be tested by comparing predicted ancient Pb-isotope compositions with those actually observed. The most meaningful such comparison is for late Archaean rocks, as it is for this particular time period that the dynamic U/Pb model predicts a rather different composition from that of the static U/Pb model (Fig. 1.3). The greenstones of the Abitibi greenstone belt of Ontario and Quebec are widely regarded to have formed from largely juvenile mantle sources. They have the most radiogenic initial Nd-isotope compositions for rocks of that age and, undoubtedly, have come from the depleted mantle (Ayer et al. 2002). Thus, Abitibi and Wabigoon greenstone belt initial Pb (conveniently preserved in ores and feldspars) can be used to test the accuracy of the dynamic U/Pb model. As is seen in Fig. 1.3, the observed Pb-isotope composition plots almost exactly on the depleted mantle-

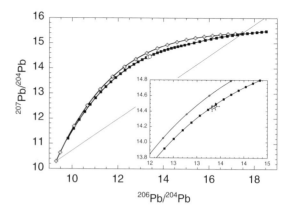

Fig. 1.3 Common Pb-isotope diagram contrasting the curve of the depleted mantle source (solid curve with black square markers) consistent with the sigmoidal continental crust volume-versus-age curve shown in Fig. 1.2(b) and a single-stage growth curve (solid curve with open-cross symbols) with a $^{238}U/^{204}Pb$ ratio of 7.91. Modified from Kramers and Tolstikhin (1997) and Kamber and Collerson (1999). Also shown for reference and in the inset are the observed initial Pb-isotope compositions of 2.72 Ga Wabigoon and 2.68 Ga Abitibi greenstone belts. Data sources: Tilton (1983); Gariépy and Allègre (1985); and Carignan et al. (1993, 1995).

evolution curve between 2.7 and 2.8 Ga, and has a much lower $^{207}Pb/^{204}Pb$ ratio than predicted by a static single-stage model for the depleted mantle.

In summary, it is not currently possible to quantify the relative contributions of a very early planet-scale depletion event versus the depletion effects of continental extraction for the mildly incompatible elements, such as the middle rare-earth elements (REE). However, for the very incompatible elements, it is clear that their inventories in the depleted mantle have changed as a function of continental extraction and recycling.

THE LOWER CRUST AS A PARTLY HIDDEN RESERVOIR OF INCOMPATIBLE ELEMENTS

The formation of the chemically-evolved continental crust, which is on average andesitic in composition, from an essentially basaltic parental melt continues to pose major mass-balance problems. The bulk density of continental crust is not permissive of the presence of ultramafic cumulates at the base of the seismically defined lower crust. Therefore, models that attempt to explain the mass imbalance resort to either placing the seismic Moho shallower than the igneous crust, or invoke the delamination of dense lower crustal cumulates, such as eclogite (e.g. Arndt and Goldstein 1989; Ellis and Maboko 1992). In the former case, lower-crustal igneous cumulates could reside below the seismic Moho where they cannot be distinguished (seismically) from residual depleted peridotite (e.g. Muentener et al. 2001). Alternatively, delaminated crustal eclogite may have foundered through the lithosphere and left the average crust with a more evolved bulk composition than the parental melt.

Regardless of this petrological issue, it is clear that the seismically-defined crust is chemically and mineralogically heterogeneous. While the composition of the upper continental crust is very well studied (e.g. Taylor and McLennan 1985), there are surprisingly few genuine lower-crustal rock sections exposed, and most of the relatively sparse information about the composition of the lower crust is actually derived from xenoliths. It is important to stress that orogenic granulites exposed in exhumed collisional mountain belt roots cannot, in most cases, be used to approximate the composition of 'typical' lower continental crust.

The key observations regarding distribution of highly incompatible elements in the lower continental crust are heat-flow measurements and determinations of geothermal gradients. The former reflect the total heat flow from a particular area of continental crust, while the latter tell us about the vertical distribution of the three main radioactive heat-producing elements – K, Th and U – which are all highly incompatible. Estimates of thermal gradients and heat flow (from deep mines and boreholes) firmly show that the heat-producing elements are strongly concentrated in the upper continental crust (e.g. Rudnick and Fountain 1995; Perry et al. 2006). The observed

mechanical strength of continents also requires a relatively stiff middle crust, which implies that temperatures there cannot exceed the brittle/plastic transition of feldspar at ca. 550°C (e.g. Pryer 1993).

There are two principal ways in which the lower crust could have ended up depleted in heat-producing elements. First, magmatic compositional stratification would necessarily lead to a feldspar-pyroxene dominated lower continental crust inherently poor in heat-producing elements. Alternatively, the lowermost continental crust originally could have harboured some heat-producing elements. The resulting build-up of heat could have led to crustal melting (e.g. Michaut et al. 2009), during the course of which the highly incompatible elements would have been removed into granitoid melts and relocated much higher in the crustal column (e.g. O'Nions et al. 1979; Whitehouse 1989).

Lead isotope compositions of lower crustal xenoliths, particularly for those that contain plagioclase, can be used to distinguish between these two scenarios. Figure 1.4(a) illustrates a theoretical model of how Pb-isotope ratios can be used to infer the U/Pb history of lower crust. Two end-member model crusts are modelled. Both are calculated to have formed at 2.50 Ga with the same isotopic composition as coeval mantle. The first curve shows the Pb-isotope evolution of an inherently low U/Pb crust and the second illustrates the curve of a crust that initially had a U/Pb ratio similar to upper continental crust, but lost 90% of its U some 500 million years later. The models show that, in the latter case, the composition plots above the depleted mantle evolution curve, but far to the left of modern continental sediment. From the Pb-isotope compositions of lower crustal xenoliths shown in Fig. 1.4(b) it is observed that almost all plot well above the mantle evolution curve and that almost none plot below. This firmly argues against an inherently U (and by analogy Th and K) -poor lower continental crust.

The distribution of observed lower crustal feldspar Pb-isotope compositions is also insightful. On account of its low (U+Th)/Pb ratio, feldspar preserves the Pb-isotope composition of the last

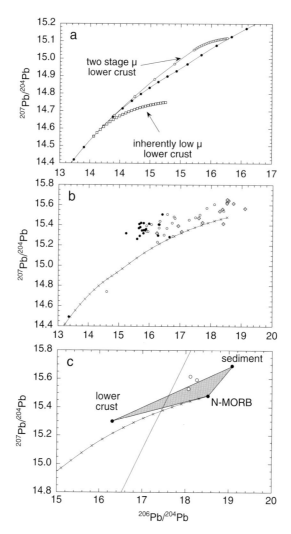

Fig. 1.4 Lead isotope models and compositions of lower crustal rocks. (a) Two contrasting models illustrating the difference in Pb-isotope evolution of an inherently low U/Pb crust (μ = 2.5) and that of a crust that initially had a U/Pb ratio (15) identical to upper continental crust but, 500 million years later, lost 90% of its U. The latter plots above the depleted mantle evolution curve (solid line connecting solid circles) but the former below; (b) compilation of average Pb-isotope compositions of world-wide lower crustal xenoliths (open crosses) after Bolhar et al. (2007). Also shown are individual Wyoming craton xenolith whole-rock (open circles) and feldspar (solid circles) compositions; (c) Positions of lower continental-crust average (from Bolhar et al. 2007), continental sediment and N-MORB (after Kramers and Tolstikhin 1997) relative to the 4568 Ma meteorite isochron. Three estimates of bulk silicate-earth Pb-isotope composition are shown as open circles. That of Galer and Goldstein (1996) plots above the triangle defined by N-MORB, upper and lower continental crust while those of Kamber and Collerson (1999) and Murphy et al. (2003) plot near and into the triangle, respectively.

time of recrystallization, at which time it will have exchanged Pb with the whole rock (Fig. 1.4b). By comparing whole-rock and feldspar Pb-isotope compositions, it is possible to get a rough idea of when U was lost and, furthermore, by approximately how much the U/Pb ratio of the whole rock was lowered. This approach was explained by Bolhar et al. (2007) using Precambrian lower crustal xenoliths from the northern Wyoming province. These authors showed that U-loss from Neoarchean lower crust occurred ca. 500 million years after igneous crystallization. Uranium loss was driven by partial melting that lowered the U/Pb ratio between 50% and 90%. Similar reductions in U/Pb were calculated for granulites from Fyfe Hills in Antarctia (dePaolo et al. 1982).

In the vertical direction, the chemical self-reorganization of the continents is driven by the geothermal gradient (e.g. Michaut et al. 2009). In the typical case of geological terrains, where only upper crustal levels are exposed, lower crustal melting is inferred from abundant late (with respect to deformation) K-rich granitoids (e.g. Sandiford and McLaren 2002). In Archaean terrains, the intrusion of these granitoids is often seen as the critical step for cratonization. The most important factors that determine the geotherm are basal heatflow, crustal thickness, and the concentration and vertical distribution of the three main heat-producing radioactive elements K, U and Th. These factors remaining equal, the delicate thermal stability of continental crust is also a strong function of time, because of the decrease in heat output, mainly from ^{235}U (Kramers et al. 2001).

Although it is difficult to predict the time period for which lower crust of a given age might have remained thermally stable before reaching the granitoid solidus, it can nonetheless be qualitatively appreciated that as radioactive heat production decays with time, the lag between crust formation and lower crustal melting will become longer. For the Neoarchean and Palaeoproterozoic examples that have been studied, a 500 million year time lag seems typical (e.g. Whitehouse 1989; Bolhar et al. 2007). Younger, Phanerozoic

crust may therefore not yet have reached its final vertical chemical stratification. This probably explains why the well-studied lower crustal xenoliths from northeastern Australia (Rudnick and Goldstein 1990) have Pb-isotope compositions similar to continental sediment. Namely, insufficient time has yet elapsed for their host lower crust to lose enough K, U and Th to eventually reach thermal stability, and for the lower (U+Th)/Pb ratio to be expressed in Pb-isotope compositions.

Partial melting of lower continental crust not only distils U, Th and K into the upper crust but also affects a host of other incompatible trace elements. Therefore, to accurately estimate the composition of the lower continental crust, it is necessary to study xenoliths from crust that have undergone the full reorganization, because average continental crust has an age in excess of 2 Ga. However, the most widely used estimate of the composition of the lower crust relies predominantly on xenoliths from Phanerozoic crust (e.g. Rudnick and Fountain 1995). Further studies are required to determine the chemistry of older xenoliths to refine current understanding of typical lower crustal composition, and, by inference, the average composition of continental crust (e.g. Taylor and McLennan 1985).

Radiogenic isotope studies of the lower continental crust demonstrate that the transfer of heat-producing elements via partial melting strengthens primary vertical chemical stratification. In the simplest case, this is achieved by a single partial melting event, but, at least for old crustal sections, multiple melting events are more plausible. The net result for Pb-isotope composition is the multi-stage lowering of the U/Pb ratio and the retardation of Pb-isotope evolution. This has an important implication for global mass balance. Figure 1.4(c) illustrates that both continental sediment and N-MORB both plot well to the right of the 4568 Ma meteorite isochron. This forms part of the first terrestrial Pb-isotope paradox (Allègre 1969) and implies the existence of hidden or rarely tapped Pb-reservoirs at depth that plot to the left of the meteorite isochron. Estimates for the bulk silicate-earth Pb-isotope composition that take into account the effect of core formation and volatility related Pb-loss are shown in Fig. 1.4(c). It is obvious in common Pb-isotope space that old lower crust (as sampled by xenoliths) could help to pull the combined compositions of upper crustal sediment and N-MORB closer towards bulk-silicate Earth estimates.

ENRICHED MANTLE RESERVOIRS AND THEIR CONNECTION TO CONTINENTAL CRUST AND THE DEPLETED MANTLE

The study of the geochemistry and radiogenic isotope systematics of basaltic rocks from enriched mantle sources has yielded a vast body of evidence that cannot be summarized here, not least because fundamental controversies about the meaning of isotopic signatures remain. Thus, only those aspects of enriched mantle reservoirs that are directly relevant to the two reservoirs already discussed – the depleted mantle and the lower continental crust are considered.

Of all possible mechanisms that could have led to mantle enrichment, as sampled by modern mantle melts, the early silicate differentiation event can be excluded on present evidence. Regardless of whether this differentiation, which apparently increased the Sm/Nd of the accessible mantle, was caused by formation and subsequent deep recycling of a protocrust (e.g. Tolstikhin and Hofmann 2005; Boyet and Carlson 2005) or by cumulates from an early magma ocean, it appears that none of the modern mantle melts is tapping into it. The lack of low ^{142}Nd/^{144}Nd mantle melts on Earth (Boyet et al. 2005) complementary to the widely accessible reservoirs could also imply a terrestrial Sm/Nd ratio of the terrestrial planets (e.g. Caro et al. 2008) that is different from that of typical chondrites, such as might be caused by incomplete mixing of products of different nucleosynthetic processes in the feeding zone of the early Solar nebula (e.g. Ranen and Jacobsen 2006). This possibility may question altogether whether a very early silicate differentiation event occurred on Earth. Alternatively, the early reservoir could

be completely hidden. Regardless, neither solution offers insight into the significance of the enriched reservoirs actually observed.

A second generally valid observation is that the nomenclature of the enriched mantle reservoirs (e.g. Zindler and Hart 1986) was developed largely on the basis of Pb-isotope compositions, for they show by far the largest dispersion compared to Sr-, Nd-, Hf- and Os- radiogenic isotope systems. The term 'enriched' in this context is somewhat ambiguous. Namely, the principal three enriched reservoirs, enriched mantle 1 (EM-1), enriched mantle 2 (EM-2) and the 'high μ' reservoir (HIMU) are enriched in U/Pb relative to depleted mantle. But by analogy with all other isotope systems, enrichment should ideally be defined relative to bulk-silicate Earth. It is noted that since the seminal work by Zindler and Hart (1986), estimates of the Pb-isotope composition of the bulk-silicate Earth have evolved (Fig. 1.4(c)) and the idea needs to be entertained that there may be two principal ways of arriving at the ^{207}Pb/^{206}Pb composition of EM-1. This issue will be discussed next, before commenting on the possible significance of the HIMU reservoir and the OIB Pb-isotope array as a whole.

Figure 1.5(a) shows the position of OIB of EM-1 character in common Pb-isotope space alongside the bulk-silicate Earth estimates already discussed in Fig. 1.4(c). It can be seen that, with the exception of Kerguelen, most EM-1 basalts have ^{207}Pb/^{204}Pb ratios no higher than estimates of undepleted (primitive) but not enriched mantle. This illustrates the point that the term 'enriched', with regard to elevated ^{207}Pb/^{204}Pb relative to depleted mantle, is not necessarily accurate. Melts from the undepleted mantle would, of course, be expected also to have high ^{3}He/^{4}He ratios, such as those found in certain Hawaiian (Loihi) and early Iceland plume picrites (e.g. Hilton et al. 2000). The alternative explanation for the Pb-isotope composition of EM-1 is that its mantle source contains a component of recycled continental material. The possibility of continental sediment recycling was proposed to account for the peculiarly high ^{208}Pb/^{206}Pb ratios (Fig. 1.5b) of seamounts of the Pitcairn chain (Woodhead and

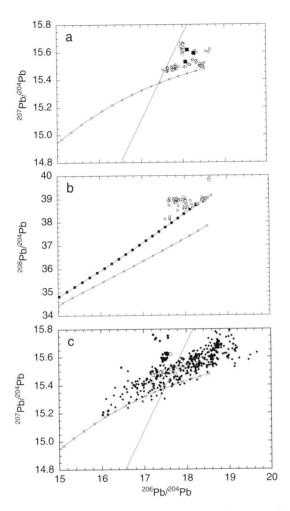

Fig. 1.5 Pb-isotope systematics of enriched mantle sources. (a) Common Pb-isotope diagram showing a selection of EM-1 type ocean-island basalts from Walvis Ridge, Pitcairn, Gough and South Atlantic Ridge seamounts (all as open small circles) and Kerguelen (open diamond symbols) compared to bulk-silicate Earth estimates from Fig. 1.4(c) (solid squares), meteorite isochron and depleted mantle curve (Kramers and Tolstikhin, 1997). Basalt data are from Eiler et al. (1995); Eisele et al. (2002); Gast et al. (1964); Ingle et al. (2002); Regelous et al. (2009); Richardson et al. (1982); Weis et al. (2001) and Woodhead and McCulloch (1989). (b) The same data in thorogenic Pb-isotope space. Note that this comparison only shows one bulk silicate-earth evolution (after Kamber and Collerson (1999), in 100 Ma time steps). (c) Common Pb-isotope diagram with lamproite (solid circles), selected kimberlite (open circles) and alkali basalt (solid small diamonds) data relative to 4568 Ma meteorite isochron (adapted from Murphy et al. 2003) and depleted mantle curve.

Devey 1993). These authors argued that preferential loss of U>Pb>Th during subduction dehydration of pelagic sediment leads to relatively low U/Pb and high Th/U ratios, which, over time, express as low $^{206}Pb/^{204}Pb$ and high $^{208}Pb/^{206}Pb$ ratios. The similarity between lower continental crust and certain EM-1 ocean island basalts in Pb-isotope composition and pertinent trace element geochemical signatures also has led to the proposal that there might be recycled lower continental crust in the shallow astenosphere (e.g. Weis et al. 2001) and that it might be the source of at least the South Atlantic EM-1 signature (e.g. Willbold and Stracke 2006; Regelous et al. 2009). It can be seen from Fig. 1.5(b) that EM-1 type OIB define a considerable spread in $^{208}Pb/^{204}Pb$ at any given $^{206}Pb/^{204}Pb$. Those samples with the lowest $^{208}Pb/^{204}Pb$ ratios plot exactly onto the undepleted mantle curve of Kamber and Collerson (1999). That curve was modelled using single-stage evolution, with $^{238}U/^{204}Pb$ and $^{232}Th/^{238}U$ ratios of 8.9 and 4.21, respectively. A $^{232}Th/^{238}U$ ratio of 4.21 is at the upper end of chondritic estimates (e.g. Kramers and Tolstikhin 1997) and it appears unlikely that those EM-1 type ocean island basalts (OIB) with the highest $^{208}Pb/^{204}Pb$ ratios could be generated in undepleted mantle. In summary, combined Th-U-Pb isotope systematics can be used to subdivide EM-1 basalts into two possible origins. Those with modestly high $^{208}Pb/^{204}Pb$ and $^{207}Pb/^{204}Pb$ ratios could come from the least U- and Th-depleted mantle, while those with much higher $^{208}Pb/^{204}Pb$ and $^{207}Pb/^{204}Pb$ ratios, as well as a very low $^{206}Pb/^{204}Pb$ ratio, would have continental material in their source. The continental material can either be subduction zone-modified sediment with low $^{238}U/^{204}Pb$ and high $^{232}Th/^{238}U$ ratios, owing to preferential loss of U over Pb and Pb over Th, or delaminated segments of old lower continental crust that preferentially had lost U>Th>Pb during earlier episodes of intracrustal differentiation. Proterozoic granulite xenoliths have been recovered in Kerguelen Plateau boreholes (e.g. Weis et al. 2001), proving that entrainment of granulite facies continental crust is feasible, at least in that particular setting. It was further suggested that such crust could have

affected the radiogenic isotope, and particularly the Pb-isotope, systematics of the entire Indian Ocean basin mantle (e.g. Weis et al. 2001).

Experimental studies of element behavior during sediment dehydration (Rapp et al. 2008) have confirmed that the proposal of a sediment component in the high $^{208}Pb/^{206}Pb$ EM-1 type source is compatible with trace-element systematics. The experimental findings by Rapp et al. (2008) further showed that lamproites (i.e. ultrapotassic magnesian volcanic rocks) could represent a more extreme approximation to melt from subducted continental sediment as originally proposed by Nelson (1992) and further elaborated by Murphy et al. (2002). It is interesting to note that many examples of highly alkaline magnesian volcanic rocks – such as lamproites, minettes, carbonatites, and lamprophyres – plot well above the depleted mantle evolution curve but to the left of the meteorite isochron (Fig. 1.5(c)). Murphy et al. (2003) suggested that their mantle source (subduction modified recycled continental sediment) could help to solve the first terrestrial Pb-isotope paradox, alongside the lower continental crust.

Another widely advocated mechanism for mantle enrichment is via recycling of oceanic crust. Again, this idea has its origin in Pb-isotope systematics. In his highly influential paper on Pb-isotope systematics of OIB, Chase (1981) noted that the OIB Pb-isotope array as a whole (Fig. 1.6a) defines a slope of ca. 1.7 Ga. Further noting that the linear regression line intersected a single stage Pb-isotope curve also at ca. 1.7 Ga, he proposed that isolation of recycled oceanic crust with a 10–20% higher U/Pb ratio than typical mantle could, over time, evolve to the array of Pb-isotope compositions displayed by OIB (other than EM-1). Thus, the increase in U/Pb relative to contemporary mantle could be inherent in oceanic crust formation (as originally proposed by Chase, 1981) and caused by U-addition during sea-floor alteration (e.g. Jochum and Verma 1996) or have its origin during preferential removal of Pb during subduction dehydration (Chauvel et al. 1995). The latter process would ultimately lead to the characteristic over-enrichment in Pb of continental crust (see Fig. 1.1).

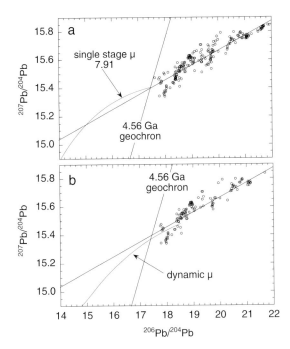

Fig. 1.6 Common Pb-isotope array of world-wide OIB modified from Kamber and Collerson (1999). In (a) a linear regression line is shown (with a slope corresponding to ca. 1.7 Ga) relative to the single stage (μ = 7.91) mantle evolution curve of Chase (1981), which is intersected at approximately 1.7 Ga; in (b) the same regression line is shown relative to the depleted mantle evolution curve of Kramers and Tolstikhin (1997), which is not intersected by the OIB linear regression line. See text for explanation.

The idea of initially increasing the U/Pb of oceanic crust via U-addition during seafloor metamorphism is attractive from the point of view of also linking this explanation to the solution to the second terrestrial Pb-paradox, as explained in the opening section of this chapter. Thus, the low Th/U ratio of N-MORB; and its position to the right of the meteorite isochron would be related to the more extreme position of the HIMU source, in particular, and the OIB Pb-isotope array, in general. The proposal that the OIB Pb-isotope systematics could have been influenced additionally by preferential (relative to Th and U) Pb-loss from oceanic crust during prograde subduction dehy-

dration (Chauvel et al. 1995) is also attractive, as it could explain the surprisingly low continental U/Pb ratio. Compared to typical lunar rocks (e.g. Premo et al. 1999), terrestrial arc rocks have much lower U/Pb ratios, clearly demonstrating the role of aqueous fluids in shaping terrestrial geochemical reservoirs.

Despite the pleasing internal consistency emerging from these models, new observations challenge the simplicity (and elegance) of the elemental cycles described above. The key problem is the highly dynamic U/Pb evolution of the depleted mantle that was discussed earlier. It has led to a Pb-isotope curve very different from that produced by a single stage, constant U/Pb model (Fig. 1.3). This has significant ramifications for the idea proposed by Chase (1981). While a regression line through the general OIB array (with a slope corresponding to ca. 1.7 Ga) indeed intersects the constant U/Pb growth curve at roughly 1.7 Ga, it does not intersect the depleted mantle curve of Kramers and Tolstikhin (1997) at all (Fig. 1.6(b)), plotting instead at too high a $^{207}Pb/^{204}Pb$ ratio for a given $^{206}Pb/^{204}Pb$ ratio. Since the dynamic U/Pb model is successful at predicting not only modern N-MORB Pb-isotope compositions but also juvenile Pb of late Archaean age, it seems difficult to derive OIB Pb-isotope ratios via storage of oceanic slabs in the mantle. This problem was originally discussed by Kamber and Collerson (1999) and has not yet been satisfactorily solved.

The second potential problem arises from considerations of Nb/Th and Nb/U ratios. If, as is widely thought, the continental Nb/Th ratio is low because Nb is preferentially retained in the metamorphosed oceanic slab, then it follows that partial melts from such slabs should have higher Nb/Th ratios than contemporaneous N-MORB, but most modern OIB have Nb/Th ratios of ca. 16. This is lower than modern N-MORB with Nb/Th of 18. However, a solution is still possible, if these OIB were derived from ancient oceanic lithosphere. For example, the observed Nb/Th ratio for N-MORB at 2.0 Ga is about 14 (Fig. 1.2(a)). Subducted oceanic crust of that age would be expected to have lost more Th than Nb, and therefore its Nb/Th ratio would be expected to be higher than 14. Modern OIB do indeed show a somewhat higher ratio, but only marginally so (Hofmann 1997).

It is now well established that continental crust has a lower Nb/Ta ratio (11–12) than chondrites (18) and N-MORB (14), leading to an unresolved mass imbalance that requires a missing high Nb/Ta reservoir (e.g. Rudnick et al. 2000). If the low continental Nb/Ta ratio is acquired during the modification of oceanic crust to eclogite that also imprints the low continental Nb/Th, it follows that melts from subducted oceanic slabs (i.e. OIB) should have high Nb/Ta ratios (Kamber and Collerson 2000). However, they do not (e.g. Pfander et al. 2007). These observations could be used to argue that most OIB are not the melting products of ancient recycled oceanic crust (e.g. Kamber and Collerson 2000; Pilet et al. 2005; Niu and O'Hara 2009) or that oceanic slabs do not have a high Nb/Ta ratio (Pfander et al. 2007).

SUMMARY

The origin of the chemical enrichment that characterizes the mantle source regions that generate OIB is still a major area of on-going research. It is not possible at present to satisfy all mass-balance constraints with a single model of mantle depletion, continent formation and recycling, and evolution of OIB source regions in the mantle. The idea of metasomatic enrichment of the deep lithosphere (e.g. Pilet et al. 2005) is gaining momentum as an alternative source for many OIB, but this proposal carries a large number of implications that fundamentally challenge widely accepted ideas regarding formation of the continents. Some of these have been discussed by Niu and O'Hara (2009). Compared to the first generation models for large-scale differentiation of the planet, current ideas place even more emphasis on the highly dynamic nature of elemental and isotopic evolution within the major terrestrial geochemical reservoirs. The extraction history of continental crust and its recycling into the mantle, modified and possibly influenced by the

change in the composition of the atmosphere, remain as very significant controls over mantle evolution. The successful combination of information from radiogenic isotopes with highly incompatible trace elements will be key in further refining our understanding of the fascinating history of terrestrial differentiation.

REFERENCES

Allègre, CJ. (1969) Comportement des systemes U-Th-Pb dans le manteau superieur et modele d'èvolution de ce dernier au cours des temps geologiques. *Earth and Planetary Science Letters*, **5**: 261–9.

Arndt, NT and Goldstein, SL. (1989) An Open Boundary between Lower Continental-Crust and Mantle – Its Role in Crust Formation and Crustal Recycling. *Tectonophysics*, **161**: 201–12.

Ayer, J, Amelin, Y, Corfu, F, Kamo, S, Ketchum, J, Kwok, K and Trowell, N. (2002) Evolution of the southern Abitibi greenstone belt based on U-Pb geochronology: autochthonous volcanic construction followed by plutonism, regional deformation and sedimentation. *Precambrian Research*, **115**: 63–95.

Bekker, A, Holland, HD, Wang, PL, Rumble, DI, Stein, HJ, Hannah, JL, Coetzee, LL and Beukes, NJ. (2004) Dating the rise of atmospheric oxygen. *Nature*, **427**: 117–120.

Blenkinsop, TG, Fedo, CM, Eriksson, KA, Martin, A, Nisbet, GE and Wilson, JF. (1993) Ensialic origin for the Ngezi Group, Belingwe greenstone belt, Zimbabwe. *Geology*, **21**: 1135–8.

Bolhar, R, Kamber, BS and Collerson, KD. (2007) U-Th-Pb fractionation in Archaean lower continental crust: Implications for terrestrial Pb isotope systematics. *Earth and Planetary Science Letters*, **254**: 127–45.

Boyet, M and Carlson, RW. (2005) 142Nd isotope evidence for early (>4.53 Ga) global differentiation of the silicate Earth. *Science*, **309**: 576–8.

Boyet, M, Garcia, MO, Pik, R and Albarede, F. (2005) A search for Nd-142 evidence of primordial mantle heterogeneities in plume basalts. *Geophysical Research Letters*, **32** (4).

Campbell, IH and Griffiths, RW. (1993) The evolution of the mantle's chemical-structure. *Lithos*, **30**: 389–99.

Carignan, J, Gariepy, C, Machado, N and Rive, M. (1993) Pb isotopic geochemistry of granitoids and gneisses from the late Archean Pontiac and Abitibi Subprovinces of Canada. *Chemical Geology*, **106**: 299–316.

Carignan, J, Machado, N and Gariépy, C. (1995) Initial lead composition of silicate minerals from the Mulcahy layered intrusion: Implications for the nature of the Archean mantle and the evolution of greenstone belts in the Superior Province, Canada. *Geochim. Cosmochim. Acta*, **59**: 97–105.

Caro, G, Bourdon, B, Halliday, AN and Quitte, G. (2008)Super-chondritic Sm/Nd ratios in Mars, the Earth and the Moon. *Nature*, **452**: 336–9.

Chan, LH, Leeman, WP and You, CF. (1999) Lithium isotopic composition of Central American volcnic arc lavas: Implications for modification of the sub-arc mantle by slab-derived fluids. *Chemical Geology*, **160**: 255–80.

Chase, CG. (1981) Oceanic island Pb: two-stage histories and mantle evolution. *Earth and Planetary Science Letters*, **52**: 277–84.

Chauvel, C, Goldstein, SL and Hofmann, AW. (1995) Hydration and dehydration of oceanic crust controls Pb evolution of the mantle. *Chemical Geology*, **126**: 65–75.

Collerson, KD and Kamber, BS. (1999) Evolution of the continents and the atmosphere inferred from Th-U-Nb systematics of the depleted mantle. *Science*, **283**: 1519–22.

DePaolo, DJ, Manton, WI, Grew, ES and Halpern, M. (1982) Sm-Nd, Rb-Sr and U-Th-Pb Systematics of Granulite Facies Rocks from Fyfe-Hills, Enderby Land, Antarctica. *Nature*, **298**: 614–18.

DePaolo, DJ and Wasserburg, GJ. (1976) Nd Isotopic Variations and Petrogenetic Models. *Geophysical Research Letters*, **3**: 249–52.

Eiler, JM, Farley, KA, Valley, JW, Stolper, EM, Hauri, EH and Craig, H. (1995) Oxygen-Isotope Evidence against Bulk Recycled Sediment in the Mantle Sources of Pitcairn Island Lavas. *Nature*, **377**: 138–41.

Eisele, J, Sharma, M, Galer, SJG, Blichert-Toft, J, Devey, CW and Hofmann, AW. (2002) The role of sediment recycling in EM-1 inferred from Os, Pb, Hf, Nd, Sr isotope and trace element systematics of the Pitcairn hotspot. *Earth and Planetary Science Letters*, **196**: 197–212.

Elliott, T, Zindler, A and Bourdon, B. (1999) Exploring the kappa conundrum: the role of recycling in the lead isotope evolution of the mantle. *Earth and Planetary Science Letters*, **169**: 129–45.

Ellis, DJ and Maboko, MAH. (1992) Precambrian tectonics and the physicochemical evolution of the conti-nental crust. I. The gabbro-eclogite transition revisited. *Precambrian Research*, **55**: 491–506.

Galer, SJG and O'Nions, RK. (1985) Residence time of uranium, thorium and lead in the mantle with implications for mantle convection. *Nature*, **316**: 778–82.

Galer, SJG and Goldstein, SL. (1996) Influence of accretion on lead in the Earth. In: A. Basu and S. Hart (Editors), *Earth processes: Reading the isotopic code*. AGU, Washington, pp. 75–98.

Gariepy, C and Allegre , CJ. (1985) The lead isotope geochemistry and geochronology of late-kinematic intrusives from the Abitibi Greenstone-Belt, and the implications for late Archean crustal evolution. *Geochimica Et Cosmochimica Acta*, **49**: 2371–83.

Gast, PW, Hedge, C and Tilton, GR. (1964) Isotopic Composition of Lead + Strontium from Ascension + Gough Islands. *Science*, **145**: 1181–5.

Hilton, DR, Thirlwall, MF, Taylor, RN, Murton, BJ and Nichols, A. (2000) Controls on magmatic degassing along the Reykjanes Ridge with implications for the helium paradox. *Earth and Planetary Science Letters*, **183**: 43–50.

Hofmann, AW. (1988) Chemical differentiation of the Earth: the relationship between mantle, continental crust, and oceanic crust. *Earth and Planetary Science Letters*, **90**: 297–314.

Hofmann, AW. (1997) Mantle geochemistry: the message from oceanic volcanism. *Nature*, **385**: 219–29.

Ingle, S, Weis, D, Scoates, JS and Frey, FA. (2002) Relationship between the early Kerguelen plume and continental flood basalts of the paleo-Eastern Gondwanan margins. *Earth and Planetary Science Letters*, **197**: 35–50.

Jochum, KP, Arndt, NT and Hofmann, AW. (1991) Nb-Th-La in komatiites and basalts: constraints on komatiite petrogenesis and mantle evolution. *Earth and Planetary Science Letters*, **107**: 272–89.

Jochum, KP and Verma, SP. (1996) Extreme enrichment of Sb, Tl and other trace elements in altered MORB. *Chemical Geology*, **130**: 289–99.

Kamber, BS and Collerson, KD. (1999) Origin of ocean-island basalts: A new model based on lead and helium isotope systematics. *J. Geophys. Res.*, **104**: 25'479–25'491.

Kamber, BS and Collerson, KD. (2000) Role of 'hidden' deeply subducted slabs in mantle depletion. *Chemical Geology*, **166**: 241–54.

Kamber, BS, Greig, A and Collerson, KD. (2005) A new estimate for the composition of weathered young upper continental crust from alluvial sediments,

Queensland, Australia. *Geochim. Cosmochim. Acta*, **69**: 1041–58.

Kamber, BS, Greig, A, Schoenberg, R and Collerson, KD. (2003) A refined solution to Earth's hidden niobium: Implications for evolution of continental crust and depth of core formation. *Precambrian Research*, **126**: 289–308.

König, S, Münker, C, Schuth, S and Garbe-Schönberg, D. (2008) Mobility of tungsten in subduction zones. *Earth and Planetary Science Letters*, **274**: 82–92.

Kramers, JD, Kreissig, K and Jones, MQW. (2001) Crustal heat production and style of metamorphism: a comparison between two Archaean high grade provinces in the Limpopo Belt, southern Africa. *Precambrian Research*, **112**: 149–63.

Kramers, JD and Tolstikhin, IN. (1997) Two terrestrial lead isotope paradoxes, forward transport modelling, core formation and the history of the continental crust. *Chemical Geology*, **139**: 75–110.

Leybourne, MI, van Wagoner, NA and Ayres, LD. (1997) Chemical stratigraphy and petrogenesis of the Early Proterozoic Amisk Lake Volcanic Sequence, Flin-Flon-Snow Lake Greenstone Belt, Canada. *J. Petrol.*, **38**: 1541–64.

McCulloch, MT, 1993. The role of subducted slabs in an evolving earth. *Earth and Planetary Science Letters*, **115**: 89–100.

Michaut, C, Jaupart, C and Mareschal, JC. (2009) Thermal evolution of cratonic roots. *Lithos*, **109**: 47–60.

Miller, DM, Goldstein, SL and Langmuir, CH. (1994) Cerium/lead and lead isotope ratios in arc magmas and the enrichment of lead in the continents. *Nature*, **368**: 514–20.

Mohan, MR, Kamber, BS and Piercey, SJ. (2008) Boron and arsenic in highly evolved Archean felsic rocks: Implications for Archean subduction processes. *Earth and Planetary Science Letters*, **274**: 479–88.

Muentener, O, Keleman, PB and Grove, TL. (2001) The role of H_2O during crytsallization of primitive arc magmas under uppermost mantle conditions and genesis of igneous pyroxenites: an experimental study. *Contrib. Mineral. Petrol.*, **141**: 643–58.

Murphy, DT, Collerson, KD and Kamber, BS. (2002) Lamproites from Gaussberg: Possible transition zone melts of Archaean subducted sediments. *J. Petrol.*, **43**: 981–1001.

Murphy, DT, Kamber, BS and Collerson, KD. (2003) A refined solution to the first terrestrial Pb-isotope paradox. *J. Petrol*, **44**: 39–53.

Nägler, TF and Kramers, JD. (1998) Nd isotopic evolution of the upper mantle during the Precambrian:

Models, data and the uncertainty of both. *Precambrian Research*, **91**: 233–52.

Nelson, DR. (1992) Isotopic Characteristics of Potassic Rocks – Evidence for the Involvement of Subducted Sediments in Magma Genesis. *Lithos*, **28**: 403–20.

Niu, YL and O'Hara, MJ. (2009) MORB mantle hosts the missing Eu (Sr, Nb, Ta and Ti) in the continental crust: New perspectives on crustal growth, crust-mantle differentiation and chemical structure of oceanic upper mantle. *Lithos*, **112**: 1–17.

O'Nions, RK, Evenson, NM and Hamilton, PJ. (1979) Geochemical modeling of mantle differentiation and crustal growth. *J. Geophys. Res.*, **84**: 6091–6101.

Peltonen, P, Kontinen, A and Huhma, H. (1996) Petrology and geochemistry of metabasalts from the 1.95 Ga Jormua Ophiolite, Northeastern Finland. *J. Petrol*, **37**: 1359–83.

Perry, HKC, Jaupart, C, Mareschal, JC and Bienfait, G. (2006) Crustal heat production in the Superior Province, Canadian Shield, and in North America inferred from heat flow data. *Journal of Geophysical Research-Solid Earth*, **111** (B4).

Pfander, JA, Münker, C, Stracke, A and Mezger, K. (2007) Nb/Ta and Zr/Hf in ocean island basalts – Implications for crust-mantle differentiation and the fate of Niobium. *Earth and Planetary Science Letters*, **254**: 158–72.

Pilet, S, Hernandez, J, Sylvester, P and Poujol, M. (2005) The metasomatic alternative for ocean island basalt chemical heterogeneity. *Earth and Planetary Science Letters*, **236**: 148–66.

Premo, WR, Tatsumoto, M, Misawa, K, Nakamura, N and Kita, NI. (1999) Pb-isotopic systematics of lunar highland rocks (>3.9 Ga): Constraints on early lunar evolution. In: GA Snyder, CR Neal and WG Ernst (Editors), *Planetary petrology and geochemistry*. GSA, Bellwether Publishing Ltd.

Pryer, LL. (1993) Microstructures in feldspars from a major crustal thrust zone: the Grenville Front, Ontario, *Canada J. of Structural Geology*, **15**: 21–36.

Ranen, MC and Jacobsen, SB. (2006) Barium isotopes in chondritic meteorites: Implications for planetary reservoir models. *Science*, **314**: 809–12.

Rapp, RP, Irifune, T, Shimizu, N, Nishiyama, N, Norman, MD and Inoue, J. (2008) Subduction recycling of continental sediments and the origin of geochemically enriched reservoirs in the deep mantle. *Earth and Planetary Science Letters*, **271**: 14–23.

Regelous, M, Niu, YL, Abouchami, W and Castillo, PR. (2009) Shallow origin for South Atlantic Dupal Anomaly from lower continental crust: Geochemical evidence from the Mid-Atlantic Ridge at 26 degrees S. *Lithos*, **112**: 57–72.

Richardson, SH, Erlank, AJ, Reid, DL and Duncan, AR. (1982) Major and trace element and Nd and Sr isotope geochemistry of basalts from the DSDP Leg 74 Walvis Ridge transsect. In: TC Moore (ed.), *Initial Reports on the Deep Sea Drilling Project 74*. US Government Printing Office, Washington DC, pp. 739–54.

Rudnick, RL, Barth, M, Horn, I and McDonough, WF. (2000) Rutile-bearing refractory eclogites: missing link between continents and depleted mantle. *Science*, **287**: 278–81.

Rudnick, RL and Fountain, DM. (1995) Nature and composition of the continental crust: a lower crustal perspective. *Rev. Geophys.*, **33**: 267–309.

Rudnick, RL and Goldstein, SL. (1990) The Pb Isotopic Compositions of Lower Crustal Xenoliths and the Evolution of Lower Crustal Pb. *Earth and Planetary Science Letters*, **98**: 192–207.

Ryan, JG and Langmuir, CH. (1993) The systematics of boron abundances in young volcanic rocks. *Geochim. Cosmochim. Acta*, **57**: 1489–98.

Sandiford, M and McLaren, S. (2002) Tectonic feedback and the ordering of heat producing elements within the continental lithosphere. *Earth and Planetary Science Letters*, **204**: 133–50.

Stacey, JS and Kramers, JD. (1975) Approximation of terrestrial lead isotope evolution by a two-stage model. *Earth and Planetary Science Letters*, **26**: 207–21.

Sylvester, PJ, Campbell, IH and Bowyer, DA. (1997) Niobium/uranium evidence for early formation of the continental crust. *Science*, **275**: 521–3.

Taylor, SR and McLennan, SM. (1985) The continental crust: its composition and evolution. *Geoscience Texts*. Blackwell, Oxford.

Tilton, GR. (1983) Evolution of depleted mantle – the lead perspective. *Geochimica Et Cosmochimica Acta*, **47**: 1191–1197.

Tolstikhin, I and Hofmann, AW. (2005) Early crust on top of the Earth's core. *Physics of the Earth and Planetary Interiors*, **148**: 109–30.

Wade, J and Wood, BJ. (2001) The Earth's 'missing' niobium may be in the core. *Nature*, **409**: 75–8.

Weis, D, Ingle, S, Damasceno, D, Frey, FA, Nicolaysen, K and Barling, J. (2001) Origin of continental components in Indian Ocean basalts: Evidence from Elan Bank (Kerguelen Plateau, ODP Leg 183, Site 1137). *Geology*, **29**: 147–50.

Whitehouse, MJ. (1989) Pb-Isotopic Evidence for U-Th-Pb Behavior in a Prograde Amphibolite to Granulite Facies Transition from the Lewisian Complex of Northwest Scotland – Implications for Pb-Pb Dating. *Geochimica Et Cosmochimica Acta*, **53**: 717–24.

Willbold, M and Stracke, A. (2006) Trace element composition of mantle end-members: Implications for recycling of oceanic and upper and lower continental crust. *Geochemistry Geophysics Geosystems*, 7.

Woodhead, JD and McCulloch, MT. (1989) Ancient seafloor signals in Pitcairn island lavas and evidence for large amplitude, small-scale mantle heterogeneities. *Earth and Planetary Science Letters*, **94**: 257–73.

Woodhead, JD and Devey, CW. (1993) Geochemistry of the Pitcairn seamounts, I: source character and temporal trends. *Earth and Planetary Science Letters*, **116**: 81–99.

Yin, Q, Jacobsen, S.B, Yamashita, K, Blichert-Toft, J, Télouk, P and Albarede, F. (2002) A short timescale for terrestrial planet formation from Hf-W chronometry of meteorites. *Nature*, **418**: 949–52.

Zimmer, M, Kröner, A, Jochum, KP, Reischmann, T and Todt, W. (1995) The Gabal Gerf complex: A Precambrian N-MORB ophiolite in the Nubian Shield, NE Africa. *Chemical Geology*, **123**: 29–51.

Zindler, A and Hart, S. (1986) Chemical geodynamics. *Annu. Rev. Earth Planet. Sci.*, **14**: 493–571.

2 Crustal Evolution – A Mineral Archive Perspective

C.J. HAWKESWORTH[1], A.I.S. KEMP[2], B. DHUIME[1, 2, 3]
AND C.D. STOREY[4]

[1]University of Bristol, Bristol, UK
[2]James Cook University, Townsville, Australia
[3]University of Bristol, Bristol, UK
[4]University of Portsmouth, Portsmouth, UK

ABSTRACT

The continental crust is the principal record of conditions on the Earth for the last 4.4 Ga. Less than 10% of the crustal rocks exposed are older than 2.5 Ga, and yet 50% of the continental crust may have stabilized by that time. A key archive is minerals like zircon which can be precisely dated and preserve robust isotope and trace element signals. Much of the early crust was mafic in composition, and the late Archaean marks the transition from a period of uniformly poor preservation potential to one in which the geological record appears to be biased by the tectonic setting in which the rocks were formed.

INTRODUCTION

A striking feature of the surface elevation of the Earth is that it is bimodal. Almost approximately 40% is 5 km higher than the rest, and most of the

Frontiers in Geochemistry: Contribution of Geochemistry to the Study of the Earth, First edition. Edited by Russell S. Harmon and Andrew Parker.
© 2011 Blackwell Publishing Ltd. Published 2011 by Blackwell Publishing Ltd.

elevated material is above sea level. These continental areas are made of relatively buoyant material, so they are difficult to destroy and, therefore, preserve the record of the evolution of the silicate Earth. The oldest known rock formed four billion years ago (Bowring and Williams 1999), and the Earth is gradually cooling, and so most of the continental crust had been generated by 2.5 Ga. However, less than 10% of the surface geology is of rocks older than 2.5 Ga (Hurley and Rand 1969). A major challenge remains as to how to investigate the early history of the Earth, and of its crust, when there is relatively little material available for study.

The continental crust differs in composition from the crust of other planets in our Solar System. Its formation modified the composition of the mantle and the atmosphere, it supports life and it remains a sink for CO_2 through weathering and erosion. The continental crust therefore has had a key role in the evolution of this planet, and yet when and how it formed remain the topic of considerable debate.

During the past two decades, the study of rocks and minerals has been revolutionized by the development of in-situ analytical techniques for the high-precision analysis of isotope ratios and trace-element abundances on the scale of tens of microns (e.g. Hinthorne et al. 1979; Compston et al. 1984; Jackson et al. 1992; Feng et al. 1993; Fryer et al. 1993, 1995; Hirata and Nesbitt 1995; Wiechert and Hoefs 1995; Horn et al. 2000; Cavosie et al. 2005; Hawkesworth and Kemp 2006c). As a consequence, materials can be characterized at micron spatial scales with much greater confidence than is possible when analysing bulk rock samples. For example, in a number of instances, igneous rocks have been shown to contain minerals of quite different histories, brought together shortly before emplacement of the host rock (e.g. Vazquez and Reid 2004; Kemp et al. 2007a). The accessory mineral zircon is widely used to obtain accurate and high-precision age determinations that provide a foundation for understanding the timing of particular events in the geological record. Such studies have shown that detrital zircons in ancient sediments preserve ages up to 400 million years older than the oldest known rock (Compston and Pidgeon 1986; Amelin et al. 1999; Wilde et al. 2001; Harrison et al. 2005). Thus, these tiny minerals offer exceptional and sometimes unique insight into the processes involved in the generation of the continental crust, and they highlight the importance of having robust archives that can also be dated precisely. Other such archives include the minerals apatite, monazite, titanite, allanite and perovskite; all can yield precise ages and radiogenic isotope tracer information (e.g. Foster and Carter 2007; McFarlane and McCulloch 2007), they may exist under very different geological conditions and they have yet to be fully exploited.

THE COMPOSITION AND DIFFERENTIATION OF THE CONTINENTAL CRUST

The broad compositional features of the continental crust are now well established. These have been derived using a number of approaches and have been summarized by Rudnick and Gao (2003). The continental crust is compositionally evolved, i.e. enriched in Si and incompatible elements, and it dominates the Earth's geochemical budget for those incompatible elements that preferentially partition into silicate liquid during partial melting of the mantle. The continental crust represents only 0.57% of the mass of the Earth's mantle, but it contains, for example, over 40% of the Earth's potassium. The bulk continental crust has 60.6% SiO_2 and 4.7% MgO (Table 2.1), a composition not likely to have been in equilibrium with the upper mantle. Most models for the generation of the continental crust therefore involve at least two stages of differentiation, the extraction of basaltic magma from the mantle and remelting or fractional crystallization of that basalt (Kuno 1968; Ellam and Hawkesworth 1988; Arndt and Goldstein 1989; Kay and Kay 1991; Rudnick 1995; Arculus 1999; Kemp and Hawkesworth 2003; Zandt et al. 2004; Plank 2005; Hawkesworth and Kemp 2006a).

The continental crust has high contents of incompatible elements, including U, Th and K, and hence elevated heat production, and negative mantle-normalized anomalies for Nb and Ta, and high Pb contents (Fig. 2.1). Thus, it is characterized by low Nb/La and Ce/Pb ratios relative to the oceanic crust and the upper mantle. These are a feature of magmas related to subduction (Hofmann et al. 1986), and so it is widely assumed that similar processes were responsible for the average composition of the continental crust (e.g. Arculus 1999; Davidson and Arculus 2006). However, there are also significant volumes of intraplate magmatism, and a number of attempts have sought to assess the balance of intraplate and subduction-related magmatism in the generation of continental crust (e.g. Rudnick 1995; Barth et al. 2000). Typically the bulk crust is estimated to contain less than 10% of material generated in intraplate settings.

The geological record is dominated by the granitic and sedimentary rocks of the upper continental crust (e.g. Taylor and McLennan 1985). Models for the formation of new crust require an understanding of how the composition of the upper crust is related to that of new continental crust. Hawkesworth and Kemp (2006a, b) argued that differentiation of the continental crust is dominated by igneous processes, and that the

Table 2.1 Major and trace element composition estimates of the lower, upper and bulk continental crust after Rudnick and Gao (2003). Major elements in weight percent and trace elements in ppm.

	Bulk	Lower	Upper
SiO_2	60.6	53.4	66.6
TiO_2	0.72	0.82	0.64
Al_2O_3	15.9	16.9	15.4
FeO_T	6.71	8.57	5.04
MnO	0.10	0.10	0.10
MgO	4.66	7.24	2.48
CaO	6.41	9.59	3.59
Na_2O	3.07	2.65	3.27
K_2O	1.81	0.61	2.80
P_2O_5	0.13	0.10	0.15
Mg#	55.3	60.1	46.7
Rb	49	11	82
Sr	320	348	320
Y	19	16	21
Zr	132	68	193
Nb	8	5	12
Cs	2	0.3	4.9
Ba	456	259	628
La	20	8	31
Ce	43	20	63
Pr	4.9	2.4	7.1
Nd	20	11	27
Sm	3.9	2.8	4.7
Eu	1.1	1.1	1.0
Gd	3.7	3.1	4.0
Tb	0.6	0.48	0.7
Dy	3.6	3.1	3.9
Ho	0.77	0.68	0.83
Er	2.1	1.9	2.3
Tm	0.28	0.24	0.30
Yb	1.9	1.5	2.0
Lu	0.3	0.25	0.31
Hf	3.7	1.9	5.3
Ta	0.7	0.6	0.9
Pb	11	4	17
Th	5.6	1.2	10.5
U	1.3	0.2	2.7
Lu/Hf	0.081	0.132	0.058
Sm/Nd	0.195	0.255	0.174
Rb/Sr	0.153	0.032	0.256

Table 1 (Hawkesworth et al.)

composition of the upper crust has not been modified significantly by erosion and sedimentation. They estimated the average composition of new continental crust, and of the mafic material derived from the mantle from which the continental crust was differentiated, and calculated trace element abundances that are broadly similar to estimates of the average lower continental crust (Rudnick and Gao 2003). The implication is that the average composition of the lower crust has not been markedly depleted by intracrustal processes, even though it locally contains residual and cumulate lithologies (Rudnick and Fountain 1995).

Together, the estimated compositions of model new crust and the upper continental crust can constrain models for the generation of the upper crust (e.g. Hawkesworth and Kemp 2006a, b). These comparisons indicate that the upper crust reflects approximately 14% partial melting, or the analogous amount of fractional crystallization, of average new, mantle-derived basaltic crust. Given this proportion of new and upper crustal material, the generation of the upper crust should result in a large volume of residual material. The Earth's upper crust is approximately 12.5 km thick, with an average thickness of 40 km (Rudnick and Gao 2003). If the upper crust is the product of 14% melting, the corresponding residue would be 77 km thick, resulting in a total crustal thickness of approximately 100 km, including the middle crust. This is thicker than estimates of the continental crust, and it implies that the volumetrically dominant residue of upper crust formation has largely foundered into the mantle (e.g. Ellam and Hawkesworth 1988; Arndt and Goldstein 1989; Kay and Kay 1991; Kempton and Harmon 1992; Arculus 1999; Kemp and Hawkesworth 2003; Zandt et al. 2004; Hawkesworth and Kemp 2006a, b). The preferred explanation is that the residence time of material in the lower crust is approximately five to six times shorter than in the upper crust, and it reinforces arguments that delamination of residual lower crustal material is critical for establishing the andesitic composition of the average continental crust. The residual material returned to

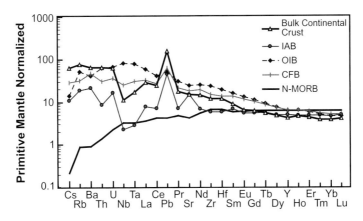

Fig. 2.1 A mantle-normalized diagram illustrating the minor and trace element abundances of the bulk continental crust (Rudnick and Gao 2003), average N-MORB (Sun and McDonough 1989), and average values for island arc basalts (IAB, $n = 644$), ocean island basalts (OIB, $n = 1520$) and continental flood basalts (CFB, $n = 2514$) compiled from the GERM database (http://earthref.org/GERM/). The elements are plotted in order of increasing relative partition coefficients during partial melting of the upper mantle from left to right. Thus ocean island basalts and MORB tend to have smoother mantle-normalized patterns, without the anomalies that characterize subduction-related magmas and the bulk continental crust.

the mantle would have had relatively high Sr/Nd and low Rb/Sr, and perhaps U/Pb ratios, and it has been invoked as the cause of trace-element enriched material in the source of some OIB (Lustrino 2005).

For a simple box model, the rates at which new crust is generated is constrained by the residence times of elements in the crust. For the upper crust, the residence time for at least the rare earth elements (REE), can be taken to be its average model Nd age of 2–2.5 Ga (e.g. Allègre 1982; O'Nions et al. 1983; Allègre and Rousseau 1984). Given that the rate of generation of the upper crust is estimated to be approximately 1/6th that of the bulk crust (Hawkesworth and Kemp 2006a, b), a residence time of 2 Ga in the upper crust requires an average crust-generation rate of 8 km^3 a^{-1}. This is two to five times greater than estimates of current rates at which new crust is generated (1.65 km^3 a^{-1} to 3.7 km^3 a^{-1}; Reymer and Schubert 1984; Clift and Vannuchi 2004; Scholl and von Huene 2007, 2009), consistent with models in which the Earth has cooled, and the rates of crust generation have decreased with time (Taylor and McLennan 1985).

THE ZIRCON ARCHIVE

In the geologically recent past, i.e. over the past 500 million years, new crust has been generated along convergent margins (e.g. Taylor and McLennan 1985; Rudnick 1995; Condie and Chomiak 1996; Condie 1998; Davidson and Arculus 2006). This, coupled with the subduction associated trace-element signatures in estimates of the average continental crust (Fig. 2.1), has encouraged models in which the crust has been very largely generated in subduction-related settings. However, questions have been raised over how different these crustal generation processes might have been in the Archaean, and even over the preservation potential of new crust generated above subduction zones (Gurnis and Davies 1985, 1986; Hawkesworth et al. 2009). One way to address these issues is through an examination of the distribution of crystallization ages, and of model ages, in crustal rocks. This is increasingly done by U–Pb and Hf isotope analysis of zircon crystals that occur within igneous rocks or as detrital grains in sediments and sedimentary rocks. Figure 2.2 summarizes U–Pb crystallization and

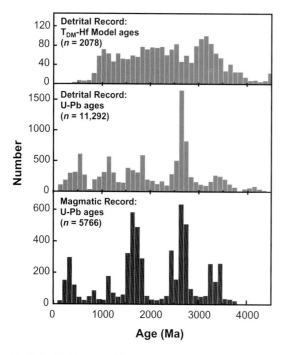

Fig. 2.2 U–Pb crystallization ages on magmatic and detrital zircons, and Hf model ages on detrital zircons in Australia (after Hawkesworth et al. 2010). It is striking that the crystallization ages preserve marked peaks in ages, reflecting discrete magmatic episodes, whereas the Hf model ages do not. The latter reflect the times when new crust was generated (Fig. 2.4), and the lack of peaks of model ages is attributed to the mixing of crust generated at different times during erosion and sedimentation, and in the generation of granitic rocks.

Hf model ages on zircons from Australia in both magmatic rocks and as detrital grains in sediments. Just over 17,000 zircons have been dated, and there are Hf model ages on over 2000 of them. The striking observation is that there are marked peaks in the ages of crystallization, but not in the Hf model ages.

Zircon has the unique combination of physiochemical resilience and high concentrations of important trace elements that include two radiogenic isotope systems of geochronological importance (namely U–Pb, Th–Pb), and another (Lu–Hf) that is widely used as a crustal evolutionary

tracer. Zircons typically crystallize from high-silica melts and they are almost ubiquitous in upper-crustal rocks. So, it follows that zircons should offer an excellent record of the evolution of differentiated compositions in the continental crust, but that they may be less useful in charting the evolution of the depleted mantle, for example. Diffusion rates for the isotopes in question are extremely low (Cherniak et al. 1999), and so zircons can retain their isotopic integrity through multiple episodes of sedimentary and magmatic recycling. Remarkably, they even appear to survive transient entrainment into the mantle via lower crustal delamination and sediment subduction (e.g. Gao et al. 2004). Given their low solubility in silicic melts (Watson and Harrison 1983), zircons can persist as refractory relics in granitic magmas, and potentially carry chemical and isotopic information about the deep crust that is otherwise inaccessible. Similarly, because weathering and erosion average large tracts of continental crust (e.g. Nance and Taylor 1976, 1977; Taylor and MacLennan 1991; Condie 1993), detrital zircons in clastic sediments may preserve a more complete temporal record of igneous and crustal growth episodes than exposed basement. Of enormous benefit is that the chemical and isotope information encoded within the complex growth structures developed within many zircons can now be extracted by micro-analytical techniques capable of high precision and spatial resolution. As the ages of discrete growth phases within single grains can be determined by in-situ U–Pb isotope analysis (Fig. 2.3), zircons provide an unparalleled time-series of changing magmatic compositions during crystal growth.

Hafnium has six naturally occurring isotopes, of which radiogenic ^{176}Hf is produced by the β^- decay of ^{176}Lu, with a half-life of 37.2 Ga in terrestrial samples. Hf is more incompatible than Lu during melting of spinel and garnet peridotite, and so the long-term enrichment of Hf relative to Lu in the continental crust results in ^{176}Hf/^{177}Hf ratios in crustal and depleted mantle reservoirs that are respectively relatively unradiogenic and radiogenic (Patchett et al. 1981). In this respect, the Lu-Hf system is analogous to Sm-Nd, and

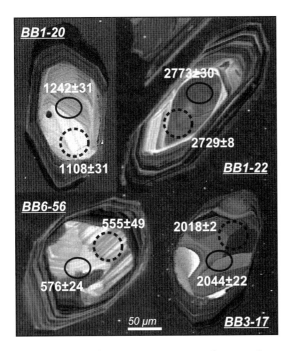

Fig. 2.3 Cathodoluminescence images of zircons from 430 Ma cordierite granites from eastern Australia (Kemp et al. 2009a). The rims have the ages of crystallization of the host granites, but they have older inherited cores that have been dated using U–Pb isotopes by ion mircoprobe (SIMS, small ellipses) and by laser ablation ICP-MS (LA-ICP-MS, dotted circles).

initial ^{176}Hf/^{177}Hf ratio inherited from the magma at the time of crystallization. In principle Hf-isotope ratios of crustal rocks are a measure of the crustal residence age, or the average time since the sources of the magmas from which the zircons crystallized were extracted from a specified mantle reservoir, usually depleted mantle. These Hf 'model ages' are calculated as shown in Fig. 2.4(a). The crystallization ages of zircons are calculated from their U–Pb isotopes, and the Hf isotopes at the time of crystallization are calculated from the measured Hf-isotope and Lu/Hf ratio of the zircon. The Hf model age is then calculated using the average Lu/Hf ratio for the continental crust to calculate when the crustal source rocks for the magmas that crystallized the zircons were themselves derived from the mantle (Fig. 2.4a). In detail the Lu/Hf ratios of igneous rocks decreases with increasing silica (Fig. 2.4b). Thus, if the slope of the isotope-evolution line for the crustal source rocks can be determined, it should be possible to evaluate whether the zircons analysed crystallized from magmas that were derived from source rocks that were broadly mafic or granitic in composition.

Overall the model age approach requires that the Lu/Hf ratio, and hence the Hf isotope ratio, of the depleted mantle is known as a function of time (Vervoort and Blichert-Toft 1999). It also presumes that, as with Sm/Nd ratios, the Lu/Hf ratio of a protolith has not been subsequently modified during intracrustal reworking. Taken together, Hf and U–Pb isotopes, therefore, offer a means of investigating the links between the ages of the major episodes of igneous activity (zircon crystallization ages) and the formation of new crust (Hf 'model ages') (see Fig. 2.2).

A complementary approach is to interrogate the trace-element contents of zircons, and to consider the implications of the mineral inclusions preserved within them. Initially, temperatures of zircon crystallization were calculated using the estimated saturation temperature of zircon in the melt (Watson and Harrison 1983), and subsequently using Ti contents in the zircon (Watson and Harrison 2005). More recently, such approaches have been extended to evaluate

indeed Hf-Nd isotopes form coherent arrays for most mantle-derived rocks (Vervoort et al. 1999). However, since the fractionation of Lu/Hf during mantle melting is approximately twice that of Sm/Nd, and the half-life of ^{176}Lu is shorter than that of ^{147}Sm, Hf isotopes may have greater resolution in identifying discrete mantle-source domains (Patchett and Tatsumoto 1980; Patchett 1983).

The great advantage of Hf isotopes for crustal studies is that the element Hf is concentrated and bound in the zircon crystal lattice, whereas the REE are far less compatible. Zircons therefore have very low Lu/Hf ratios (typically less than 0.001), so that Hf isotope corrections due to in-situ radiogenic growth are typically small (Patchett 1983). Thus, even at the present day, zircons preserve ^{176}Hf/^{177}Hf ratios close to the

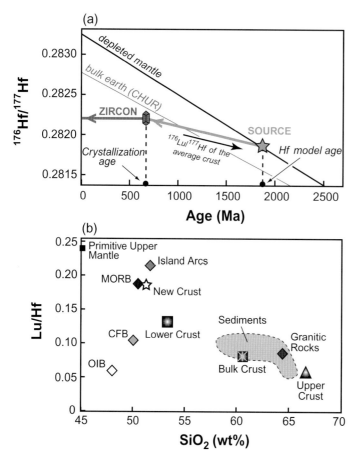

Fig. 2.4 (a) A plot showing how Hf isotope ratios vary with time for a hypothetical sample. The analysis of the zircon yields its crystallization age (from U–Pb isotopes) and its Hf isotope ratio at the time of crystallization. The Hf model age is the time that the crustal precursor to the magma that crystallized the zircon was itself derived from the upper mantle. The model age is therefore calculated using the $^{176}Lu/^{177}Hf$ ratio of the likely crustal precursors.

(b) A plot of Lu/Hf ratios and SiO_2 contents for different average compositions of common rock types and crustal compositions. The New Crust is the model composition of magmas from which the average bulk continental crust was derived from Hawkesworth and Kemp (2006a, b). The objective is in part to estimate the Lu/Hf ratios of primitive crust from variations in Hf isotopes through time, and then to use Lu/Hf- SiO_2 to estimate the SiO_2 content of that crust. Primitive mantle after Sun and McDonough (1989), average value for the granitic rocks ($SiO_2 \geq 55\%$; $n = 7371$) compiled from the GERM database (http://earthref.org/GERM/), other references as Fig. 2.1.

pressures of crystallization using Si in muscovite inclusions and Al in hornblende inclusions (Hopkins et al. 2008). For Hadean (greater than 4.0 Ga) zircons from the Jack Hills, Western Australia, these suggest temperatures of approximately 700°C at pressures of 0.7 ± 0.2 GPa, and, hence, a relatively low geotherm, with 30 K km^{-1} inferred to be a maximum value. This translates into a near-surface heat flow of approximately 75 mW m^{-2}, much lower than is typically predicted for the Hadean. Assuming that the estimates of pressure and temperature are robust, Hopkins et al. (2008) argue that the simplest mechanism for developing relatively low heat-flow values is by subduction of cold material back into the mantle. This implies that a plate tectonic

regime similar to that operating today was established in the early Earth.

More generally, there have been discussions over the implication of the distributions of temperatures of crystallization of zircon, and in particular of those in detrital and inherited zircons (Nuttman 2006; Harrison et al. 2007). The preservation of zircon is less common for those with high U and Th contents because of the effects of radiation damage, and U and Th contents are higher in more differentiated, lower-temperature magmas. Thus, the peak of temperatures for the population of zircons preserved may be higher than it would have been for the zircons at the time of crystallization. However, given that zircons tend to crystallize from relatively high-silica melts, the principal concern is that the assemblages of mineral inclusions in zircon, and the estimated crystallization temperatures, will converge at low temperatures. For zircons that crystallized from low-temperature differentiates, it remains difficult to establish the extent to which they retain significant information on the less-evolved parental magmas, which are likely to be more diagnostic of the source regions, and even the tectonic setting, in which they were generated.

THE IGNEOUS AND SEDIMENTARY RECORDS

Sediments and sedimentary rocks, and igneous rocks, offer different perspectives on the evolution of the continental crust. For the most part, igneous rocks are generated in magmatic provinces that are restricted in space and time, and they reflect the episodes in which new crust may have been generated. For example, destructive plate margins are characterized by linear belts of igneous rocks, and the present orogenic cycle in the Andes appears to have persisted for the last 180 Ma. Continental flood basalts occur in more equi-dimensional provinces, they are generated in relatively short periods of time, (less than 10 Ma), and yet new continental crust may be generated in both settings. Sediments, in contrast, are mixtures of the material available from their source regions, eroded, and subsequently deposited. They provide average compositions, and particularly for insoluble elements they have yielded robust estimates of the average trace-element contents of the upper continental crust. (Garrels and Mackenzie 1971; Nance and Taylor 1976, 1977; O'Nions et al. 1983; Condie 1993) Whole-rock analyses of sediments provide Nd and even Hf model ages that reflect when the average of their source rocks was derived from the mantle, and they have been used to constrain the evolution paths for average continental crust (Nance and Taylor 1976, 1977; O'Nions et al. 1983; Allègre and Rousseau 1984; Michard et al. 1985; Condie 1993). However, such sediments are mixtures, and so the model ages are

necessarily hybrids that, in most cases, do not reflect discrete crust-forming events (see Fig. 2.10(a), inset). Such hybrid ages are difficult to interpret, because the ages, and the relative contributions of different source terrains, cannot be independently determined from whole-rock samples. In practice, two ways to address this dilemma been developed: (i) to use O isotopes to screen out zircons that crystallized from magmas that may contain a contribution from sedimentary sources regions (Kemp et al. 2006, 2007a), and may therefore yield hybrid model ages, and (ii) to use the crystallization and model ages of zircons to constrain the source regions, and their relative contributions, represented in whole rock sediments (Dhuime et al. 2009).

The $^{18}O/^{16}O$ ratio, expressed as $\delta^{18}O$, stated as permil relative to the SMOW standard, is most readily changed by low-temperature and surficial processes, and so the $\delta^{18}O$ of mantle-derived magmas (5.37 to 5.81‰ in fresh mid-ocean-ridge basalt (MORB) glass; Eiler et al. 2000) contrasts with those from rocks that have experienced a sedimentary cycle which typically have elevated $\delta^{18}O$. This is reflected in the high $\delta^{18}O$ of the crystallizing zircons and is a 'fingerprint' for a reworked component in granite genesis. Empirical studies have established that oxygen diffusion in zircon is sufficiently sluggish that the original igneous $\delta^{18}O$ remains intact, even through protracted metamorphism and crustal fusion (King et al. 1998; Peck et al. 2003). Significantly, $\delta^{18}O$ can be measured in-situ in zircon with excellent precision (less than 0.5‰) by large-radius ion microprobes that have multi-collector capability (Valley 2003).

In principle, the coupling of radiogenic and stable isotopes can uniquely reveal whether zircon crystallized from a juvenile magma during crustal generation, or from magma derived by reworking of pre-existing igneous or metasedimentary rocks. This provides greater insight into crustal evolution than was previously possible. The combined Hf and O isotope studies of zircons from the Lachlan Fold Belt granites of eastern Australia highlighted how those in strongly peraluminous 'S-type' granites tended to have higher

$\delta^{18}O$ and lower Hf isotope ratios than those in metaluminous 'I-types' (Kemp et al. 2007a). However, they also show that I-type granites contain contributions from sedimentary source rocks, and that they are not therefore simple melts of pre-existing igneous rocks, as previously thought (Chappell and Stephens 1988).

CRUSTAL GROWTH PROCESSES FROM THE IGNEOUS RECORD USING ZIRCONS

Tectonic activity regulates the availability and composition of magma sources. Thus, provided that the different source regions have distinct isotope ratios, the changes in isotope ratios through time in magmatic rocks should be different in orogens developed in different tectonic settings. Kemp et al. (2009b) have demonstrated

striking secular trends defined by bulk rock εNd and zircon εHf-$\delta^{18}O$ data from granitic rocks of the eastern Australian Tasmanides, which formed behind a long-lived, dominantly west-dipping subduction zone (Fig. 2.5). These rocks were generated during three cycles of tectonic activity between 520 and 230 Ma, and the cyclic trends are attributed to variation in the incorporation of older, dominantly metasedimentary material, during alternating crustal thickening and thinning episodes induced by compressive and extensional dynamics at the subduction zone. In this model, the negative segments of the isotope trends, i.e. those marked by greater crustal input, are associated with compressional deformation (transient flat-slab subduction) and reflect enhanced metasedimentary input into magmas from thickened crust. The positive isotope-time trends, those that reflect increasing mantle contribution through time, are associated with sub-

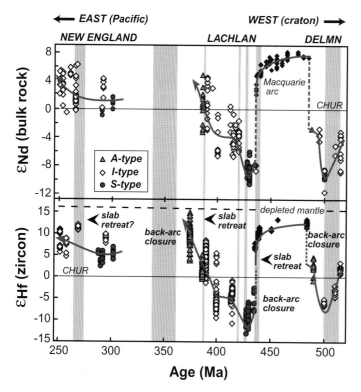

Fig. 2.5 Isotope and tectono-magmatic evolution of the Tasmanides defined by the ε_{Nd} values of igneous rocks (top) and ε_{Hf} (bottom) of granite-hosted zircons (sources for the isotope data and the timing of events are given in Kemp et al. 2009b). The shaded time slices correspond to major contractional episodes.

sequent extension caused by the re-initiation of subduction-zone retreat. They therefore indicate declining rates of crustal reworking. Trace element evidence suggests a decreasing slab-flux component within the basaltic component of granitic magmas that formed during these extensional periods.

These orogens in Eastern Australia are distinctive in that they are dominated by turbiditic sedimentary assemblages and granites, often with relatively short time periods between the deposition of the turbidites and the generation of the granites. They were generated in back-arc extensional settings behind retreating-arc systems (Gray and Foster 2004) and they involved the generation of a third of the present Australian crust in 300 Ma (Kemp et al. 2009b). It is argued below that the preservation potential of igneous rocks generated in subduction zones is very poor, whereas, in contrast, crust generated in back-arc settings seems to have been stabilized, in part by the influx of sediment that is melted to form granite, and in part by the development of granulites that are also difficult to deform. Rapid crustal growth in eastern Australia was promoted by the progressive oceanward retreat of the subduction-accretionary system, and such systems also may have had a significant role during rapid crustal generation in the Precambrian, especially for turbidite-granite dominated Proterozoic orogens.

THE LACHLAN CASE STUDY

There are now literally tens of thousands of Hf and U–Pb isotope analyses of zircons in the literature (see recent review by Condie et al. 2008). They have been used to identify segments of crust of similar histories, to investigate the nature of source rocks in river catchments and how they contribute to the zircon populations in the mouth of major rivers, to unravel granite petrogenesis and to investigate the ages of crust formation events in the Hadean and early Archaean. In most cases, it has been difficult to distinguish those samples that may yield hybrid ages, because the magmas contain a contribution from sedimentary material. As discussed above, however, oxygen isotopes offer a means to identify sedimentary input into felsic magmas. The first study to combine in-situ O and Hf isotopes in zircon was conducted on zircons from igneous and sedimentary rocks from the Lachlan Fold Belt in southeast Australia (Kemp et al. 2006).

The Lachlan Fold Belt is a classic region for granite petrogenesis, as the area where the I- and S-type granite scheme was initially developed (Chappell and White 1974; 1992). The zircons discussed here are inherited zircons in the granites that yield crystallization ages older than the ages of the granites, and detrital zircons in the Ordovician turbidites. The crystallization ages of these pre-granite zircons range from up to 3400 Ma, and there are peaks in the age distribution for zircons (Fig. 2.6) thought to reflect major magmatic episodes at approximately 1000 and 500 Ma. The Hf model ages are all much older and, therefore, it can be concluded that little new crust was generated during the 1000 and 500 Ma magmatic episodes in the provenance region of the zircons. The Hf model ages range from approximately 3.9 to approximately 1.3 Ga (Fig. 2.6), but those from zircons with low $\delta^{18}O$, which are considered to be from magmas derived from igneous sources, yield more restricted Hf model ages that cluster around two peaks at 3.3 and 1.9 Ga. The Hf model ages of these grains therefore indicate periods of new crustal growth at 3.3 and 1.9 Ga. In contrast, the zircons with high $\delta^{18}O$ crystallized from magmas with a sedimentary component, and thus have mixed-age source rocks. The Hf model ages of these zircons peak at 2.6 to 2.2 Ga represent a mixture of the rocks previously generated at 3.3 and 1.9 Ga. O-isotope data were critical in establishing that the 2.2 to 2.1 Ga age peak represented mixing in the sedimentary environment, rather than a real crust-forming event.

More generally, with this approach we can now contrast the respective evolutions of the igneous and sedimentary reservoirs in the continental crust, and compare the information from zircons with the models for the evolution of the continental crust based on Nd-isotope ratios in shales

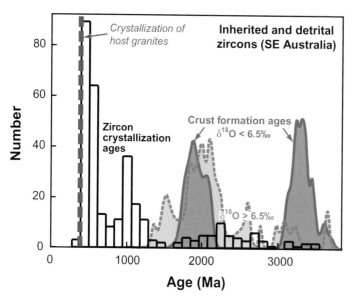

Fig. 2.6 A histogram of ages obtained from inherited and detrital zircons from approximately 430–380 Ma old granites and Ordovician turbidites in the Lachlan Fold Belt (Kemp et al. 2006). The crystallization ages have peaks at approximately 500 and 1000 Ma, whereas the Hf model ages (see Fig. 2.4a), here termed crust formation ages, are much older with peaks at approximately 1.9 and 3.3 Ga. These peaks of model ages in zircons with $\delta^{18}O$ less than 6.5‰ indicate when new crust in the areas sampled by these zircons was generated (for discussion, see text).

(Allègre and Rousseau 1984, and 'The sedimentary record and erosion models' below).

THE CONTINENTAL RECORD – PEAKS OF CRUST GENERATION OR A FUNCTION OF PRESERVATION?

It has long been known that the age distribution of the rocks at the Earth's surface is uneven, with more rocks of some ages exposed than others. There are late Archaean (2.7 to 2.5 Ga) terrains in every continent, and yet there are relatively few rocks preserved that formed in the periods before or shortly thereafter. McCulloch and Bennett (1994) and Condie (1998) summarized the ages of rocks in the continental crust that were considered to have been derived from the mantle and, therefore, represented new crust. The age distributions had peaks of crystallization times at approximately 2.7, 1.9 and 1.2 Ga that were thought to represent periods of relatively rapid crust generation (Fig. 2.7). Such peaks contrasted with the relatively smooth curves developed for the increase in the volume of stable continental crust through time (e.g. Taylor and McLennan

1985), a number of which were based on radiogenic isotopes in continental sediments (see Hawkesworth et al. 2010 for a review). Once again, it seems important to be able to distinguish the records of the generation and evolution of the continental crust preserved in sediments and in igneous rocks.

The major question posed by the peaks of crust formation is whether these times truly represent periods of rapid growth, or instead are some artefact of differential preservation. If they do reflect periods of rapid growth, they would imply pulses of new crustal growth that are difficult to attribute to plate-tectonic processes, and thus these times have been attributed to episodes of thermal instabilities within the Earth and some form of super-plume event (e.g. Stein and Hofmann 1994; Albarède 1998, Condie 1998). It is striking that no marked peaks of crust generation were recognized in the last billion years, and this is readily linked to an increasing dominance of subduction-related magmatism later in Earth history. However, the bulk composition of the continental crust has trace-element features more typically attributed to convergent margin magmatism (e.g. high Pb/Ce, low Nb/La; see discussion of Fig. 2.1) and dif-

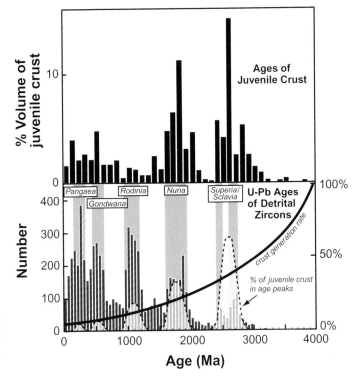

Fig. 2.7 Histograms of estimated relative volumes of rocks of different ages that reflect juvenile crust (upper panel, Condie 2005), and of approximately 7,000 U–Pb crystallization ages from detrital zircons from recent rivers and beach sands (lower panel, Campbell and Allen 2008). The peaks of the crystallization ages in particular appear to match up with the periods when supercontinents were developed on Earth. The broad curve illustrates a model in which the volume of new continental crust generated decreased exponentially with time, and the sketched-in peaks illustrate how the relative volume of new crust generated in the periods of the peaks of crystallization ages will also have decreased with time.

ficult to reconcile with plume-related crustal growth.

More recently there have been a number of studies that have used Hf- and O- isotope analyses of zircons to investigate whether there are peaks of model Hf ages, peaks when new crust was formed, and to compare the timing of those peaks with the ages of the peaks of crystallization ages as summarized in Fig. 2.7. The data from the Lachlan Fold belt yielded peaks of Hf model ages at 3.3 and 1.9 Ga (Kemp et al. 2006), and so the 1.9 Ga peak is observed in both the distribution of crystallization and model Hf ages. The 3.3 Ga peak dates from a time for which the volume of preserved rock is relatively small, and so peaks of crust generation are best investigated using Hf model ages in zircons that preserve mantle-like $\delta^{18}O$ values. Pietranik et al. (2008) reported Hf and O isotope analyses on detrital zircons from 2.8 Ga sediments in the Slave Province, Canada. These indicate that there were three periods of crust

formation at approximately 4.4, 3.8 and 3.4 Ga, and highlighted some of the difficulties in demonstrating that peaks in Hf model ages in different continents were, or were not, coeval. Condie et al. (2008) reviewed the crystallization ages for more than 25,000 zircons and concluded that there may have been more periods of relatively rapid crust generation than indicated in Fig. 2.7. They suggested there were prolonged peaks of granitic magmatism in the late Archaean and early Proterozoic that cannot easily be attributed to short-lived plume events at 2.7 and 1.9 Ga.

A number of studies have refocused attention on the extent to which the distribution of crustal formation ages might be a function of preservation in the geological record. Kemp et al. (2007b) demonstrated that high-temperature, low-pressure granulites in North Japan were generated in a back-arc setting. They noted that this provides a link between granulite facies metamorphism and the generation of new crust, and

pointed out the link between the peaks of crust
generation and the ages of granulite metamorphic
events. Brown (2007) categorized high-grade orogenic belts into high-temperature belts that
formed under high-, intermediate-, or low-pressure
conditions. He noted that the high-pressure belts
were restricted to the last 600 Ma, and concluded
that they reflect cold subduction as observed at
present along convergent margins. Intermediate-
to low-pressure rocks are preserved into the late
Archaean, and Brown (2007) further demonstrated
that their ages are grouped in clusters similar to
the peaks of crust generation summarized in Fig.
2.7. The periods of granulite facies metamorphism would appear to be linked to the processes
of crust generation, and/or the peaks of the ages
of crust generation and granulite metamorphism
are themselves a function of the unevenness of
the continental record.

Campbell and Allen (2008) investigated the
build-up of atmospheric oxygen, which they
linked to the development of supercontinents.
They compiled the crystallization ages of around
7000 detrital zircons from Australian dunes and
major rivers world-wide demonstrating that their
distributions yield peaks of ages that coincide
with the times of supercontinent formation (Fig.
2.7). One implication is that the distribution of
crystallization ages in geological record of the
continental crust may be biased by periods of
supercontinent formation, and so interpretations
of the apparent peaks in crust formation may
need to take account of the processes of preservation, rather than just those of crust generation.

Figure 2.7 compares the histograms of U–Pb
crystallization ages in detrital zircons from
Campbell and Allen (2008), and the ages of rocks
that reflect new continental crust from Condie
(1998, 2005). It is striking that in the last billion
years BP there are marked peaks in the zircon
crystallization ages, and no analogous peaks of
crust formation. In older rocks there appears to be
a better match between the peaks of magmatic
activity (zircon ages) and the peaks of crust generation. The geological record is dominated by
younger events, when less new crust was
generated.

Just how might the development of supMerconti-
nents bias the geological record? The development of supercontinents at different stages in
Earth history is thought to be a predictable consequence of plate movements (e.g. Dalziel 1992).
The cycle of supercontinent formation and
destruction involves periods of subduction-related
magmatism, collisional orogens and extensional
magmatism, and the preservation potential of
rocks generated in these different settings may be
different. Hawkesworth et al. (2009) illustrated
the relative volumes of magma generated, and
their preservation potential through the development, stabilization and destruction of a supercontinent (Fig. 2.8). Given that the bulk composition

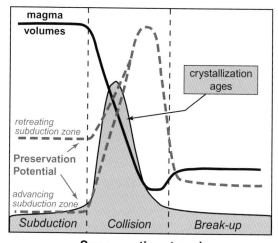

Supercontinent cycle

Fig. 2.8 A schematic illustration of how the volumes
of magma generated, and their likely preservation
potential, may vary in the three stages associated with
the amalgamation and break-up of a supercontinent
(Hawkesworth et al. 2009). The preservation potential
in the first stage is greater at margins in which the
subduction zone retreats oceanwards and results in
extensional basins than at margins in which the subduction zone advances towards the continent. It is inferred
that the peaks in the ages of crystallization that are
preserved reflects the balance between the volumes of
magmatic rocks generated in the three stages, and their
preservation potential.

of the continental crust is dominated by geochemical signatures similar to those in magmas generated above subduction zones, it is surprising that the preservation potential of rocks generated in this setting is so low, apart, perhaps, from those generated behind retreating subduction zones. There is increasing evidence that the global rates of removal of continental and island-arc crust through subduction into the mantle are similar to the rates at which crust is generated at modern magmatic arcs (approximately $3\,km^3\,y^{-1}$, Clift and Vannuchi 2004; Scholl and von Huene 2007, 2009). Mineral deposits that form predominantly in convergent-margin settings, such as epithermal and porphyry copper deposits and orogenic gold deposits, are also generally less than a hundred million years old, as a consequence of rapid exhumation and erosion (Gastil 1960; Kesler and Wilkinson 2006). Instead, there is increasing evidence that subduction-related magmatic rocks are better preserved in extensional basins that lie inboard from the subduction zone, as discussed in below.

Magmas associated with collisional mountain-building are dominated by partial melting of pre-existing crust. They are granitic in composition, and although the volumes generated may be small relative to other tectonic settings, they will tend to be protected within the enveloping supercontinent (Fig. 2.8). These igneous rocks thus have good preservation potential in the geological record. The extensional phase is dominated by mafic magmatism with subordinate rhyolite, as seen in the continental-flood basalt sequences associated with the break-up of Gondwana (Storey 1995; Hawkesworth et al. 1999). It is unlikely to result in large volumes of zircons, and the rocks may in any case be relatively prone to erosion and hence have relatively poor preservation potential.

In summary, the record of continental development and evolution is likely to be dominated by periods when supercontinents assembled, not because this is necessarily a major stage of crust generation, but because it provides a setting for the selective preservation of crustal rocks. The rocks thus preserved tend to be from the end of the subduction-related and the collisional mountain building stages (Fig. 2.8), because they are insulated in the cratonic interior and thus preserved from convergent plate-margin erosion. Therefore, peaks in age distribution reflect the interplay between zircon crystallization and preservation bias related to tectonic setting, and do not require accelerated pulses of crust generation or magmatic activity (Hawkesworth et al. 2009).

Earlier in Earth history the peaks of ages of crystallization also match up with episodes of crust generation (Fig. 2.7). The peaks of ages are not then dominated by periods of crustal remelting, and so other processes must have dominated, and shaped the geological record. The implication is that the products of crust formation are better preserved in the late Archaean and early Proterozoic than they have been subsequently. The striking feature of the end of the Archaean (2.7 to 2.5 Ga) is that it was the time of the stabilization of the Archaean cratons as currently preserved. This presumably reflects the thermal evolution of the Earth (Davies 1995), and a predictable consequence would be a marked peak of magmatic activity preserved from that time (Fig. 2.7). Records of older events are much less complete and, as argued above, they may be best recorded in the crystallization and the Hf model ages of zircons. One implication is that these ages may more closely mirror the events that took place, less biased by the controls on preservation that mark post-Archaean magmatic processes.

THE COMPOSITION OF THE EARLY PROTO CONTINENTAL CRUST

The earlier discussion of the Lu/Hf ratios in igneous rocks highlighted that mafic rocks typically have higher Lu/Hf ratios than granitic rocks (Fig. 2.4). Different Lu/Hf ratios result in divergent Hf-isotope evolution paths on plots of Hf isotopes versus time, and the evolution of mafic and granitic crust is shown in Fig. 2.9. The key is to obtain Hf isotope data on zircons with a range of crystallization ages from rocks that may

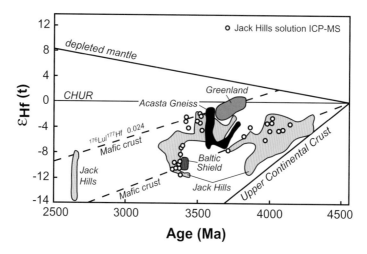

Fig. 2.9 A plot of initial Hf isotope ratios, expressed in the epsilon notation, against crystallization ages for selected zircons ranging in age from 4.3 to 2.6 Ga. The solution ICP-MS detrital Jack Hills zircons (open circles) are from Amelin et al. (1999) and the Acasta Gneiss data (black field) are from Amelin et al. (2000). The remaining fields are for in-situ analyses of Pb and Hf isotopes (Kemp et al. 2009a; Kemp et al. 2010). The evolution of closed systems takes place along straight lines and their slopes depend on the parent/daughter ratio $^{176}Lu/^{177}Hf$. Felsic rocks, and therefore the upper continental crust, have lower $^{176}Lu/^{177}Hf$ ratios than mafic rocks (Fig. 2.4(b)), and hence they evolve along lines with steeper slopes. Most of the zircon data are consistent with crystallization from granitic magmas that were themselves derived from mafic source regions, and just some of the older zircons from Jack Hills may have crystallized from magmas that were derived from upper-crustal source rocks.

reasonably be thought to have been derived from similar crustal source rocks. Figure 2.9 summarizes both dissolution and laser ICP-MS data on zircons that crystallized between 4.25 and 2.7 Ga from Western Australia, the Baltic Shield, Canada and Greenland. In some cases Pb and Hf isotopes have been determined together from the same laser spot, and this tends to result in tighter arrays than when Pb and Hf are analysed on different instruments.

The oldest known zircons are from the Jack Hills and Mount Narryer supracrustal belts in the Yilgarn craton in Western Australia (Amelin et al. 1999; Wilde et al. 2001; Harrison et al. 2005). The data define a reasonable linear array, and whereas the best-fit line has a slope similar to that of mafic crust, it can also be argued that the lower limit is consistent with the presence of a granitic source component. It both cases the Hf model ages are approximately 4.5 to 4.4 Ga (Fig. 2.9). Considering all data in Fig. 2.9, many points scatter around

slopes consistent with high Lu/Hf ratios that are a feature of mafic crust (Fig. 2.4b). The implication of this observation is that most of the crust that was the source of the magmas from which the zircons crystallized was mafic in composition. Similar conclusions have been obtained from detrital zircons in the Slave Province (Pietranik et al. 2008), and by Kramers (2007) using published data on detrital zircons from Jack Hills in Western Australia. It also can be inferred from the higher Cr/V, Ni/Co and Ni/V ratios of Archaean sediments (Taylor and McLennan 1985). Kamber et al. (2005) explored the constraints available from the thermal conditions likely to be present in the Hadean. Zircons tend to crystallize from high-silica magmas and these will also have elevated U, Th and K contents. The radioactive heat production of all rock types was three to four-and-a-half times higher in the Hadean than at present. Even with low basal heat-flow values, mafic crust is likely to have melted at depths of

20 to 40 km, and felsic crust is likely to have been at temperatures in excess of 1000°C at mid-crustal depths. Thus felsic crust, which is the dominant source of zircon, is unlikely to have been much more than 10 km thick in the Hadean, and most of the continental crust at depth is likely to have been mafic in composition – consistent with the Hf isotope data in Fig. 2.9.

THE SEDIMENTARY RECORD AND EROSION MODELS

Detrital sediments sample the continental crust formed at different times and in different places (e.g. Garrels and Mackenzie 1971; Nance and Taylor 1976, 1977; O'Nions et al. 1983; Allègre and Rousseau 1984; Michard et al. 1985; Condie 1993). Some lithologies are more susceptible to erosion than others, and so a central issue is to understand how the compositions of the different source rocks are then recorded in the sediments. The relative contributions of different source terrains are usually expressed through an erosion factor 'K', or an equivalent erosion parameter. In crustal-evolution studies, K relates the model age of the bulk sediments to the average model age of the continental blocks from which they were derived (Allègre and Rousseau 1984). The determination of K, and the extent to which it varies in different erosion systems, has fundamental implications for models of continental growth based on radiogenic isotopes in continental sediments. Typically the values of K are usually assumed in these models, because it remains difficult independently to assess the relative contributions of sources of different ages in a sample of sediment (Allègre and Rousseau 1984; Goldstein and Jacobsen 1988; Jacobsen 1988; Kramers and Tolstikhin 1997; Kramers 2002; Tolstikhin and Kramers 2008).

Dhuime et al. (2009) recently reported the first estimates of K from integrated Hf and U–Pb isotopes in detrital zircons, and Nd isotope ratios of bulk recent sediments along the Frankland River in SW Australia. The Frankland River constitutes one of a series of southward flowing rivers that developed along the southwest coast of Western Australia following the break-up of Australia and Antarctica at around 65 Ma. It has a length of approximately 320 km and the catchment area is 4630 km². It offers an opportunity to link sediments to their source rocks, because it drains just two crustal blocks with distinctive age components, the Archaean Yilgarn craton and the Proterozoic Albany-Fraser mobile belt (Cawood et al. 2003).

The Hf model age and the U–Pb crystallization age data on detrital zircons sampled along the Frankland River can be used to calculate the contribution from the Yilgarn and the Albany-Fraser belt in each sample, and to compare that with the proportion of those two units in the catchment again for each sample (Fig. 2.10, Dhuime et al. 2009). Although there is a large range of Hf model ages in any group of zircons with similar U–Pb ages, it is striking that the trend of the average value of the Hf model ages through time is broadly similar to that for Nd model ages in Australian shales (Allègre and Rousseau 1984) (Fig. 2.10a). This suggests that the distribution of zircon data may be used to constrain values of K for different sediment samples.

Overall there is a downstream decrease in the proportion of Yilgarn material in the recent sediments (Fig. 2.10(b), grey curve). K is not constant and it increases by a factor of two to three downstream, and with the gradient of the river (K ~4–6 to K ~15–17, Fig. 2.10b). Samples with K approximately 9–10 and K approximately 15–17 are from below the inflection that reflects a weak escarpment at approximately 80 km from the coast associated with Miocene-Pliocene uplift (Cawood et al. 2003) (Fig. 2.10b). The steepening of the river's profile has resulted in preferential erosion of material from the Albany-Fraser Belt, and hence an increase in the calculated K values. In turn the 'stable' segment of the Frankland River is best sampled above the escarpment at distances of 100–150 km from the coast. It is concluded that values of K = 4–6 are representative of mature river systems that sample large source areas. These values should therefore be used to re-evaluate models of crust formation and evolution

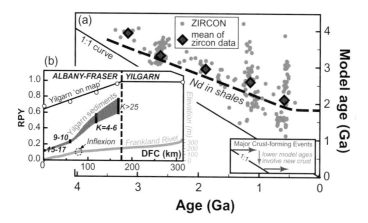

Fig. 2.10 (a) Model ages versus crystallization ages of detrital zircons from five recent sediments from the Frankland River in southwest Australia (Dhuime et al. 2009). The diamonds are the means of the zircon data (small dots) grouped into five main periods of zircon crystallization. The Nd model ages in Australian shales (dashed curve) are from Allègre and Rousseau (1984) who used them to model the evolution of the continental crust using different values of the erosion factor 'K'. Inset: on a plot of the model age of the sample against its sedimentation age, new crust that is then reworked in subsequent events results in horizontal arrays. Samples that have younger model ages in younger sediments, as seen for Nd isotopes in shales (Allègre and Rousseau 1984), indicate that new crust was generated in younger events and it has then been sampled by the younger sediments. As the slope of the data array flattens out, it indicates that with time, less and less new crust has been generated.
(b) Relative contribution of material from the Archaean Yilgarn block in fine-grained recent sediments (dark grey envelope curve), plotted against distance along the Frankland River. These are compared with the proportion of Archaean source rock available in the catchment for each sediment sample (open squares). The calculated K values for each sediment, and the change in elevation down the river, are also shown. The high values of K close to the contact with the Yilgarn are subject to large errors, and are poorly understood. Values of K = 4–6 appear representative of mature river systems that sample large areas of continental crust. RPY: relative proportion of Yilgarn component (e.g. 0.8 means 80% of Yilgarn and 20% of Albany-Fraser); DFC: distance from the coast.

through time, since many previous models used values of K = 2–3 (Allègre and Rousseau 1984; Goldstein and Jacobsen 1988; Jacobsen 1988; Kramers and Tolstikhin 1997; Kramers 2002), which biases the outcomes to younger ages of crust formation.

SUMMARY

The continental crust is a unique reservoir, and it retains the principal record of conditions on the Earth for at least the last 4.4 Ga. The oldest known rocks are approximately 4 Ga, and the record of the previous 500 Ma is isolated in tiny grains of detrital zircons. Thus, the development of in-situ

analytical techniques for high-precision analyses of isotope ratios and trace element abundances has revolutionized crustal-growth studies. Zircons tend to crystallize from relatively differentiated magmas, and that in turn biases the record as perceived through the zircon archive. Heat production considerations indicate that in the first 500 Ma of Earth history, mafic crust may remelt to form more felsic crust once it is 20 to 40 km thick, and that the latter may only reach thicknesses of approximately 10 km before it too is unstable (Kamber et al. 2005). Thus most of the early crust is likely to have been mafic in composition, consistent with estimates of the Lu/Hf ratios of the early proto-crust from the Hf-isotope data (Fig. 2.9).

Sediments and igneous rocks offer different perspectives on the evolution of the continental crust. Igneous rocks reflect the episodes in which new crust may have been generated, whereas sediments are mixtures from different source regions available at the time of deposition. Whole-rock analyses of sediments provide hybrid model ages that are difficult to interpret. However, the relative contributions of different source terrains can be independently determined using zircons to constrain the source regions, and their relative contributions, in individual sediment samples. The combination of Nd and Hf isotopes allows investigation of the links between bulk sediments and their source rocks, and values for the erosion factor 'K' can now be determined directly from river systems. This in turn allows more quantitative models to be developed linking the geological records of igneous and sedimentary rocks of the continental crust.

A striking feature of the continental crust is that its history seems to be marked by distinct episodes of strong magmatic activity, reflected in the peaks of zircon crystallization ages (Fig. 2.7). In some cases, these peaks are of the ages of rocks that represent new continental crust (Condie 1998). Such peaks would be difficult to ascribe to changes in plate tectonic regimes, and so they have been taken as evidence that in such periods crust generation was associated with the emplacement of deep-seated mantle plumes. An alternative interpretation is that the peaks in crystallization ages are themselves an artefact of preservation. They are associated with periods of supercontinent formation (Campbell and Allen 2008), when magmatic rocks may be preserved preferentially enveloped within a supercontinent. New continental crust generated along destructive plate margins is generated at similar rates to which material is returned to the mantle by subduction (Scholl and von Huene 2007, 2009), and so the overall preservation potential of destructive plate-margin magmas appears to be low. Reconciling this observation with the composition of the bulk continental crust, which is dominated by subduction-related minor and trace-element characteristics, is the subject of considerable current debate. The bulk composition displays features that are associated with subduction zones (fluid release from the subducted slab), and others that reflect fractionation within the crust that is then eroded and returned to subduction zones in subducted sediment (Plank 2005). These are manifest in different trace-element ratios. The role of sediments may be critical, as they can also be melted in settings with better preservation potential for the resultant granitic magmas than many destructive plate margins.

Less than 10% of the crustal rocks exposed are older than 2.5 Ga, and yet 50% of the continental crust may have stabilized by that time. The late Archaean marks the transition from a period of uniformly poor preservation potential to one in which the geological record appears to be biased by the tectonic setting in which the rocks were formed. The episodic patterns of crystallization ages are largely based on the records of igneous and sedimentary rocks selectively preserved in 'stable' areas and are therefore biased by the formation of supercontinents. The key now is to explore the geological records from stages with 'poor preservation' potential.

ACKNOWLEDGEMENTS

We thank Peter Cawood, Tony Prave, Anna Pietranik and Horst Marschall for many discussions on the generation and evolution of the continental crust, and Russell Harmon and Andrew Parker for their helpful comments on the manuscript. This work was supported by the NERC (NE/E005225/1), a Royal Society Merit Award to CJH, an NERC Fellowship (NE/D008891/1) to CS, and an Australian Research Council Fellowship (DP077094) to AISK.

FURTHER READING

Armstrong, RL. (1991) The persistent myth of crustal growth. *Australian Journal of Earth Sciences*, **38**: 613–30.

Brown, M and Rushmer, T. (2006) Evolution and Differentiation of the Continental Crust Cambridge University Press. ISBN-13: 9780521066068 550pp.

DePaolo, DJ. (1980) Crustal growth and mantle evolution: inferences from models of element transport and Nd and Sr isotopes. *Geochimica et Cosmochimica Acta*, **44**: 1185–96.

Hamilton, WB. (2007) Earth's first two billion years The era of internally mobile crust. In: R.D. Hatcher, Jr, Carlson, MP, McBride, JH and Martínez Catalán, JR. (eds.). 4-D Framework of Continental Crust. *Geological Society of America Memoir* **200**, 233–296.

Rollinson, H. (2008) Secular evolution of the continental crust: Implications for crust evolution models. *Geochemistry Geophysics Geosystems*, **9**, doi:10.1029/2008GC002262.

Stern, RJ. (2008) Neoproterozoic crustal growth: The solid Earth system during a critical episode of Earth history. *Gondwana Research*, **14**: 33–50.

Taylor, SR and McLennan, SM. (1985) The continental crust: its composition and evolution. Blackwell, Malden.

REFERENCES

Albarède, F. (1998) The growth of continental crust. *Tectonophysics*, **296**: 1–14.

Allègre, CJ. (1982) Chemical geodynamics. *Tectonophysics*, **81**: 109–32.

Allègre, CJ and Rousseau, D. (1984) The growth of continents through geological time studied by Nd isotope analysis of shales. *Earth and Planetary Science Letters*, **67**: 19–34.

Amelin, Y, Lee, D-C, Halliday, AN and Pidgeon, RT. (1999) Nature of the Earth's earliest crust from hafnium isotopes in single detrital zircons. *Nature*, **399**: 252–55.

Amelin, Y, Lee, D-C and Halliday, AN. (2000) Early middle Archaean crustal evolution deduced from Lu–Hf and U–Pb isotopic of single zircon grains. *Geochimica et Cosmochimica Acta*, **64**: 4205–4225.

Arculus, RJ. (1999) Origins of the continental crust. *Journal and Proceedings of the Royal Society of New South Wales*, **132**: 83–110.

Arndt, NT and Goldstein, SL. (1989) An open boundary between lower continental crust and mantle: its role in crust formation and crustal cycling. *Tectonophysics*, **161**: 201–212.

Barth, M, McDonough, WF and Rudnick, RL. (2000) Tracking the budget of Nb and Ta in the continental crust. *Chemical Geology*, **165**: 197–213.

Bowring, SA and Williams, IS. (1999) Priscoan (4.00–4.03) orthogneisses from northwestern Canada. *Contributions to Mineralogy and Petrology*, **134**: 3–16.

Brown, M. (2007) Metamorphic conditions in orogenic belts: A record of secular change. *International Geology Review*, **49**: 193–234.

Cawood, PA, Nemchin, AA, Freeman, M and Sircombe, K. (2003) Linking source and sedimentary basin: Detrital record of sediment flux along a modern river system and implications for provenance studies. *Earth and Planetary Science Letters*, **210**: 259–68.

Campbell, IH and Allen, CM. (2008) Formation of supercontinents linked to increases in atmospheric oxygen. *Nature Geoscience*, **1**: 554–8.

Cavosie, AJ, Valley, JW and Wilde, SA. (2005) Magmatic $\delta^{18}O$ in 4400–3900 Ma detrital zircons: a record of the alteration and recycling of crust in the Early Archaean. *Earth and Planetary Science Letters*, **235**: 663–81.

Chappell, BW and White, AJR. (1974) Two contrasting granite types. *Pacific Geology*, **8**: 173–4.

Chappell, BW and Stephens, WE. (1988) Origin of infracrustal (I-type) granite magmas. *Transactions of the Royal Society of Edinburgh: Earth Sciences*, **79**: 71–86.

Chappell, BW and White, AJR. (1992) I- and S-type granites in the Lachlan Fold Belt. *Transactions of the Royal Society of Edinburgh: Earth Sciences*, **83**: 1–26.

Cherniak, DJ, Hanchar, JM and Watson, EB. (1999) Diffusion of tetravalent cations in zircon. *Contributions to Mineralogy and Petrology*, **127**: 383–90.

Clift, P and Vannuchi, P. (2004) Controls on tectonic accretion versus erosion in subduction zones: Implications for the origin and recycling of the continental crust. *Review of Geophysics*, **42**: 1–31.

Compston, W and Pidgeon, RT. (1986) Jack Hills, evidence of more very old detrital zircons in Western Australia. *Nature*, **321**: 766–9.

Compston, W, Williams, IS and Meyer, C. (1984) U–Pb geochronology of zircons from Lunar breccia 73217 using a sensitive high mass-resolution ion microprobe. *Journal of Geophysical Research*, **89**: Supplement: B525–B534.

Condie, KC. (1993) Chemical composition and evolution of the upper continental crust: Contrasting results from surface samples and shales. *Chemical Geology*, **104**: 1–37.

Condie, KC. (1998) Episodic continental growth and supercontinents: a mantle avalanche connection? *Earth and Planetary Science Letters*, **163**: 97–108.

Condie, KC. (2005) Earth as an Evolving Planetary System. Elsevier, Amsterdam.

Condie, KC and Chomiak, B. (1996) Continental accretion: Contrasting Mesozoic and Early Proterozoic tectonic regimes in North America. *Tectonophysics*, **265**: 101–26.

Condie, KC, Belousova, E, Griffin, WL and Sircombe, KN. (2008) Granitoid events in space and time: Constraints from igneous and detrital zircon age spectra. *Gondwana Research*, **15**: 228–42.

Dalziel, IWD. (1992) On the organization of American plates in the Neoproterozoic and the breakout of Laurentia. *GSA Today*, **2**: 238–41.

Davidson, JP and Arculus, RJ. (2006) The significance of Phanerozoic arc magmatism in generating continental crust. In: M Brown and T Rushmer (eds.) Evolution and Differentiation of the Continental Crust, Cambridge University Press, pp. 135–72.

Davies, GF. (1995) Punctuated tectonic evolution of the Earth. *Earth and Planetary Science Letters*, **136**: 363–79.

Dhuime, B, Hawkesworth, CJ, Storey, C and Cawood, PA. From sediments to their source rocks:Hf and Nd isotopes in recent river sediments *Geology (in press)*.

Eiler, JM, Schiano, P, Kitchen, N and Stolper, EM. (2000) Oxygen-isotope evidence for recycled crust in the sources of mid-ocean-ridge basalts. *Nature*, **403**: 530–34.

Ellam, RM and Hawkesworth, CJ. (1988) Is average continental crust generated at subduction zones? *Geology*, **16**: 314–317.

Feng, R, Machado, N and Ludden, J. (1993) Lead geochronology of zircon by LaserProbe-Inductively Coupled Plasma Mass Spectrometry (LP-ICPMS). *Geochimica et Cosmochimica Acta*, **57**: 3479–86.

Foster, GL and Carter, A. (2007) Insights into the patterns and locations of erosion in the Himalaya – A combined fission-track and in situ Sm–Nd isotopic study of detrital apatite. *Earth and Planetary Science Letters*, **257**: 407–18.

Fryer, BJ, Jackson, SE and Longerich, HP. (1993) The application of laser ablation microprobe-inductively coupled plasma-mass spectrometry (LAM-ICP-MS) to in situ U–Pb geochronology. *Chemical Geology*, **109**: 1–8.

Fryer, BJ, Jackson, SE and Longerich, HP. (1995) The design, operation and role of the Laser-Ablation Microprobe coupled with an Indutively Coupled Plasma Mass Spectrometer (LAM-ICP-MS). *The Canadian Mineralogist*, **33**: 303–12.

Gao, S, Rudnick, RL, Yuan, H-L, Liu, X-M, Liu, Y-S, Xu, W-L, Ling, W-L, Ayers, J, Wang, X-C and Wang, Q-H. (2004) Recycling lower continental crust in the North China craton. *Nature*, **432**: 892–7.

Garrels, RM and MacKenzie, FT. (1971) Evolution of Sedimentary Rocks. Norton, WW & Co., New York.

Gastil, G. (1960) The distribution of mineral dates in time and space. *American Journal of Science*, **258**: 775–84.

Goldstein, SJ and Jacobsen, SB. (1988) Nd and Sr isotopic systematics of river water suspended material: implications for crustal evolution. *Earth and Planetary Science Letters*, **87**: 249–65.

Gray, DR and Foster, DA. (2004) Tectonic evolution of the Lachlan Orogen, southeast Australia: historical review, data synthesis and modern perspectives. *Australian Journal of Earth Sciences*, **51**: 773–818.

Gurnis, M and Davies, GF. (1985) Simple parametric models of crustal growth. *Journal of Geodynamics*, **3**: 105–135.

Gurnis, M and Davies, GF. (1986). Apparent episodic crustal growth arising from a smoothly evolving mantle. *Geology*, **14**: 396–9.

Harrison, TM, Blichert-Toft, J, Muller, W, Albarède, F, Holden, P and Mojzsis, SJ. (2005) Heterogeneous Hadean Hafnium: Evidence of Continental Crust at 4.4 to 4.5 Ga. *Science*, **310**: 1947–50.

Harrison, TM, Watson, EB and Aikman, AB. (2007) Temperature spectra of zircon crystallization in plutonic rocks. *Geology*, **35**: 635–38.

Hawkesworth, CJ and Kemp, AIS. (2006a) The Differentiation and Rates of Generation of the Continental Crust. *Chemical Geology*, **226**: 134–43.

Hawkesworth, CJ and Kemp, AIS. (2006b) Evolution of the Continental Crust. *Nature*, **443**: 811–817.

Hawkesworth, CJ and Kemp, AIS. (2006c), Using hafnium and oxygen isotopes in zircons to unravel the record of crustal evolution. *Chemical Geology*, **226**: 144–62.

Hawkesworth, C, Kelley, S, Turner, S, le Roex, A and Storey, B. (1999) Mantle processes during Gondwana break-up and dispersal. *Journal of African Earth Sciences*, **28**: 239–61.

Hawkesworth, C, Cawood, P, Kemp, T, Storey, C and Dhuime, B. (2009) A Matter of Preservation. *Science*, **323**: 49–50.

Hawkesworth, CJ, Dhuime, B, Pietranik, AB, Cawood, PA, Kemp, AIS and Storey, CD. (2010) The Generation and Evolution of the Continental Crust. *Journal of the Geological Society, London*, **167**: 228–48.

Hinthorne, JR, Andersen, CA, Conrad, RL and Lovering, JF. (1979) Single-grain [207]Pb/[206]Pb and U/Pb age determinations with a 10-μm spatial resolution using the ion microprobe mass analyzer (IMMA). *Chemical Geology*, **25**: 271–303.

Hirata, T and Nesbitt, R. (1995) U–Pb isotope geochronology of zircon: Evaluation of the laser probe-inductively coupled plasma mass spectrometry technique. *Geochimica et Cosmochimica Acta*, **59**: 2491–500.

Hofmann, AW, Jochum, KP, Seufert, M and White, WM. (1986) Nb and Pb in oceanic basalts: new constraints on mantle evolution. *Earth and Planetary Science Letters*, **79**: 33–45.

Hopkins M, Harrison, TM and Manning, CE. (2008) Low heat flow inferred from >4 Gyr zircons suggests Hadean plate boundary interactions. *Nature*, **456**: 493–6.

Horn, I, Rudnick, R and McDonough, WF. (2000) Precise elemental and isotope ratio determination by simultaneous solution nebulization and laser ablation-ICP-MS: application to U–Pb geochronology. *Chemical Geology*, **164**: 281–301.

Hurley, PM and Rand, JR. (1969) Pre-drift continental nuclei. *Science*, **164**: 1229–42.

Jacobsen, SB. (1988) Isotopic constraints on crustal growth and recycling. *Earth and Planetary Science Letters*, **90**: 315–29.

Jackson, SE, Longerich, HP, Dunning, GR and Fryer, BJ. (1992) The application of laser ablation microprobe; inductively coupled plasma mass spectrometry (LAM ICP MS) to in situ trace element determinations in minerals. *Canadian Mineralogist*, **30**: 1049–64.

Kamber, BS, Whitehouse, MJ, Bolhar, R and Moorbath, S. (2005) Volcanic resurfacing and the early terrestrial crust: Zircon U–Pb and REE constraints from the Isua Greenstone Belt, southern West Greenland. *Earth and Planetary Science Letters*, **240**: 276–90.

Kay, RW and Kay, SM. (1991) Creation and destruction of lower continental crust. *Geologische Rundschau*, **80**: 259–78.

Kemp, AIS and Hawkesworth, CJ. (2003) Granitic perspectives on the generation and secular evolution of the continental crust. In: RL Rudnick (ed.) The Crust. Treatise on Geochemistry, **3**: 349–410.

Kemp, AIS, Hawkesworth, CJ, Paterson, BA and Kinny, PD. (2006) Episodic growth of the Gondwana supercontinent from hafnium and oxygen isotope ratios. *Nature*, **439**: 580–3.

Kemp, AIS, Hawkesworth, CJ, Foster, GL, Paterson, BA, Woodhead, JD, Hergt, JM, Gray, CM and Whitehouse, MJ. (2007a) Magmatic and crustal differentiation history of granitic rocks from hafnium-oxygen isotopes in zircon. *Science*, **315**: 980–3.

Kemp, AIS, Shimura, T and Hawkesworth, CJ. (2007b) Linking granulites, silicic magmatism, and crustal growth in arcs: Ion microprobe (zircon) U–Pb ages from the Hidaka metamorphic belt, Japan. *Geology*, **35**: 807–10.

Kemp, AIS, Foster, GL, Schersten, A, Whitehouse, MJ, Darling, J and Storey, C. (2009a) Concurrent Pb-Hf isotope analysis of zircon by laser ablation multi-collector ICP-MS, with implications for the crustal evolution of Greenland and the Himalayas. *Chemical Geology*, **261**: 244–60.

Kemp, AIS, Hawkesworth, CJ, Collins, WJ, Gray, CM, Blevin, PL and EIMF (2009b) Nd, Hf and O isotope evidence for rapid continental growth during accretionary orogenesis in the Tasmanides, eastern Australia. *Earth and Planetary Science Letters*, **284**: 455–66.

Kemp, AIS, Wilde, SA, Hawkesworth, CJ, Coath, CD, Nemchin, A, Pidgeon, RT, Vervoort, JD and DuFrane A. (2010) Hadean crustal evolution revisited: new constraints from Pb-Hf isotope systematics of the Jack Hills zircons. *Earth and Planetary Science Letters*, **296**: 45–56.

Kempton, PD and Harmon, RS. (1992) Oxygen isotope evidence for large-scale hybridization of the lower crust during magmatic underplating. *Geochimica et Cosmochimica Acta*, **56**: 971–86.

Kesler, SE and Wilkinson, BH. (2006) The Role of Exhumation in the Temporal Distribution of Ore Deposits. *Economic Geology*, **101**: 919–22.

King, EW, Valley, JW, Davis, DW and Edwards, GR. (1998) Oxygen isotope ratios of Archaean plutonic zircons from granite-greenstone belts of the Superior Province: indicator of magmatic source. *Precambrian Research*, **92**: 47–67.

Kramers, JD. (2002) Global modeling of continent formation and destruction through geological time and implications for CO_2 drawdown in the Archaean. In: Fowler, CMR, Ebinger, CJ and Hawkesworth, CJ. (eds.). The early Earth: physical, chemical and biological development. Geological Society of London, Special Publications, **199**: 259–74.

Kramers, JD. (2007) Hierarchical Earth accretion and the Hadean Eon. *Journal of the Geological Society*, **164**: 3–17.

Kramers, JD and Tolstikhin, IN. (1997) Two terrestrial lead isotope paradoxes, forward transport modelling, core formation and the history of the continental crust. *Chemical Geology*, **139**: 75–110.

Kuno, H. (1968) Origin of andesite and its bearing on the Island arc structure. *Bulletin of Volcanology*, **32**: 141–76.

Lustrino, M. (2005) How the delamination and detachment of lower crust can influence basaltic magmatism. *Earth-Science Reviews*, **72**: 21–38.

McCulloch, MT and Bennett, VC. (1994) Progressive growth of the Earth's continental crust and depleted mantle: geochemical constraints. *Geochimica et Cosmochimica Acta*, **58**: 4717–38.

McFarlane, CRM and McCulloch, MT. (2007) Coupling of in-situ Sm–Nd systematics and U–Pb dating of monazite and allanite with applications to crustal evolution studies. *Chemical Geology*, **245**: 45–60.

Michard, A, Gurriet, P, Soudant, M and Albarède, F. (1985) Nd isotopes in French Phanerozoic shales: external vs. internal aspects of crustal evolution. *Geochimica et Cosmochimica Acta*, **49**: 601–10.

Nance, WB and Taylor, SR. (1976) Rare earth element patterns and crustal evolution-I. Australian post-Archaean sedimentary rocks. *Geochimica et Cosmochimica Acta*, **40**: 1539–51.

Nance, WB and Taylor, SR. (1977) Rare earth element patterns and crustal evolution-II. Archaean sedimentary rocks from Kalgoorlie, Australia. *Geochimica et Cosmochimica Acta*, **41**: 225–31.

Nuttman, AP. (2006) Comments on 'Zircon thermometer reveals minimum melting conditions on earliest Earth' *Science*, **311**: 779b.

O'Nions, RK, Hamilton, PJ and Hooker, PJ. (1983) A Nd isotope investigation of sediment related to crustal development in the British Isles. *Earth and Planetary Science Letters*, **63**: 229–40.

Patchett, JP. (1983) Importance of the Lu-Hf isotopic system in studies of planetary chronology and chemical evolution. *Geochimica et Cosmochimica Acta*, **47**: 81–91.

Patchett, PJ and Tatsumoto, M. (1980) Hafnium isotope variations in oceanic basalts. *Geophysical Research Letters*, **7**: 1077–80.

Patchett, JP, Kouvo, O, Hedge, CE and Tatsumoto, M. (1981) Evolution of continental crust and mantle heterogeneity: evidence from Hf isotopes. *Contributions to Mineralogy and Petrology*, **78**: 279–97.

Peck, WH, Valley, JW and Graham, CM. (2003) Slow diffusion rates of O isotopes in igneous zircons from metamorphic rocks. *American Mineralogist*, **88**: 1003–14.

Pietranik, AB, Hawkesworth, CJ, Storey, CD, Kemp, AIS, Sircombe, KN, Whitehouse, MJ and Bleeker, W. (2008) Episodic, mafic crust formation from 4.5 to 2.8 Ga: New evidence from detrital zircons, Slave craton, Canada. *Geology*, **36**: 875–8.

Plank T. (2005) Constraints from thorium/lanthanum on sediment recycling at subduction zones and the evolution of continents. *Journal of Petrology*, **46**: 921–44.

Reymer, A and Schubert, G. (1984) Phanerozoic addition rates to the continental crust and crustal growth. *Tectonics*, **3**: 63–77.

Rudnick, RL. (1995) Making continental crust. *Nature*, **378**: 573–8.

Rudnick, RL and Fountain, DM. (1995) Nature and composition of the continental crust: a lower crustal perspective. *Reviews of Geophysics*, **33**: 267–309.

Rudnick, RL and Gao, S. (2003) Composition of the continental crust. In: RL Rudnick (ed.) The Crust. Treatise on Geochemistry, **3**: 1–64.

Scholl, DW and von Huene, R. (2007) Crustal recycling at modern subduction zones applied to the past-issues of growth and preservation of continental basement crust, mantle geochemistry, and supercontinent reconstruction. In: RD Hatcher, MP Carlson, JH McBride and JRM Catalán (eds.) 4-D Framework of Continental Crust. *Geological Society of America Memoir* **200**: 9–32.

Scholl, DW and von Huene, R. (2009) Implications of estimated magmatic additions and recycling losses at the subduction zones of accretionary (non-collisional) and collisional (suturing) orogens. In: Cawood, PA and Kröner, A. (eds.) Earth Accretionary Systems in Space and Time. Geological Society, London, Special Publications **318**: 105–25.

Stein, M and Hofmann, AW. (1994) Mantle plumes and episodic crustal growth. *Nature*, **372**: 63–8.

Storey, BC. (1995) The role of mantle plumes in continental break-up: case histories from Gondwanaland. *Nature*, **377**: 301–8.

Sun, S-s and McDonough, WF. (1989) Chemical and isotopic systematics of oceanic basalts: implications for mantle composition and processes. In: Saunders, AD. & Norry, MJ. (eds.) Magmatism in the Ocean Basins. Geological Society Special Publication, **42**: 313–45.

Taylor, SR and McLennan, SM. (1985) The continental crust: its composition and evolution. Blackwell, Malden.

Taylor, SR, McLennan, SM. (1991) Sedimentary rocks and crustal evolution: Tectonic setting and secular trends. *Journal of Geology*, **99**: 1–21.

Tolstikhin, I and Kramers, JD. (2008) The Evolution of Matter. Cambridge University Press, New York.

Valley, JW. (2003) Oxygen isotopes in zircon. *Reviews in Mineralogy and Geochemistry*, **53**: 343–80.

Vazquez, JA and Reid, MR (2004) Probing the Accumulation History of the Voluminous Toba Magma. *Science*, **305**: 991–4.

Vervoort, JD and Blichert-Toft, J. (1999) Evolution of the depleted mantle: Hf isotope evidence from juvenile rocks through time. *Geochimica et Cosmochimica Acta*, **63**: 533–66.

Vervoort, JD, Patchett, PJ, Blichert-Toft, J and Albarède, F. (1999) Relationships between Lu-Hf and Sm-Nd isotopic systems in the global sedimentary system. *Earth and Planetary Science Letters*, **168**: 79–99.

Watson, EB and Harrison, TM. (1983) Zircon saturation revisited: temperature and composition effects in a variety of crustal magma types. *Earth and Planetary Science Letters*, **64**: 295–304.

Watson, EB and Harrison, TM. (2005) Zircon thermometer reveals minimum melting conditions on earliest Earth. *Science*, **308**: 841–4.

Wiechert, U and Hoefs, J. (1995) An excimer laser-based micro analytical preparation technique for in-situ oxygen isotope analysis of silicate and oxide minerals, *Geochimica et Cosmochimica Acta*, **59**: 4093–101.

Wilde, SA, Valley, JW, Peck, WH and Graham, CM. (2001) Evidence from detrital zircons for the existence of continental crust and oceans on the Earth 4.4 Gyr ago. *Nature*, **409**: 175–8.

Zandt, G, Gilbert, H, Owens, TJ, Ducea, M, Saleeby, J and Jones, CH. (2004) Active foundering of a continental arc root beneath the southern Sierra Nevada in California. *Nature*, **431**: 41–6.

3 Discovering the History of Atmospheric Oxygen

University of Pennsylvania, Philadelphia, PA, USA

ABSTRACT

In the summer of 1771 Joseph Priestley put a sprig of mint into a quantity of air and found that the 'goodness' of the air had been improved. He had discovered photosynthesis. Antoine Lavoisier identified the gas that Priestley had made, and named it oxygen. The heroic decade of the 1770s closed with the demonstration by Jean Senebier and Jean Ingenhousz' that light is required for the generation of oxygen by plant photosynthesis. In the 1840s Jean-Jacques Ebelmen documented much of the geochemical cycle of oxygen. However, writings about the geologic history of oxygen remained highly speculative until 1927, when AM Macgregor linked the history of atmospheric oxygen to the geologic rock record. He concluded that atmospheric oxygen was either absent or present at very low levels during the Archaean. This proposition has been confirmed by a large number and variety of observations. It is now generally, though not universally, accepted that significant quantities of oxygen first appeared in the atmosphere ca. 2.4 Ga, and that oxygen levels have risen somewhat irregularly since then. The progressive oxygenation of the atmosphere is due to the evolution of green-plant photosynthesis, and probably also to gradual changes in the composition of volcanic and hydrothermal inputs to the atmosphere.

Frontiers in Geochemistry: Contribution of Geochemistry to the Study of the Earth, First edition. Edited by Russell S. Harmon and Andrew Parker. © 2011 Blackwell Publishing Ltd. Published 2011 by Blackwell Publishing Ltd.

INTRODUCTION

The search for the history of atmospheric oxygen has long occupied astronomers, biologists, palae-ontologists, geologists and, particularly, geochemists. There is now general, though not universal, agreement on the broad outlines of the history of O_2. However, many important aspects of the redox state of the ancient atmosphere remain to be clarified, and there is no consensus regarding the processes that have driven the progressive oxidation of the atmosphere. The search for answers to the outstanding questions is an active field of research, because atmospheric oxygen is central to so many aspects of Earth history. This chapter offers an abbreviated account of the history of the field. It ends with some ideas and observations which may help to shape its future.

JOSEPH PRIESTLEY'S SPRIG OF MINT

Joseph Priestley (1733–1804) was most remarkable. He was a Dissenting minister, a steadfast champion of that hypothesis respecting the Divine nature which is termed Unitarianism by its friends, and Socialism by its foes. 'Regardless of all odds he was ready to do battle with all-comers in that cause, and if no adversaries entered the lists, he would sally forth to meet them' (Huxley 1881). In addition to his duties as a minister, he carried forward innumerable scientific experiments. Most of these were directed toward finding and characterizing new 'airs.' Many of the results of his researches were published in six large volumes. Their contents are described in Robert Schofield's biography (1997, 2004), and by Steven Johnson (2008) in a more popular vein. Priestley was also politically active, a good friend of Benjamin Franklin, a champion of American liberties and a defender of the French Revolution.

Priestley spent the years 1767–73 in Leeds as minister to one of the Dissenting congregations in that bustling town. In his scientific endeavours he was concerned with the consumption of air by 'fires of all kinds, volcanoes, etc.' and wondered 'what provision there is in nature for remedying the injury which the atmosphere receives by this means' (Priestley 1781: 43–4). He found the restorative remedy in plants (Priestley 1781, pp. 52–4):

Accordingly, on the 17th of August 1771, I put a sprig of mint into a quantity of air, in which a wax candle had burned out, and found that, on the 27th of the same month, another candle burned perfectly well in it. This experiment I repeated, with the least variation in the event, not less than eight or ten times in the remainder of the summer. Experiments made in the year 1772 abundantly confirmed my conclusion concerning the restoration of air, in which candles had burned out by plants growing in it. The first of these experiments was made in the month of May; and they were frequently repeated in that and the two following

months, without a single failure. This remarkable effect does not depend upon any thing peculiar to mint, which was the plant that I always made use of till July 1772; for on the 16th of that month, I found a quantity of this kind of air to be perfectly restored by sprigs of balm, which had grown in it from the 7th of the same month. That this restoration of air was not owing to any aromatic effluvia of these two plants, not only appeared by the essential oil of mint having not sensible effect of this kind; but from the equally complete restoration of this vitiated air by the plant called groundsel, which is usually ranked among the weeds, and has an offensive smell. This was the result of an experiment made the 16th of July, which the plant had been growing in the burned air from the 8th of the same month. Besides, the plant which I have found to be the most effectual of any that I have tried for this purpose is spinach, which is of quick growth, but will seldom thrive long in water. One jar of burned air was perfectly restored by this plant in four days, and another in two days. This last was observed on the 22d of July.

Priestley had discovered photosynthesis. He did not realize the importance of sunlight in this process, and he did not understand the mechanism by which plants improve the 'goodness of the air.' The role of sunlight was discovered later in the 1770s by J Ingenhousz (1780) and by J Senebier (1782). Antoine Lavoisier identified oxygen as the product of photosynthesis and the source of the 'goodness of the air.' This discovery marked the beginning of modern chemistry (Lavoisier 1789).

Near the end of Priestley's first volume of *Experiments and Observations on Different Kinds of Air* (1781), there is a section entitled 'Queries, Speculations, and Hints' (pp. 258–86). Priestley comments at the outset of this section that he is apprehensive lest, after being considered a dry experimenter, he should now be considered a visionary theorist. Nevertheless, he launched into a long theoretical discourse. This contains some fine theorizing. He proposed that volcanoes may be responsible for the origin of our atmosphere as well as of all solid land, that the fixed air (carbon dioxide) of the atmosphere may be 'supplied by volcanos from the vast masses of calcareous matter lodged in the earth, together with inflammable air (hydrogen).' He ventured that a part of the carbon dioxide in the atmosphere may be supplied from the fermentation of vegetables, and that at present 'as fast as it is precipitated and imbibed by one process, it may be set loose by others.' He was not sure 'whether there be, on the whole, an increase or a decrease in the general mass of the atmosphere ...' but 'imagines that it rather increases.' He had launched the study of the history of the atmosphere.

JACQUES JOSEPH EBELMEN: THE GEOCHEMICAL CYCLE OF OXYGEN

Jacques Joseph Ebelmen (1814–52) is not a geochemical household name. His papers on the decomposition (weathering) of silicate minerals and rocks (Ebelmen 1845, 1847) were cited frequently during the 19th century, sparsely during the first decade of the 20th century, and apparently not at all thereafter, until they were discovered a few years ago by Berner and Maasch (1996). Ebelmen did a great deal of work on the chemistry of weathering, and used his results to define all the major processes by which CO_2 and O_2 are added to and lost from the atmosphere. Table 3.1 summarizes his findings. They were rediscovered independently during the second half of the 20th century to the great satisfaction of Holland (1978), Garrels and Lerman (1984) and other members of the geochemical fraternity.

Ebelmen was a French chemist and mining engineer. He became a Professor at the École des Mines and the Director of the Sèvres Porcelain

Table 3.1 Processes controlling the levels of carbon dioxide and oxygen in the atmosphere according to Ebelmen. Carbon dioxide is substituted here in place of the term carbonic acid as used by Ebelmen for atmospheric CO_2 (translated from the summary table of Ebelmen 1845; addition in italics by Berner and Maasch 1996).

I Causes which tend to raise the proportion of carbon dioxide contained in air
   (a) Without diminution of the proportion of oxygen
   (b) With diminution of the proportion of oxygen
      (i) Destruction of organic matter contained in sedimentary rocks (oil shales, lignites, coal)
      (ii) Weathering of iron pyrites and some iron minerals.

II Causes which tend to diminish the proportion of carbon dioxide
   (a) With the liberation of oxygen
      (i) Formation of iron pyrites
      (ii) Formation of mineralized combustibles (coal, etc.) and the conservation (burial) of organic debris
   (b) With or without the absorption of oxygen
      (i) Weathering of silicates of igneous rocks.

III Causes which tend to raise the proportion of oxygen contained in the air
   (a) Without diminishing the proportion of carbon dioxide
      (i) None
   (b) In lowering the proportion of carbon dioxide
      (i) Formation of mineralized combustibles (coal, etc.)
      (ii) Formation of iron pyrites.

IV Causes that produce a diminution of oxygen contained in the air
   (a) With formation of carbon dioxide
      (i) The destruction (oxidation) of organic matter contained in the ground
      (ii) The weathering of iron pyrites and some iron minerals
   (b) With the absorption of carbon dioxide
      (i) Weathering of silicates in igneous rocks.

works. He was well known for his work on met-
allurgy, ceramics, mineral synthesis and the
chemistry of blast furnaces. His brilliant career
was cut short by illness and death when he was
only 37 years old. His geochemical papers deal
only peripherally with the history of the atmosphere. He concluded his first paper (Ebelmen
1845) with the following propositions [author's
translation]:

*Nevertheless, several circumstances tend to prove
that in ancient geologic periods the atmosphere was
denser and richer in carbon dioxide, and perhaps in
oxygen, than in the present age. A heavier gaseous
envelope is associated with a greater condensation of
solar heat and with atmospheric phenomena of greatly
enhanced intensity. The changes in the nature of the
atmosphere were undoubtedly in constant rapport
with the organisms which lived in each of these
periods. Has the composition of our atmosphere arrived
at a permanent equilibrium? It will undoubtedly
require several centuries to solve this problem. The
analytical techniques which we now possess are sufficiently precise to allow us to bequeath to future gen-
erations some certain parts of the answer to that
important question.*

These propositions go a step beyond Priestley, but
only a small one. In his now classic monograph
The Data of Geochemistry, FW Clarke (1911)
assembled a large amount of geochemical information, most of it dating from the second half of
the 19th century. He rarely deviated from his
major aim, and he apologized mildly for discours-
ing about the primitive atmosphere, for positing
that 'a few words as to the origin of our atmosphere may not be out of place.' (Clarke 1911: 52)
He pointed out that:

*Upon this subject much has been written, especially in
recent years; but none of the widely variant theories so
far advanced can be regarded as conclusive. The
problem, indeed, is one of cosmology, and chemical
data supply only a single line of attack. Physical, astro-
nomical, mathematical, and geological evidence must
be brought to bear upon the question, before anything
like an intelligent conclusion can be reached. Even
then, with every precaution taken, we can hardly be
sure that our fundamental premises are sound.*

He was surely right. The first major steps toward
defining the evolution of the atmosphere were
still nearly two decades in the future.

ALEXANDER MIERS MACGREGOR: A LOOK AT THE ROCK RECORD

In 1927 AM Macgregor, a distinguished geologist
of the Southern Rhodesia Geological Survey, began
his justly famous paper on the *Problem of the
Precambrian Atmosphere* by pointing out that
'Certain types of rocks which occur among the
early Precambrian Basement Schists of Southern
Rhodesia appear to the writer to be most readily
explained by the supposition that they were
formed beneath an atmosphere deficient in oxygen,
and otherwise radically different from that of the
present day' (Macgregor 1927). He observed that
many of his Archaean rocks were unusually car-
bonatized, and concluded that this observation
was best explained if the atmosphere was 'relatively enriched in carbon dioxide.' Three sets of
observations convinced him that the Archaean
atmosphere was 'relatively or completely deficient in free oxygen.' All three have been confirmed and extended by more recent observations.
He had found that 'the Archaean sediments in
Rhodesia are characterized by an abundance of
ferrous carbonate in the form of siderite (chalybite)
or of iron-bearing calcite' in areas where reduction
by organic matter was out of the question. Analyses
of a greyish Archaean arkose typical of the
Ndutiana grits, a dark grey quartzose greywacke,
and arkoses at the Shamna mine, showed a heavy
preponderance of ferrous oxide over ferric oxide.
'Not one of the analyses gave indication of any
oxidation having taken place during the formation
of these rocks. On the other hand, reduction is
suggested in the case of some.'

These observations have been confirmed abundantly by analyses of pre-2.3 Ga shales. The data
in Fig. 3.1 show that the weight ratio of Fe_2O_3 to
FeO in all the analyses is less than or equal to 1.0;
their average Fe_2O_3/FeO ratio is well below that
of average igneous rocks, perhaps because, as
Macgregor suggested, 'magnetite may have been

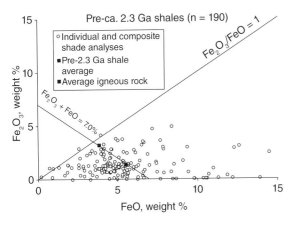

Fig. 3.1 The FeO and Fe_2O_3 content of pre-2.3 Ga shales (after Bekker et al., 2003a, b, in prep.).

separated from the other detritus' during transport and sedimentation 'while ferrous iron precipitated from solution has been added to that present in the silt.' (Macgregor 1927). Macgregor then addressed the knotty problem of the transport of iron that accumulated in the banded iron formations (BIF) that are 'beyond question the most characteristic of the Precambrian rocks. Under present conditions silica is only slightly soluble, and iron is thrown out of solution by oxidation almost as soon as it is exposed to the atmosphere or to oxygenated water, and there is no accepted explanation as to how these substances were ever carried in solution from one place to another' in the Archaean. He pointed out that Gruner's proposition of iron transport in 'peaty waters' was unlikely to be correct, but that 'if it is assumed that the atmosphere was devoid or nearly devoid of oxygen and was rich in carbon dioxide, the difficulties of the problem become almost insignificant.' That is surely true. However, research on banded iron formations since 1927 has shown that much of the iron was delivered to the oceans via hydrothermal black smoker vents (Beukes and Gutzmer 2008) and that the fate of the iron was determined by the oxidation state of the oceans rather than by that of the atmosphere.

The deeper parts of the ocean can be anoxic even when the atmosphere is oxygenated. This observation explains the formation of banded iron formations during the late Proterozoic. These BIFs are highly oxidized, because Fe^{2+} that was introduced into the anoxic deep ocean was oxidized to Fe^{3+} in shallow waters and was precipitated there dominantly as haematite (Klein and Beukes 1993; Klein and Ladeira 2004). However, Macgregor's comments still apply to the mineralogy of Archaean shallow-water banded iron formations. The shallow-water facies of the 2.49 ± 0.03 Ga Griquatown IF were deposited on a ca. 800×800 km shelf (Beukes and Klein 1990). Several of the members of the Griquatown IF can be traced across this shallow platform into its lagoonal, near-shore part. There the mineralogy of the iron formation is dominated by Fe^{2+} minerals. This can only be explained if the O_2 content of the ambient atmosphere was extremely low (Holland 2004).

The third line of evidence cited by Macgregor as indicative of little or no oxygen in the Archaean atmosphere was based on the observation that some of the pebbles in Rhodesian Archaean conglomerates contain pyrite. He pointed out that this is true although 'that mineral is absent from the matrix of the rock, nor is there any iron oxide which might be derived from pyrite by the process of weathering' (Macgregor 1927). The pyrite may, of course, be due to subsequent differential mineralization, but the specimens do not suggest this, and it seems unlikely to have happened in the three distinct cases which he cited. 'If, however, the pyrite was present in the pebbles, which in all cases are well-rounded and must have travelled considerable distances before their deposition, it is clear evidence of relative freedom from oxidizing conditions.' This observation has turned out to be particularly fruitful.

HAROLD UREY, STANLEY MILLER AND THE ORIGIN OF LIFE

Speculations regarding the origin of life go back at least as far as Biblical times. The first modern approach was A.E. Oparin's (1894–1980) book *Origin of Life* (Oparin 1924). Oparin started his story with a very hot Earth. On cooling, masses of lava containing iron and carbides reacted with superheated steam in the primordial atmosphere to produce hydrocarbons. Unsaturated hydrocarbons formed first. These reacted with oxygen to produce alcohols, aldehydes, ketones and organic acids, compounds which were capable of further transformations. Nitrogen, combined with metals as nitrides, reacted with superheated steam to form ammonia, a compound that could also have formed by the condensation of hydrogen and nitrogen in the upper layers of the atmosphere.

When the temperature of the atmosphere dropped below 100°C, water rained out, and the atmospheric organic compounds dissolved in this early ocean. Hydroxy and amino derivatives of hydrocarbons combined to create ever-longer and more complicated compounds, including carbohydrates and proteins. These became the foundation of life. The continued growth of complicated particles led to the formation of colloidal solutions in water, and then to precipitates, coagula and gels. The precipitated gels were the first substances to acquire structure, and organic compounds were transformed into individual organic bodies. 'With certain reservations we can even consider that first piece of organic slime ... as being the first organism' (Oparin 1924). The bits of slime grew by absorbing nutrients dissolved in the oceans. They were frequently broken up by waves or in the surf, and gave rise to varieties of new 'primitive organisms.' With time, the physicochemical structure of the gels improved, and new properties arose, among them the ability to metabolize. Among gels that developed this ability a 'fierce struggle for existence, a fight to the death' ensued for the supply of nutrient organic material. Ultimately, the only organisms that survived were those that ate their weaker comrades or that could nourish themselves on very simple inorganic compounds.

Oparin offered several disclaimers. He pointed out that his explanation was not unique, that little was known about the structure of colloidal gels, and that even less was known about the physicochemical structure of protoplasm. But he was sure that 'this ignorance of ours is certainly only temporary. ... Very soon the last barrier between the living and the dead will crumble under the attack of patient work and powerful scientific thought.' He had essentially outlined the research program which occupied him for the rest of his life. *Origin of Life* went unnoticed internationally. In Russia it was not mentioned in Vladimir Vernadsky's important book *The Biosphere* (1926). *Origin of Life* was translated into English in 1938, but the reception of Oparin's views was lukewarm even then, in part because they were not backed by experimental data.

The next important chapter was written by Harold Urey (1893–1981) and Stanley Miller (1930–2007). Miller entered the University of Chicago as a graduate student in 1951. He was enormously impressed by Harold Urey's lectures on the early history of the Earth, particularly with the proposition that the early atmosphere was highly reducing, that it consisted largely of methane, ammonia (or nitrogen), hydrogen and water vapour (Urey 1952), and that these gases could have reacted together to produce organic compounds. Urey pointed out that few experiments had been conducted to test the reality of such potential prebiotic syntheses. Miller approached Urey with the proposal to do prebiotic synthesis experiments, sparking mixtures of methane, ammonia and hydrogen in an apparatus with circulating water vapour. Permission was given somewhat grudgingly, but the results were spectacular. Glycine and several other amino-acids were synthesized within a week (Miller 1953, 1955). Subsequent experiments yielded a score of different amino-acids as well as hydroxy-acids and urea. The synthesis of life seemed to be within reach, perhaps only a few years away.

It has not turned out that way. The notion of a highly-reduced early atmosphere ran into difficulties almost immediately. W.W. Rubey (1955) made a strong case for the view that the atmosphere is the result of the gradual degassing of the Earth, potentially modified by the recycling of its components. He rejected the notion of a dense, early, reduced atmosphere, because he saw no evidence for such an atmosphere in the geologic record. I provided somewhat of a reprieve for a Miller–Urey atmosphere by pointing out that a highly reduced atmosphere could have existed very early in Earth history when metallic iron was still present in the upper mantle. Volcanic gases would then have been H_2-rich, and CH_4 and NH_3 could have been present in the atmosphere (Holland 1962). However, this gave only slight comfort. Metallic iron probably left the upper mantle to form the core within a few tens of millions years. It may have been replenished by the infall of meteorites, but if the origin of life required a highly reduced atmosphere, life probably began very early, during a very inhospitable period of Earth history. A good case can now be made for a mildly reducing atmosphere during much of the Archaean (see below). The synthesis of organic compounds in such an atmosphere is much less efficient than in a Miller–Urey mixture, but it may have been sufficient to provide the basic building blocks of life.

PAUL RAMDOHR AND MANFRED SCHIDLOWSKI'S URANINITE MUFFINS

The Witwatersrand Basin of South Africa has produced more than 48,000 tons of gold, some 35% of all the gold ever mined (Robb 2005), and worth ca. $1400 billion at 2008 gold prices. The bankets of the Basin also contain uraninite (UO_2). Although the region is much less famous for its uranium than for its gold, the production of uranium is by no means negligible; it accounts for about 10% of the known world production of uranium. The origin of the gold, the uraninite, and the accompanying pyrite has been warmly debated since their discovery. Are they placer accumulations, were they precipitated from late hydrothermal solutions or are they, as seems likely today, placer accumulations that were redistributed by later hydrothermal solutions (see, for instance, Robb 2005). The classic 1958 paper by P Ramdohr (1890–1985) (Ramdohr, 1958) on the shape, size, textures and distribution of these minerals in the

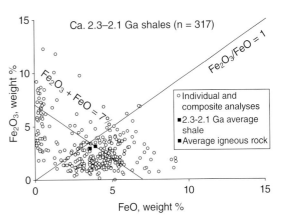

Fig. 3.2 The FeO and Fe_2O_3 content of shales deposited between 2.3 and 2.1 Ga (after Bekker et al., 2003a, b, in prep.).

Witwatersrand ores showed that much of the uraninite and pyrite is detrital. His observations were confirmed and extended by M Schidlowski (1966). Many of his uraninite grains had been shaped into small 'muffins' (Fig. 3.3) by mechanical abrasion during transport and deposition.

The Witwatersrand sediments were deposited between approximately 2.7 and 2.9 Ga. The youngest conglomeratic uranium ores containing placer uraninite are the 2.45 Ga sediments of the Elliot Lake-Blind River district in Southern Canada. The occurrence of detrital uraninite in sediments older than 2.4 Ga is of interest in reconstructing the oxygen content of the atmosphere, because UO_2 is oxidized rapidly in today's atmosphere and occurs only sporadically in modern stream sediments where erosion is extraordinarily rapid (Holland 1984). D Grandstaff's experiments (Grandstaff 1976) on the kinetics of the dissolution of uraninite have been confirmed by S Ono's (2002) measurements and have been used to set an approximate limit on the O_2 content of the atmosphere before 2.45 Ga (Grandstaff 1980; Holland 1984). These calculations are uncertain, because the assumptions that have to be made are not firmly based. The best estimate of the maximum O_2 pressure based on the available uraninite dissolution data is $10^{-2.4}$ atm (Holland 1984). This is nearly two orders of magnitude lower than today, but the calculations indicate that the preservation of uraninite in the Witwatersrand bankets did not require an oxygen-free atmosphere.

It would seem strange if Archaean sandstones did not also preserve other oxygen-sensitive minerals. Rasmussen and Buick (1999) have shown that the expected preservation did indeed occur. Heavy minerals in 3250–2750 Ma fluvial siliciclastic sediments from the Pilbara Craton of Australia that have never undergone significant alteration include pyrite, uraninite, gersdorffite (NiAsS) and siderite(FeCO$_3$). Locally, siderite is a major constituent (to 90%) of the heavy-mineral population, with grains displaying evidence for several episodes of erosion, rounding and subsequent authigenic overgrowth. Siderite is rare in Phanerozoic sandstones, because it is easily oxidized. Its frequent survival of prolonged transport in well-mixed and therefore well-aerated Archean river waters that contained little organic matter imply that the contemporary atmosphere was much less oxidizing than at present. Ohmoto's (1999) criticisms of the Rasmussen and Buick (1999) paper were shown to be baseless and irrelevant by Rasmussen et al. (2002). The presence of readily oxidized minerals as detrital constituents of Archean sandstones and conglomerates offers excellent evidence in support of the proposition that the Archaean atmosphere was much less oxidizing than today's.

PRESTON CLOUD'S RED BEDS

Preston Cloud (1912–91) was one of a small group of extraordinary individuals who established the field of Archaean and Proterozoic palaeontology. He also made landmark contributions to our knowledge of the pre-Phanerozoic Earth. He insisted that traditional palaeontology must be married to geology and geochemistry, because early biological history needed to be understood in the context of the developing planetary surface (Knoll 1990). In one of his seminal papers Cloud pointed out that 'available records indicate that the oldest thick and extensive red beds are about 1.8 to 2 aeons (Ga) old – a little younger than or overlapping slightly with the youngest of the major banded-iron formations. The evidence of

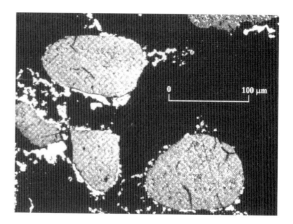

Fig. 3.3 Photomicrograph of detrital grains of uraninite, UO$_2$, (oil immersion at 375x magnification) with characteristic dusting of galena, PbS, partly surrounded by lead sulphide overgrowths from the basal reef footwall in the Loraine Gold Mines, South Africa (source Schidlowski, 1966). The large grain displays a typical 'muffin shape'.

lithospheric evolution suggests that this may mark the time of atmospheric evolution when free O$_2$ began to accumulate' (Cloud 1968). The suggestion was important. It defined the end of the low or no-O$_2$ period of atmospheric history and the beginning of the high-O$_2$ period. The timing of the transition has been revised somewhat. The oldest known red beds have an age of 2.3 Ga. The major increase in the Fe$_2$O$_3$/FeO ratio of shales occurred between 2.3 and 2.1 Ga (see Fig. 3.2).

My early estimate of the transition time of the atmosphere to a highly oxygenated state (Holland 1962) was based on the persistence of detrital uraninite in conglomerates. This date was poorly constrained until the discovery of the uranium ore deposits at Oklo in Gabon, which set an upper age limit on the end of the period of low atmospheric O$_2$ period (Gauthier-Lafaye and Weber 1989). These ores occur in a series of deltaic plain sediments, which are directly overlain by organic-rich shales. The ores are clearly not of placer origin. The highest-grade ores occur where fracturing is well developed. Uranium in the ores was released by oxidizing solutions from minerals in the Francevillian sands and conglomerates. It was reduced by contact with organic matter, and was precipitated as uraninite. The ores are well-dated at 2.05 Ga (Gancarz 1978). They are the oldest of the type of hydrothermal ore deposits that contain the majority of the world's uranium reserves. The evolution of the geology of uranium ores therefore brackets the rise of atmospheric O_2 between 2.45 Ga and 2.05 Ga, a range that is consistent with the evidence from the appearance of major red beds and the increase in the Fe_2O_3/FeO ratio in shales.

PALAEOSOLS

The large changes in the oxidation state of the atmosphere that were inferred above should have left their mark in the history of soils. Intense oxidation, particularly in present-day tropical soils should have been absent in Archaean and earliest Palaeoproterozoic soils. This has turned out to be the case. The summary diagram of the evolution of the oxidation state of the atmosphere that has been extracted from ancient soil profiles agrees remarkably well with that derived from all of the other lines of evidence.

The path leading to Fig. 3.4 has been rather long and arduous. Few palaeosols (ancient soils) are preserved in the geologic record, and most of those that are preserved lost their top portions to erosion prior to burial. In addition, the chemical evolution of soils tends to be complicated by biological processes as well as by post-soil-forming events. The most successful studies of palaeosols in search of atmospheric evolution have been those that concentrated on palaeosols developed on basalts, because these contain a good deal of FeO and tend

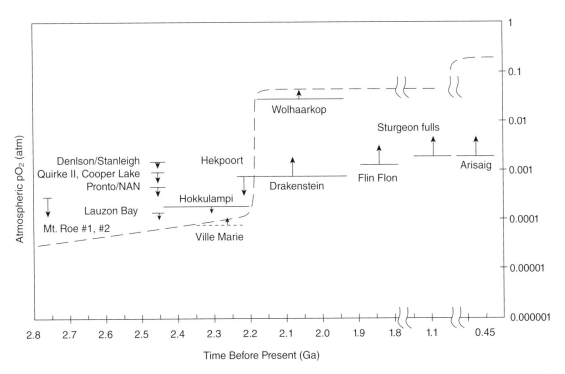

Fig. 3.4 Partial pressure of atmospheric oxygen during the last 2.8 billion years. The maximum or minimum P_{O_2} value was estimated on the basis of the mobility of Fe during weathering of the 15 definite palaeosols. Palaeosols that lost Fe during weathering have a downward-pointing arrow in the figure, indicating that atmospheric P_{O_2} was equal to or less than the value depicted. Palaeosols that retained Fe during weathering have an upward-pointing arrow, indicating that atmospheric P_{O_2} was equal to or greater than the value depicted. The dashed curve is the approximate trajectory of atmospheric P_{O_2} as constrained by a variety of data (Rye and Holland, 1998).

to be uniform in composition. Both attributes are important when using the theoretical framework developed during the last two decades for interpreting the mineralogy and the chemical composition of palaeosols (Holland and Zbinden 1988; Pinto and Holland 1988; Kasting et al. 1985).

All the palaeosols which we have studied that are younger than 2.0 Ga are highly oxidized (see Fig. 3.4). Their original FeO content has been oxidized to Fe_2O_3 and has been largely or entirely retained in the palaeosols during and since their formation. The palaeosols of age equal to or greater than 2.45 Ga are reduced. FeO in their upper portions was not oxidized during soil formation. Much of their original iron content was lost, presumably to soil water. Some was used to generate Fe^{2+}-rich montmorillonite near the base of the palaeosols. This mineral was converted to high-Fe^{2+} chlorite plus quartz during the mild metamorphism to which all of these palaeosols have been subjected. In most areas where it is exposed, the chemistry of the very extensive ca. 2.25 Ga Hekpoort palaeosol in South Africa looks much like its older counterparts.

However, in Strata I, a drill core through this palaeosol in Botswana, an iron-deficient zone is overlain by a Fe^{3+}-rich zone. This is followed upwards by a sedimentary red-bed sequence. As expected from its age, the Hekpoort palaeosol developed during the transition to the strongly oxygenated atmosphere. The highly oxidized uppermost layer of the palaeosol in Strata I shows that O_2 was present in the atmosphere. The lower, iron-depleted horizon indicates that O_2 levels were still quite low. Beukes et al. (2002) have pointed out that this profile is similar to that of modern laterites, and have proposed that the palaeosol developed under present day atmospheric O_2 levels. However, a detailed analysis of this proposition (Yang and Holland 2003) has found it to be unsatisfactory.

At the time of this writing, the Archaean palaeosol record is still quite sparse. Fortunately a potential palaeosol was discovered recently in the Coonterunah Group of Western Australia (Altinook 2006). This may extend the palaeosol record back to 3.46 Ga.

METALS IN CARBONACEOUS SHALES

Carbonaceous black shales are well known for their abnormally high concentrations of metals (Vine and Tourtelot 1970). During the 1970s, measurements of the concentration of trace metals in seawater became much more precise. By the end of that decade, it was clear that the concentration of many of the metals that are concentrated in black shales is roughly proportional to their concentration in seawater, and that seawater is the major source of these metals in carbonaceous shales (Holland 1979). Rivers are the dominant source of most metals for the oceans. The river flux of many metals depends directly or indirectly on the oxidation state of the atmosphere. Elements such as Mo, V and U are essentially immobile during weathering in the absence of O_2. Elements such as Cu, Zn and Pb, which are partially present as constituents of sulphides in igneous rocks, are liberated during weathering when enough O_2 is present to oxidize sulfides to sulphates. The concentration of these metals in black shales, therefore, offers a way to estimate the O_2 content of the atmosphere in times past.

The first attempt to apply this technique (Holland 1984) relied heavily on the composition of the carbonaceous shales of the Outokumpu region in Finland. These were deposited between 2.0 and 1.9 Ga. Peltola's analyses of the shales showed that the redox-sensitive elements are present in concentrations that are considerably in excess of those in average shale (Peltola 1960; 1968). While V, Pb and Zn are only modestly enriched, Mo, U and Se are strongly enriched in these shales. The pattern of enrichment is identical to that found in Phanerozoic black shales. This observation is consistent with the palaeosol evidence and with the history of red beds: by the time the Outokumpu black shales were deposited, the atmosphere already contained significant quantities of O_2.

A large amount of analytical data for the composition of Precambrian black shales has accumulated since 1984 (Fig. 3.5). Some of the results were summarized by Yang and Holland (2002) and Holland (2004) (Fig. 3.5). A much larger data set

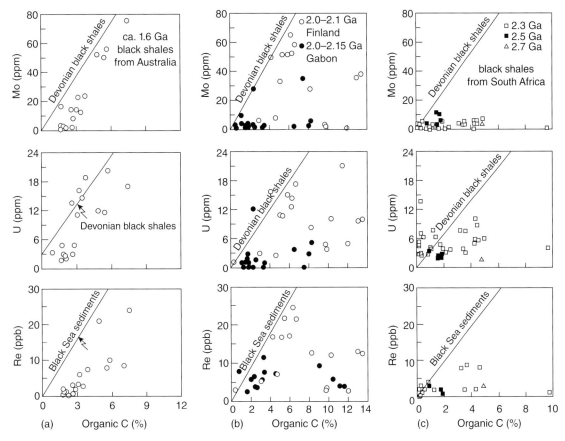

Fig. 3.5 Variations of Mo, U, and Re trace-element concentrations as a function of organic content of carbonaceous shales from South Africa deposited at 2.7 Ga (open triangles), 2.5 Ga (filled squares), and 2.3 Ga (open squares) – right column; from the Outokumpu region of Finland deposited between 2.0 and 1.9 Ga (open circles) and from Gabon deposited between 2.15 and 2.0 Ga (filled circles) – centre column; and from Australia deposited at ca. 1.6 Ga (open circles) – left column. The reference lines shown in the diagrams is the trace metal – organic carbon relationship for either Devonian black shales (Mo and U) or Black Sea sediments (Holland, 2004).

was obtained by Yamaguchi (2002). Essentially all the black-shale analyses are consistent with the proposition that atmospheric O_2 levels were very low until ~2.4 Ga. With very few exceptions, only carbonaceous shales younger than 2.4 Ga are enriched in redox-sensitive metals. The major exception is the 2.50 Ga McRae shale in the Pilbara region of Australia. Some parts of this formation are significantly enriched in Mo, though not in U (Anbar et al. 2007). This anomaly may be due to the presence of extensive oxygen oases in the oceans just before the rise of atmospheric O_2 (Kaufman et al. 2007).

MARK THIEMENS, JAMES FARQUHAR AND THE MASS INDEPENDENT FRACTIONATION OF THE SULPHUR ISOTOPES

By the end of the last century, all but one of the techniques that have proved useful for estimating past O_2 levels in the atmosphere were well-developed. A new and very powerful tool was added in 2000 AD: the mass-independent fractionation (MIF) of the sulphur isotopes. Thiemens (1999) had pointed out the widespread occurrence of MIF in oxygen and had concluded that 'there are many atmospheric oxygen and sulphur-bearing species that remain to be measured that might have their atmospheric roles clarified by isotopic analysis.' This conjecture was verified almost immediately. During the following year, Farquhar, Bao and Thiemens (2000) reported the discovery of very significant degrees of MIF in the sulphur isotopes of Archaean sulphides and sulphates. No significant MIF was found in the sulphur of Phanerozoic and Proterozoic minerals.

The source of the MIF effects is almost certainly the impact of solar ultraviolet radiation on atmospheric SO_2 in the absence or near-absence of O_2. MIF effects disappear when the O_2 content of the atmosphere is greater than a few parts per million. The presence of an MIF signal in sulphide and sulphate minerals, therefore, indicates that O_2 levels during the Archaean were less than a few ppm. The absence of MIF signals since approximately 2.45 Ga suggests that O_2 levels have been greater than a few ppm during all of the Phanerozoic and Proterozoic. These conclusions were greeted with great enthusiasm by all those who had been convinced that O_2 levels were very low in the Archaean atmosphere and that the O_2 level rose rapidly between 2.5 and 2.0 Ga. The conclusions were greeted with disbelief and dismay by those who maintained that atmospheric levels have changed very little during the past 3.5 to 4.0 Ga (Ohmoto et al. 2006).

A very large number of measurements (Fig. 3.6) have been added to those published by Farquhar et al. (2000). The new data confirm the conclusions drawn on the basis of the first paper. The difference between the pre-2.5 Ga and the post-2.5 Ga data is striking. Ohmoto et al. (2006) have pointed out that the MIF signals between 3.2 and 2.8 Ga are small, and have proposed that O_2 levels may have followed a 'yo-yo pattern' during the Archaean. However, the analysis of the MIF signal in $\delta^{36}S$ as well as in $\delta^{33}S$ suggests very strongly

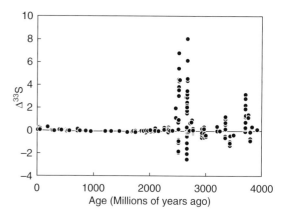

Fig. 3.6 Plot of $\Delta^{33}S$ values, indicator of mass-independent fractionation of sulphur extracted from pyrite and barite, versus sample age for the past 4 Ga (updated from Farquhar et al., 2000, Holland, 2006 and Hofmann et al., 2009). Variable values of $\Delta^{33}S$ values for samples older than ca. 2.5 Ga are considered to indicate the influence of a low-O_2 atmosphere on the sulphur cycle, whereas samples younger than 2.5 Ga with negligible $\Delta^{33}S$ values are considered to indicate a sulphur cycle dominated by oxidative weathering of continental sulphide and sulphate.

that the small MIF signals between 3.2 and 2.8 Ga are real, and that they demand very low atmospheric O_2 levels during this time period (Farquhar and Johnston 2008) unless mechanisms other than the effect of solar UV on SO_2 were responsible for their presence in the minerals deposited during this period. The possible existence of such mechanisms was explored by Lasaga et al. (2008), and Watanabe et al. (2009) have found that an MIF signal of 2.1‰ was generated during the reaction of powdered aminoacids with sulphate at 170 to 200°C. This discovery is important for understanding the aqueous processes that lead to MIF of the sulphur isotopes, but it seems unlikely that they have been geologically important. The sharp change in the geologic MIF signal close to 2.5 Gs, and the virtual absence of the signal since 2.0 Ga, are readily explained by a rise in the level of atmospheric oxygen. There is at present no viable alternative explanation.

THE HISTORY OF
ATMOSPHERIC OXYGEN

The atmospheric history of O_2 was summarized by (Kerr 2005) in a New Focus article in *Science*. He pointed out that the data of Farquhar et al. (2000) provided 'an unequivocal method for determining the presence or absence of oxygen in Earth's history. ... That discovery – now buttressed by theoretical work and studies of other rocks – pretty much clinches the case for a late "Great Oxidation Event".' However, Kerr (2009) pointed out that the experiments of Watanabe et al. (2009) 'challenge the mainstream scenario by showing how supposed signs of an early lack of oxygen could have come from unrelated geochemical reactions.' Nevertheless, after polling a number of leading geochemists, he concluded that, 'The Great Oxidation Event will likely reign a while longer.' A rough indication of the history of atmospheric oxygen since the 'Great Oxidation Event' is shown in Fig. 3.7. As described earlier, O_2 levels rose significantly between 2.4 and 2.0 Ga, red beds appeared, the chemistry of iron in palaeosols changed, detrital redox-sensitive minerals disappeared, uranium ore deposits that require the presence of oxygenated ground water appeared, and the MIF signal of sulphur in sulphides and sulphates disappeared. Evaporites containing gypsum/anhydrite made their appearance during this period, but the isotopic composition of sulphur in pyrite apparently remained close to magmatic values, indicating that the geochemical cycle of sulphur had not yet become modern, except perhaps briefly between 2.20 and 2.05 Ga.

There are still no precise O_2 barometers. The best estimate for atmospheric O_2 pressure at 2.0 Ga is between 10% and 20% PAL, i.e. between 0.02 and 0.04 atm. The long period between 2.0 and 0.8 Ga seems to have been a period of O_2 stasis, which has earned it the sobriquet the boring billion. This long period was followed by 300 Ma of intense and rapid change. Several major ice ages occurred, perhaps the most severe ever. By 500 Ma, the sulphur cycle had become modern, and the concentration of SO_4^2 in seawater had reached present-day levels. There is compelling

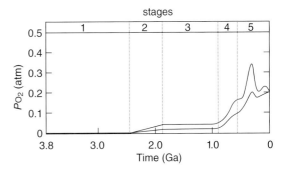

Fig. 3.7 The probable evolution of oxygen in the Earth's atmosphere (Holland, 2006) from 3.8 Ga to the present. The two curves represent the likely upper and lower bounds of the atmospheric P_{O_2} value. The four stages shown in the figure (Stage 1: 3.85–2.45 Ga, Stage 2: 2.45–1.85 Ga, Stage 3: 1.85–0.85 Ga, Stage 4: 0.85–0.54 Ga, Stage 5: the last 0.54 Ga) indicate the major phases of development of the atmosphere. During Stage 1, atmospheric O_2 was almost certainly less than a few parts per million, except possibly during the period between 3.0 and 2.8 Ga. Atmospheric O_2 rose during the 'Great Oxygenation Event'. Its value at 2.0 Ga is still poorly constrained; it was probably higher than 0.02 atm but significantly lower than 0.2 atm. Stage 3 was a time of static conditions, with no evidence for large changes in the O_2 content of the atmosphere. Stage 4 saw a steady rise in the level of atmospheric O_2 and the level of atmospheric O_2 during Stage 5 may have passed through a maximum during the Carboniferous before returning to its present value of 0.2 atm, perhaps along a rather irregular path.

evidence for a rise in atmospheric O_2 between 800 and 550 Ma, which probably played a major role in the burst of biologic evolution at the end of the Proterozoic era. The O_2 content of the atmosphere during the Phanerozoic, the last 542 Ma of Earth history, has been rather variable. The isotopic composition of carbon in limestones together with gigantism in the insect population suggests that O_2 levels reached a major peak during the Carboniferous. A more modest peak probably occurred during the Cretaceous (Berner 2004). There are no direct measurements of atmospheric P_{O_2} until the Pleistocene. The O_2 content of gas bubbles in amber offers a potential source of reasonably direct P_{O_2} measurements (Berner and Landis, 1987, 1988). However, the preliminary data have been criticized severely on the grounds that the rate of diffusion of gases through amber is so rapid that the present-day composition of air bubbles in this material is unlikely to be helpful in reconstructing their original composition. I believe that the matter is still unresolved and deserves further study.

WHY DID THE ATMOSPHERE BECOME OXYGENATED?

There is now general, though not universal, agreement regarding the major features of the history of atmospheric oxygen. However, there is no agreement about the reasons for the progressive oxygenation of the atmosphere. Kump and Barley (2007) have ascribed the change in the oxidation state of the atmosphere 2.5 Ga to an increase in subaerial volcanism. Claire et al. (2006) have proposed an increase in the oxygen fugacity of gases evolved during the metamorphism of crustal rocks. Sleep et al. (2004) have invoked the effects of the decrease in the proportion of ultramafic rocks. Bjerrum and Canfield (2004) have proposed that the fraction of volcanic CO_2 which was reduced to organic matter was much less than previously proposed. These explanations tend to be incomplete, and they do not seem to account for the full sweep of the oxygenation history of the atmosphere as satisfactorily as a recent theory advanced by the author (Holland 2009).

The oxidation state of the atmosphere depends quite strongly on the fate of hydrogen in volcanic gases (Holland 2002). Today, a part of this hydrogen is used (indirectly) to reduce volcanic CO_2 to organic matter. The remainder of the volcanic H_2 is used (again indirectly) to reduce volcanic SO_2 to FeS_2 (pyrite). Not enough H_2 is available in present-day volcanic gases to reduce the entire inventory of volcanic SO_2 to FeS_2; the remaining SO_2 is oxidized to SO_3 and is removed from the atmosphere-ocean system, largely as a constituent of gypsum ($CaSO_4 \cdot 2H_2O$) and anhydrite ($CaSO_4$). The highly oxygenated state of the

present atmosphere is due in large part to the intense photosynthetic activity of the biosphere and to the requirements of mass balance between the volcanic and hydrothermal gas inputs to the atmosphere and their removal into the solid Earth Fig. 3.8.

The atmosphere could be made reducing by increasing the flux of volcanic H_2, while maintaining the flux of the carbon and sulphur gases and the fraction of CO_2 that is reduced to organic matter at their present level. If the H_2 input were raised, so that it exceeded the level at which all of the volcanic SO_2 could be reduced to FeS_2, then the excess H_2 would accumulate in the atmosphere until its rate of escape into interplanetary space balanced the excess H_2, flux. This observation can be made into the basis of a model of atmospheric evolution. If the oxygen fugacity of volcanic gases were somewhat lower during the Archaean than today, their H_2/H_2O ratio was then higher, and the atmosphere could have contained a significant quantity of H_2. If volcanic gases have gradually become more oxygenated, the excess H_2 would have disappeared, and O_2 would have taken its place at the time of the Great Oxidation Event. This explanation (Holland 2002) became untenable when Canil (2002) and Lee et al. (2003) showed that the redox state of volcanic gases has changed very little since the early Archaean.

Fortunately there is another change in the composition of volcanic gases that has the same effect as a progressive decrease in the H_2 content: an increase in the content of CO_2 and SO_2. If the H_2/H_2O ratio of Archaean volcanic gases was the same as today, but the CO_2/H_2O and SO_2/H_2O ratios were sufficiently lower, there would have been enough H_2 to reduce 20% of the CO_2 to CH_2O, to reduce all the SO_2 to FeS_2, and for a remainder to supply the atmosphere with excess H_2. The rationale for proposing a gradual increase in the CO_2/H_2O and SO_2/H_2O ratio of volcanic gases is illustrated in the highly simplified representation of the Earth system in Fig. 3.8. Recycled gases dominate the CO_2, SO_2 and H_2O outputs of volcanoes today. Carbon and sulphur compounds in volcanic gases are removed from the atmosphere in large part as solids: carbon as a constituent of carbonates – mainly $CaCO_3$ and $CaMg(CO_3)_2$ – and organic matter (CH_2O), sulphur as a constituent of sulphides (mainly FeS_2) and sulphates (mainly $CaSO_4$ and $CaSO_4 \cdot 2H_2O$). The rate of recycling of these phases by metamorphism and magmatism during subduction depends on their concentration in the subducted materials. Their concentration in subducted materials was almost certainly low early in Earth history, and has increased with time as their quantity in the crust has increased.

The recycling rate of water follows a different pattern. The water content of subducted material depends largely on the water content of the subducted sediments and on the water content of the subducted oceanic crust. The total water-content of subducted materials has probably depended only weakly on the volume of the contemporary oceans. The progressive increase in the carbon and sulphur content of subducted material, coupled with a near-constancy of its water content, must have led to a progressive increase in the CO_2/H_2O and SO_2/H_2O ratio of volcanic gases generated by the release of these volatiles from the subducted material. It is proposed that this increase, together with the near-constancy of the H_2/H_2O ratio in volcanic gases and the continued photosynthetic reduction of CO_2 to CH_2O accounts for much of the progressive oxidation of the atmosphere.

A simple, first-order, quantitative model based on Fig. 3.8 has been developed for the cycling of carbon, sulphur, and water during Earth history (Holland 2009). The history of the CO_2/H_2O and the SO_2/H_2O ratio of volcanic gases that is predicted by this model agrees surprisingly well with the geologic record. The predicted values of the H_2 content of the atmosphere during the Archaean agree with those that seem to be required to explain Archaean climates, and the predicted CO_2 fluxes are consistent with the CO_2 fluxes that were needed to generate Archaean sediments. The timing of the crossover from an H_2-dominated to an O_2-dominated atmosphere is consistent with the record of the mass-independent fractionation of sulphur in sulphide and sulphate minerals. The changes in the geochemical cycle of Fe, C and S

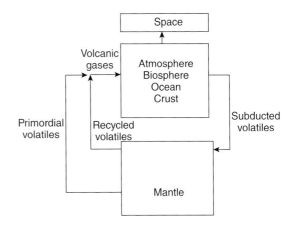

Fig. 3.8 Simple model for the cycling of carbon, sulfur, and water during Earth history.

during the early Proterozoic Great Oxidation Event are consistent with the predictions of the model, as is the modernization of the sulphur cycle by the end of the Proterozoic. However, the model is quite incomplete. It includes few of the complexities of the geologic record, but its successes encourage the hope that it can be used as the basis of a more complete explanation of the history of atmospheric oxygen.

ACKNOWLEDGEMENTS

The author wishes to record his indebtedness to Michael Mottl, James Kasting, Lee Kump, Roger Buick, Andrey Bekker, Robert Berner and David Eggler for many helpful discussions during the preparation of this paper, and to my wife Alice for all the typing.

REFERENCES

Altinook, E. (2006) Soil formation beneath the Earth's oldest known (3.46 Ga) unconformity. Abstracts with Programs, *Geol. Soc. Amer. Ann. Mtg.* **38**(7): 533.

Anbar, AD, Duan, Y, Lyons, TW, Arnold, GL, Kendall, G, Creaser, RA, Kaufman, AJ, Gordon, GW, Scott, C, Garvin, J and Buick, R. (2007) A whiff of oxygen before the Great Oxidation Event. *Science* **317**: 1903–6.

Bekker, A, Holland, HD, Young, GM and Nesbitt. HS. (2003a) Fe$_2$O$_3$/FeO ratio in average shale through time: a reflection of the stepwise oxidation of the atmosphere. Abstracts with Programs *Geol. Soc. Amer. Ann. Mtg.* **35**: 83.

Bekker, A, Holland, HD, Young, GM. and Nesbitt, HW. (2003b) The fate of oxygen during the early Paleoproterozoic carbon isotope excursion. *Astrobiology* **2**: 477.

Bekker, A, Holland, HD, Young, GM and Nesbitt, HW. The evolution of the Fe$_2$O$_3$/FeO ratio in shales, in preparation.

Berner, RA and Landis, GP. (1987) Chemical analysis of gaseous bubble inclusions in amber: the composition of ancient air. *Amer. Jour. Sci.* **287**: 757–62.

Berner, RA and Landis, GP. (1988) Gas bubbles in fossil amber as possible indicators of the major gas composition of ancient air. *Science* **239**: 1406–9.

Berner, RA and Maasch, KA. (1996) Chemical weathering and controls on atmospheric O$_2$ and CO$_2$: fundamental principles were enunciated by JJ Ebelmen in 1845. *Geochim. Cosmochim. Acta* **60**: 1633–7.

Berner, RA. (2004) The Phanerozoic Carbon Cycle: CO$_2$ and O$_2$. Oxford University Press. Oxford.

Beukes, NJ and Klein, C. (1990) Geochemistry and sedimentology of a facies transition from microbanded to granular iron-formation in the Early Proterozoic Transvaal Supergoup. South Africa. *Precambrian Res.* **47**: 99–139.

Beukes. NJ, Dorland, H, Gutzmer, J, Nedachi, M and Ohmoto. H. (2002) Tropical laterites, life on land, and the history of atmospheric oxygen in the Paleoproterozoic. *Geology* **30**: 491–4.

Beukes, NJ and Gutzmer, J. (2008) Origin and paleoenvironmental significance of major iron formations at the Archean-Paleoproterozoic boundary, *SEG Reviews* **15**: 5–47.

Bjerrum, CJ and Canfield, DE. (2004) New insights into the burial history of organic carbon on the early Earth, *Geochem. Geophys. Geosyst.* **5**: 2004 GC 000713.

Canil, D. (2002) Vanadium in peridotites, mantle redox and tectonic environments. *Earth Planet. Sci. Letters* **195**: 75–90.

Claire, MW, Catling, DC and Zahnle, KJ. (2006) Biogeochemical modelling of the rise in atmospheric oxygen. *Geobiology* **1–31**: DOI: 10. 1111/j. 1472–4669. 2006. 00084x

Clarke, FW. (1911) The Data of Geochemistry, 2nd Ed. Bull. 49, US Geological Survey, Government Printing Office, Washington DC.

Cloud, PE Jr. (1968) Atmospheric and hydrospheric evolution of the primitive Earth. *Science* **160**: 729–736.

Ebelmen, JJ. (1845) Sur les produits de la decomposition des especes minerales de la famille des silicates; *Annales des Mines, 4^c Series* **7**: 3–66.

Ebelmen, JJ. (1847) Sur la decomposition des roches; *Annales des Mines 4^c Series* **12**, 627–654.

Farquhar, J, Bao, HM and Thiemens, MM. (2000) Atmospheric influence of Earth's earliest sulfur cycle. *Science* **289**: 756–8.

Farquhar, J and Johnston, DT. (2008) The oxygen cycle of the terrestrial planets: insights into the processing and history of oxygen in surface environments. *Reviews in Mineralogy and Geochemistry* **68**: 463–92 Min. Soc. of America.

Gancarz, AJ. (1978) U-Pb age (2.05 × 10^9 years) of the Oklo deposit; pp. 513–520 in: The Natural Fission Reactors. Internat. Atomic Energy Agency, Vienna.

Garrels, RM and Lerman, A. (1984) Coupling of the sedimentary sulfur and carbon cycles: an improved model. *Amer. Jour. Sci.* **284**: 474–511.

Grandstaff, DE. (1976) A kinetic study of the dissolution of uraninite. *Econ. Geol.* **71**: 1493–1506.

Grandstaff, DE. (1980) Origin of uraniferous conglomerates at Elliott Lake, Canada and Witwatersrand, South Africa: Implications for oxygen in the Precambrian atmosphere. *Precamb. Res.* **13**: 1–26.

Gauthier-Lafaye, F. and Weber, F. (1989) The Francevillian (Lower Proterozoic) uranium ore deposits of Gabon. *Econ. Geol.* **84**: 2267–885.

Hofmann, A, Bekker, A, Rouxel, O, Rumble, D and Master, S. (2009) Multiple sulphur and iron isotope composition of detrital pyrite in Archean sedimentary rocks: A new tool for provenance analysis. *Earth and Planetary Science Letters*, **286**: 436–445.

Holland HD. (1962) Model for the evolution of the Earth's atmosphere; pp. 447–77: In Petrologic Studies: A Volume in Honor of AF Buddington. AE Engel, HL James and BF Leonard (eds.)

Holland, HD. (1978) The Chemistry of the Atmosphere and Oceans; Wiley-Interscience, New York.

Holland, HD. (1979) Metals in black shales: a reassessment. *Econ. Geol.* **74**: 1676–80.

Holland HD. (1984) The Chemical Evolution of the Atmosphere and Oceans; Princeton Univ. Press, Princeton, N.J.

Holland, HD and Zbinden, EA. (1988) Paleosols and the evolution of the atmosphere, Part I; pp. 61–82. In Physical and Chemical Weathering in Geochemical Cycles. A Lerman and M Meybeck (eds.) Kluwer Academic Publishers.

Holland, HD. (2002) Volcanic gases, black smokers and the Great Oxidation Event. *Geochim. Cosmochim. Acta* **66**: 3811–26.

Holland, HD. (2004) The geologic history of seawater; Chapter 6.21, pp. 583–625 in Vol. 6 of Treatise on Geochemistry. H Elderfield (vol. ed.), HD Holland and KK Turekian (exec. eds.) Elsevier Pergamon, Amsterdam.

Holland HD. (2006) The oxygenation of the atmosphere and oceans. *Phil. Trans. Roy. Soc. B* **361**: 903–15; doi:10. 1098/rstb. 2006. 1838.

Holland, HD. (2009) Why the atmosphere became oxygenated: A proposal. *Geochim. Cosmochim. Acta* **73**: 5241–55.

Huxley, TH. (1881) Joseph Priestley, An Address delivered on the occasion of the Presentation of a Statue of Priestley to the Town of Birmingham, on the 1st of August, 1874. Science and Culture and Other Essays, Macmillan, London.

Ingenhousz, J. (1780) Expériences Sur Les Végétaux; Didot, Paris.

Johnson, S. (2008) The Invention of Air; Riverhead Books, New York.

Kasting, JF, Holland, HD. and Pinto, JP. (1985) Oxidant abundances in rainwater and the evolution of atmospheric oxygen. *Jour. Geophys. Res.* **90**: 10, 497–10, 510.

Kaufman, AJ, Johnston, DT, Farquhar, J, Masterson, AL, Lyons, TW, Bates, S, Anbar, AD, Arnold, GL, Garvin, J. and Buick, R. (2007) Late Archean biospheric oxygenation and atmospheric evolution. *Science* **317**: 1900–1903.

Kerr, RA. (2005) The story of O$_2$. *Science* **308**: 1730–32.

Kerr, RA. (2009) Great oxidation event dethroned. *Science* **324**: 321.

Klein, C and Beukes, NJ. (1993) Sedimentology and geochemistry of the glaciogenic late Proterozoic Rapitan iron-formation in Canada. *Econ. Geol.* **88**: 542–65.

Klein, C and Ladeira, EA. (2004) Geochemistry and mineralogy of Neoproterozoic banded iron-formations and some selected, siliceous manganese formations from the Urucum District, Mato Grasso Do Sul, Brazil. *Econ. Geol.* **99**: 1233–44.

Knoll, AH. (1990) Editoral Introduction to Proterozoic Evolution and Environments; AH. Knoll and H. Ostrom. (eds) *Amer. Jour. Sci. 290-A*, V–VII.

Kump, LR and Barley, ME. (2007) Increased subaerial volcanism and the rise of atmospheric oxygen 2.5 billion years ago. *Nature* **448**: 1033–6.

Lasaga, AC, Otake, T, Watanabe, Y. and Ohmoto, H. (2008) Anomalous fractionation of sulfur isotopes during heterogeneous reactions. *Earth Planet Sci. Letters* **268**: 225–38.

Lavoisier, A. (1789) Traité Elémentaire de Chimie. Chez Cuchet, Paris.

Lee, C, Brandon, AD. and Norman, MD. (2003) Vanadium in peridotites as a proxy for paleo-f$_o$ during partial melting: Prospects, limitations and implications. *Geochim. Cosmochim. Acta* **67**: 3045–3064.

Macgregor, AM. (1927) The problem of the Precambrian atmosphere. *South Afr. Jour. Sci.* **24**: 155–72.

Miller, SL. (1953) A production of amino acids under possible primitive earth conditions. *Science* **117**: 528–9.

Miller, SL. (1955) Production of some organic compounds under possible primitive conditions. *Jour. Amer,. Chem. Soc.* **77**: 2351–61.

Ohmoto, H. (1999) Redox state of the Archean atmosphere from detrital heavy minerals in ca. 3250–2750 Ma sandstones from the Pilbara Craton, Australia; Comment. *Geology* **27**: 1151–2.

Ohmoto, H, Watanabe, Y, Ikemi, H, Poulson, SR and Taylor, BE. (2006) Sulphur isotope evidence for an oxic Archean atmosphere. *Nature* **442**: 908–911.

Ono, S. (2002) Uraninite and the early Earth's atmosphere: SIMS analyses of uraninite in the Elliot Lake District and the dissolution kinetics of natural uraninite. Unpublished Ph. D. Dissertation, Pennsylvania State University, State College, PA.

Oparin, A. (1924) Proiskhozhdenie Zhizny; (Origin of Life), Moscow.

Peltola, E. (1960) On the black schists of the Outokumpu region in eastern Finland. Bull. 192 de la Comm. Géol. de Finlande, Helsinki.

Peltola, E. (1968) On some geochemical features in the black schists of the Outokumpu area, Finland. *Bull. Geol. Soc. Finland* **40**: 39–50.

Pinto, JP and Holland, HD. (1988) Paleosols and the evolution of the atmosphere; Part II, pp. 21–34. In Paleosols and weathering through geologic time; Geological Society of America Special Paper 216, Reinhardt, J and Sigleo, WT. (eds.).

Priestley, J. (1781) Experiments and Observations on Different Kinds of Air; J. Johnson, London, Third Edition of Vol. 1.

Ramdohr, P. (1958) Die Uran-und Goldlagerstatten Witwatersrand-Blind River District-Dominion Reef-Sierra de Jacobina; Erzmikroskopische Untersuchungen und ein geologischer Vergleich; Abh. der Deutschen Akademie der Wissenschaften, *Berlin, Nr. 3*, 1–35.

Rasmussen, B and Buick, R. (1999) Redox state of the Archean atmosphere: evidence from detrital heavy minerals in ca. 3250–2750 Ma sandstones from the Pilbara Craton, Australia. *Geology* **27**: 115–118.

Rasmussen, B, Buick, R. and Holland, HD. (2002) Redox state of the Archean atmosphere: evidence from detrital heavy minerals in ca. 3250–2750 sand-stones from the Pilbara Craton, Australia. *Geology* **27**: 152.

Robb, L. (2005) Introduction to Ore-Forming Processes; Blackwell, Malden, MA., Oxford, U.K.

Rubey, WW. (1955) Development of the hydrosphere and atmosphere, with special reference to probable composition of the early atmosphere; pp. 631–650. In A. Poldervaart (ed.) Crust of the Earth. Geol. Soc. Amer. Special Paper 62.

Rye, R and Holland, HD. (1998) Paleosols and the evolution of atmospheric oxygen: A critical review. *Amer. Jour. Sci.* **298**: 621–72.

Schidlowski, M. (1966) Beitrage zur Kenntnis der radioaktiven Bestandteile der Witwatersrand-Konglomerate. I. Uranpecherz in den Konglomeraten des Oranje-Freistaat-Goldfeldes; Neues Jb. Mineral. *Abh.* **105**: 183–202.

Schofield, RE. (1997) The Enlightenment of Joseph Priestley; A Study of His Life and Work from 1733 to 1773. The Pennsylvania State University Press, University Park, PA.

Schofield, RE. (2004) The Enlightened Joseph Priestley; A Study of His Life and Work from 1773–1804; The Pennsylvania State University Press, University Park, PA.

Senebier, J. (1782) Mémoires Physico-Chimiques; Barthelemi Chirol, Geneva.

Sleep, NH, Meibom, A, Fridriksson, R, Coleman, RG and Bird, DK. (2004) H_2-rich fluids from serpentiniza-tion: geochemical and biotic implications. *Proc. Nat. Acad.Sci.* **101**: 12818–23.

Thiemens, MM. (1999) Mass-independent isotope effects in planetary atmospheres and the early solar system. *Science* **283**: 341–345.

Urey, HC. (1952) On the early chemical history of the earth and the origin of life. *Proc. Natl. Acad. Sci. (US)* **38**: 351–63.

Vernadsky, VI. (1926) Biosfera, (The Biosphere); Leningrad: Nauka.

Vine, JD and Tourtelot, EB. (1970) Geochemistry of black shale deposits: A summary report. *Econ. Geol.* **65**: 253–72.

Watanabe, Y, Farquhar, J. and Ohmoto, H. (2009) Anomalous fractionations of sulfur isotopes during thermochemical sulfate reduction. *Science* **324**: 370–3.

Yamaguchi, K. (2002) Geochemistry of Archean and Paleoproterozoic black shales: the early evolution of the atmosphere, oceans, and biosphere; Unpublished Ph. D. Thesis, The Pennsylvania State University.

Yang, W and Holland, HD. (2002) The redox-sensitive trace elements Mo, U, and Re in Precambrian carbonaceous shales: indicators of the Great Oxidation Event (abs.): Abstracts with Programs. *Geol. Soc. Amer.* **34**: 382.

Yang, W and Holland, HD. (2003) The Hekpoort paleosol in Strata 1 at Gaborone, Botswana: soil formation during the Great Oxidation Event. *Amer. Jour. Sci.* **303**: 187–220.

4 Geochemistry of the Oceanic Crust

KARSTEN M. HAASE

Universität Erlangen-Nürnberg, Erlangen, Germany

ABSTRACT

The oceanic crust has an average thickness of 7 km and forms by decompression melting of peridotitic rocks of the upper mantle at divergent plate boundaries, so-called mid-ocean ridges. The oceanic crust consists of tholeiitic basalts and gabbros, typically with low concentrations of incompatible elements like K, U and the light rare-earth elements. Evolved rocks like rhyolites or andesites and their plutonic equivalents are rare in the oceanic crust. Variations in the composition of the mafic rocks are due to variable degrees and depths of partial melting and chemical heterogeneities in the upper mantle. The chemical heterogeneities often contain higher concentrations of incompatible elements than the upper-mantle peridotites, and reflect upwelling of relatively hot deep mantle plumes causing age-progressive volcanic chains next to the mid-ocean ridges. After crystallization of the magmas, the rocks of the upper part of the oceanic crust chemically react with seawater, leading to metamorphic processes and severe alteration of the uppermost kilometre of the crust. Intrusion and extrusion of intraplate magmas also change the composition of the oceanic crust after its generation at the mid-ocean ridge.

Frontiers in Geochemistry: Contribution of Geochemistry to the Study of the Earth, First edition. Edited by Russell S. Harmon and Andrew Parker. © 2011 Blackwell Publishing Ltd. Published 2011 by Blackwell Publishing Ltd.

THE STRUCTURAL FRAMEWORK OF THE OCEAN CRUST

Although the compositional and tectonic features observed in the oceanic crust are relatively simple compared to the continental crust, the increasingly detailed study of the seafloor during the past 20 years has revealed significant complexities in the structure of the oceanic crust. One of the main parameters for the formation and evolution of the oceanic crust is the spreading rate, which ranges from few mm/yr to 180 mm/yr. Thus, spreading axes have been classified into fast- (>90 mm/yr), intermediate-, slow- (<40 mm/yr), and ultra-slow-spreading (Macdonald 1982; Perfit and Chadwick 1998; Dick et al. 2003). Spreading rate affects the crustal morphology, structure, and thickness owing to variable melting processes in the mantle (Fig. 4.1). Fast-spreading axes consist of narrow (<1 km wide) volcanic ridges where most of the active volcanism occurs (Fig. 4.1b). In contrast, slow-spreading axes typically are several kilometres wide and the volcanically-active zone occurs near the centre of a deep rift (Fig. 4.1d). Apart from spreading velocity the varying magma budget of ridge segments also affects the morphology of spreading axes (Scheirer and Macdonald 1993). High magma-supply at fast spreading rates leads to a widening of the volcanic ridge, whereas high magma-supply rates at slow-spreading segments leads to a resemblance of the ridge to intermediate or even fast-spreading axes, i.e. the deep rift valley disappears (Macdonald 2001).

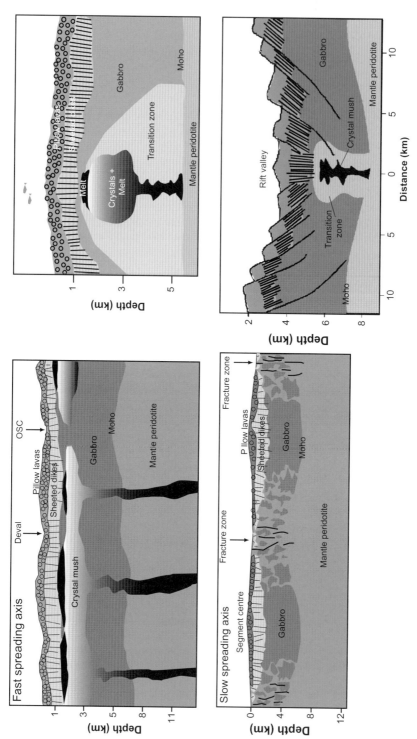

Fig. 4.1 Schematic along- and across-axis sections of fast (a & b) and slow (c & d) spreading axes redrawn after Sinton and Detrick (1992), and Cannat (1996).

The standard model of the oceanic crust structure is well-known from seismic and petrologic studies of many parts of the oceanic crust and ophiolites (Gass and Smewing 1973; Macdonald 1982; Karson 2002). This model is valid for intermediate- to fast-spreading axes (>40 mm/yr) whereas crust generated at slow-spreading mid-oceanic ridges is highly variable (Cannat 1996). Thus, mid-ocean ridges spreading faster than 40 mm/yr at full-rate have relatively constant thicknesses of 6 to 7 km, based on numerous seismic studies in all ocean basins (Chen 1992). The oceanic crust at such spreading rates typically forms three layers, based on seismic studies of oceanic crust and ophiolites (Hill 1957; Christensen and Smewing 1981; White 1984). Layer 1 consists of pelagic sediments, layer 2A of pillow lavas, layer 2B of sheeted dykes and layer 3 of gabbros (Fig. 4.1a). However, at lower spreading rates the thickness of the magmatic crust decreases and along-axis crustal thicknesses vary significantly, with the maximum thickness occurring at segment centres (Cannat 1996). A new class of ultra-slow spreading axes has been observed in the Arctic and Indian Oceans (Dick et al. 2003), which have segments where basaltic crust is absent, and mantle rocks, i.e. serpentinized peridotite forms the crust. Ultra-slow spreading also occurs in regions where either oceanic lithosphere is rifted (Vogt and Jung 2004; Beier et al. 2008) or where spreading axes become extinct owing to a jump of the spreading axis (Batiza 1989).

Spreading rates vary between segments, which range in length between about 50 and 100 km and which represent separate structural and magmatic units forming above ascending mantle diapirs (Crane 1985; Macdonald et al. 1991; Batiza 1996). Slow-spreading axes have large offsets between segments, which are less pronounced along most fast-spreading segments. Segment boundaries are divided into four categories, with transform faults or fracture zones representing the first-order boundaries, overlapping spreading centres (OSC) second-order boundaries (Fig. 4.1a) and deviations of axial linearity (devals) the third- and fourth-order boundaries (Macdonald 2001).

Continuous magma lenses have been observed along several hundreds of kilometres beneath the East Pacific Rise (Detrick et al. 1987; Hooft et al. 1997). These magma lenses appear to be stable for long periods of time and form a significant barrier for ascending magmas. The presence of these stable magma lenses explains the relatively evolved composition of MORB at fast-spreading axes (Sinton and Detrick 1992). The depth of the magma lenses varies with time (Hooft et al. 1997) but in general there appears to be a relationship between the magma-chamber depth and the spreading rate, with fast-spreading ridges having shallower magma reservoirs than slow-spreading axes (Purdy et al. 1992). In slow-spreading oceanic crust, the magma reservoirs are transient, and consist of vertical crystal-liquid mush instead of liquid magma lenses (Sinton and Detrick 1992). Seismic evidence for magma lenses at slow-spreading ridges is very scarce, and may be restricted to magmatically active segments with a thickened crust, i.e. an increased magma production (Sinha et al. 1997). Slow-spreading segments are believed to be fed by individual magma systems (Magde et al. 2000). These differences are most likely related to the much higher magma-supply rate along fast-spreading ridges, whereas slow-spreading ridges are dominated by tectonic processes for much of their time (Perfit and Chadwick 1998).

OCEANIC CRUST FORMATION: MELTING AND DIFFERENTIATION PROCESSES AT MID-OCEAN RIDGES

Chemical variations in rocks of the oceanic crust

The uppermost oceanic crust, i.e. the lavas of layer 2A, have been sampled relatively well during the past 40 years, and even remote parts of the about 65,000 km-long global spreading system, for example, beneath the Arctic ice are known reasonably well. Whereas early studies of the oceanic crust stressed compositional homogeneity of MORB, recent more detailed studies

revealed a significant geochemical variation of
rock types. The vast majority are tholeiitic basalts
with SiO_2 and K_2O contents of 50 to 52 wt.% and
less than 0.2 wt.%, respectively (Fig. 4.2) but alkaline basalts with K_2O greater than 0.8 wt.% have
been observed along spreading axes, for example,
on the Mid-Atlantic Ridge at 72°N near Jan
Mayen (Schilling et al. 1983; Haase et al. 1996).
Basalts erupted along the spreading axes vary in
light rare earth element enrichment and radiogenic isotope ratios, as observed, for example, in
the chondrite-normalized La/Sm ratios [(La/Sm)$_N$]
and Sr isotopes, respectively (Fig. 4.2). Furthermore,
the volcanic rocks range from picritic to rhyolitic
lavas, which are known from all ocean basins.
Major-element variations are due to variable
partial-melting processes and variable differentiation by fractional crystallization processes of the
magmas. Variations in degree and depth of partial
melting can be due to differences in mantle temperature and possibly volatile content (Kay et al.
1970; McKenzie and Bickle 1988; Langmuir et al.
1992). Furthermore, major-element source heterogeneity affects MORB compositions because the
mantle peridotite may be variably depleted by
partial melting, or the mantle may have become
re-enriched by small-degree melts or recycling of
subducted crustal material (Langmuir and Hanson
1980; Hirschmann and Stolper 1996). Layer 3 of
the oceanic crust at intermediate- to fast-spreading
ridges consists mainly of gabbroic rocks (Fig. 4.1)
but more evolved plutonic rocks like plagiogranites occur (Sinton and Detrick 1992). Plutonic
rocks are generally believed to represent the
intrusive counterparts of the lavas or residual
cumulates forming in magma chambers and are
thus complementary to the lavas. The crust may
be significantly thinner at slow-spreading ridges
than at faster spreading axes and the lower crust
may consist of a large part of peridotite with
gabbro intrusions (Cannat 1996).

The composition of primary MORB magmas

Early batch-melting experiments of mantle peridotites led to arguments on whether primary
MORB magmas were picritic, with MgO contents

greater than 18 wt.% (O'Hara 1968; Jaques and
Green 1980), or whether the most primitive
MORB glasses with MgO contents of 10–11 wt.%
represented primary magmas (Presnall et al. 1979;
Takahashi and Kushiro 1983). More recently,
polybaric fractional melting has been studied
experimentally, leading to primary MORB melts
with MgO contents of 10–12.5 wt.% (Fig. 4.3) or
Mg# of about 72 (Kinzler and Grove 1992; Kinzler
and Grove 1993), because the melts must be in
equilibrium with mantle olivine, which typically
has forsterite contents of 88–92 (Roeder and
Emslie 1970; Hess 1992). Another model suggests
that primary MORB may have 13 to 16% MgO
(Green et al. 2001), comparable to the MgO contents of Hawaiian primitive lavas (Clague et al.
1991). However, primitive magmas arc rare at
spreading axes, but MORB glasses with MgO up
to 10 wt.% or Mg# up to 72 (Fig. 4.3) have been
found occasionally (Presnall and Hoover 1984)
and appear to be relatively abundant close to fracture zones (Perfit et al. 1996; Wendt et al. 1999).
Picritic magmas are relatively dense and thus normally stagnate in a crustal magma chamber until
the densities decrease by crystal fractionation
processes (Stolper and Walker 1980; Ryan 1993).
Consequently, most lavas along the mid-ocean
ridges have Mg# between 65 and 50. Compared to
most other magmas on Earth, MORB magmas
contain low concentrations of volatiles. Melt
inclusions of primitive basalts yield water contents of about 0.1 wt.% and CO_2 contents of less
than 250 ppm (Saal et al. 2002).

MORB melting and temperature of the upper mantle

The basaltic magmas of the oceanic crust form
by polybaric near-fractional adiabatic partial
melting processes of the upper mantle, which
ascends beneath the mid-ocean ridge (Ahern and
Turcotte 1979; McKenzie and Bickle 1988). The
degree and depth of partial melting depend on
temperature and composition, i.e. variable fertility and water content of the mantle. Higher temperature, water content and/or fertility of the
mantle, lead to deeper partial melting and a

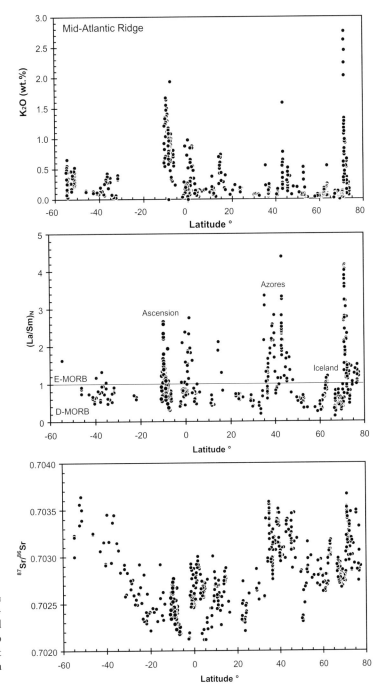

Fig. 4.2 Variation of K$_2$O, (La/Sm)$_N$ and Sr-isotope ratios along the Mid-Atlantic Ridge. Note that enriched MORB are abundant not only close to off-axis volcanoes like the Azores but also at the Equator and at 25°N. Data from the PetDB.

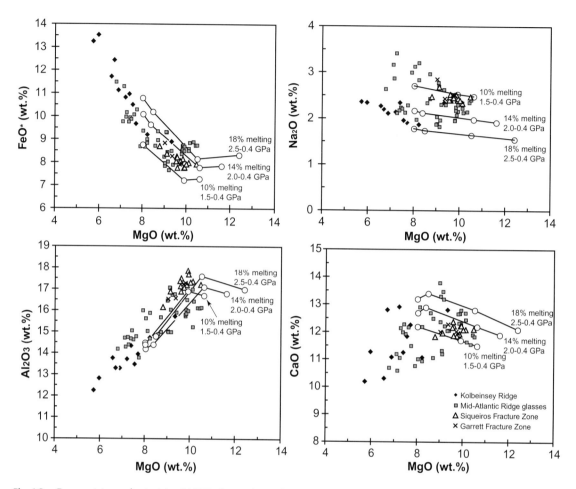

Fig. 4.3 Compositions of primitive MORB glasses (Sigurdsson 1981; Presnall and Hoover 1987; Hékinian et al. 1995; Perfit et al. 1996) compared to the calculated accumulated melt compositions from experiments and low-pressure crystal fractionation trends of these magmas (Kinzler and Grove 1992; Kinzler and Grove 1993).

longer melting column. The partial melts at different depths segregate from their residual mantle once the threshold of melt migration is reached, i.e. a permeability develops within the ascending mantle (McKenzie 1985). This threshold will be relatively low because of the deformational stress in the ascending mantle (Kohlstedt 1992). Thermodynamic models suggest that melting of a peridotite source starting at about 80 km depth would produce a basaltic crust thicker than 10 km rather than the observed 6–7 km (McKenzie

and Bickle 1988; Iwamori et al. 1995). The results of polybaric melting models suggest that MORB primary magmas form by 10 to 20% partial melting of peridotite at pressures of 2 to 0.5 GPa (Niu and Batiza 1991; Kinzler and Grove 1992; Kinzler and Grove 1993; Kushiro 2001). However, because the depth of the end of the melting column and the amount of melt retained in the peridotite are unknown, there is considerable uncertainty in the model of melt production and segregation. The onset of melting critically depends on temperature, fertility, and volatile content, where the latter two parameters determine the solidus temperature of the mantle. The potential temperature (the temperature of the mantle if it would reach the surface without melting) of the average upper mantle ranges between 1280°C (McKenzie and Bickle 1988) and 1430°C (Green et al. 2001) and the estimates depend on the MgO content of the assumed primary magma. The fertility, and particularly the alkali element content of the mantle, has a strong effect on the solidus temperature. For example, the solidus temperature at 3 GPa varies between 1450°C for mantle with Mg# 87 and Na_2O+K_2O of 0.5 wt.% to 1490°C for mantle with Mg# 90 and 0.3 wt.% alkalies (Hirschmann 2000).

In order to compare the compositions of primitive MORB along the global spreading system, the lavas can be fractionation-corrected to primitive MORB. These compositions show correlations between the water depth and fractionation-corrected FeO^T ($Fe_{8.0}$) and Na_2O ($Na_{8.0}$) contents. These variations were explained by variable degrees and depths of partial melting beneath spreading axes and the crustal thickness, which in turn affect the water depth of the spreading segments. Following Klein and Langmuir (1987), MORB with low Na8.0 and high Fe8.0 formed by high degrees of partial melting, which starts deep in the mantle and leads to thick basaltic crust forming shallow bathymetric anomalies. Deep-ridge segments are underlain by a thin basaltic crust and MORB of these segments have high Na8.0 and low Fe8.0 due to low-degree melting relatively shallow in the mantle. These global correlations were explained by mantle temperature variations of about 250°C along the ridge system, with the increased temperatures representing inflow of hot mantle plumes into the ridge axis (Klein and Langmuir 1987). An alternative model uses a different fractionation correction, yielding smaller variations of FeO^T and Na_2O in primitive MORB. In this model the temperature variations in the upper mantle are less than 50°C, and the variations in partial melting are explained by variations of mantle fertility (Niu and O'Hara 2008).

Whereas early studies concentrated on the role of peridotite in the upper mantle beneath the spreading axes (Ringwood 1975), more recent models stress the importance of other rock types and volatiles in the mantle. Data of Lu/Hf and U-series isotope systems suggested the presence of residual garnet in the melting column of MORB (Condomines et al. 1981; Salters and Hart 1989) which would imply a deep onset of partial melting of garnet peridotite (Robinson and Wood 1998), or the presence of garnet pyroxenite which is stable at relatively shallow depths in the mantle (Hirschmann et al. 1995). Pyroxenite has been observed in many peridotite bodies and may have formed either as basaltic intrusions or cumulates in lithospheric mantle portions (Bodinier et al. 1987; Suen and Frey 1987) or as recycled basaltic oceanic crust (Allègre and Turcotte 1986). The presence of pyroxenite in the upper mantle would significantly affect the onset of partial melting as well as melt productivity, yielding more melt than peridotite (Pertermann and Hirschmann 2003).

Evolution of MORB magmas and the nature of the lower oceanic crust

Mid-oceanic-ridge basalts are relatively evolved and have an average Mg# of about 55 (Klein 2003). Ascending magmas resided for a significant period of time in crustal magma reservoirs where they fractionated olivine, plagioclase, spinel and more rarely clinopyroxene (Fig. 4.4). The magmas probably stagnate at a level of neutral buoyancy where their density is similar to that of the surrounding crustal rocks (Ryan 1987). The crystallization and fractionation of olivine reduces the density of the magma until its buoyancy allows further ascent to the surface (Stolper and Walker 1980). The extrusives of the oceanic crust are erupted through a shallow dike system that forms the sheeted dike complex known from ophiolites and some deep fractures of the oceanic crust (Stewart et al. 2002). Most dikes are one to a few metres thick and studies have shown that the dikes have homogeneous compositions (Staudigel et al. 1999; Stewart et al. 2002). Below the sheeted dike complex at 1.5 to 2.5 km depth, crustal magma lenses have

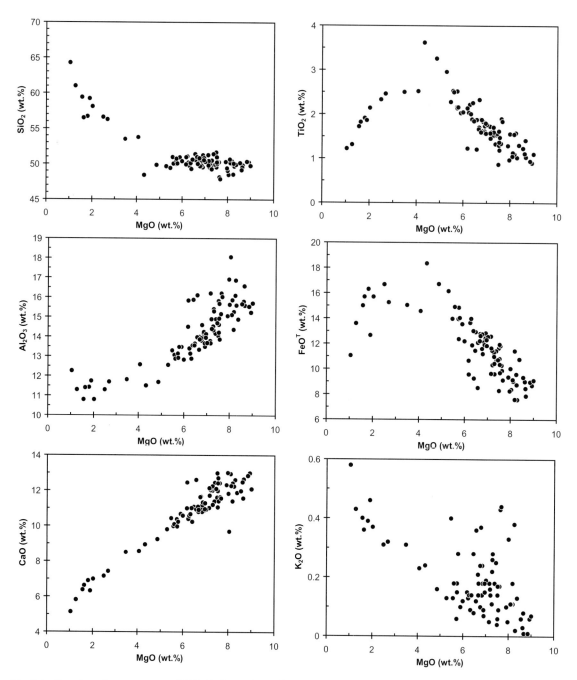

Fig. 4.4 Fractionation processes of MORB from the Galapagos spreading centre, leading to andesitic and dacitic magmas (Schilling et al. 1982; Perfit et al. 1983; Puchelt and Emmermann 1983; Perfit et al. 1999).

been observed in seismic data from fast-spreading ridges and occasionally on slow-spreading segments (Detrick et al. 1987). In fact, there appears to be a variable depth of the magma lens with spreading rate, where increasingly deep magma bodies occur at slower spreading ridges. The depth of the magma lens also seems to fluctuate with magma intrusion into the crust, i.e. during stages of high magma production the melt lens lies shallower than at times when magma production is lower (Hooft et al. 1997). It is debatable how the lower oceanic crust forms, and models suggest sill intrusion at different depths (Henstock et al. 1993) or vertical and horizontal flow of crystal mush from a shallow magma chamber (Nicolas et al. 1988). Portions of the lower crust show signs of deformation. The largest part of the oceanic crust consists of gabbros of layer 3 from crystallization of mafic magmas. The lowermost part of the crust consists of ultramafic cumulates and layered gabbros which have high seismic velocities similar to the mantle, and thus there is a distinction between a petrological and a seismic Moho. The composition of the mafic rocks of the lower oceanic crust ranges from troctolites and norites to gabbros and plagiogranites. Estimates of bulk lower-crustal sections indicate a mafic composition with a Mg# of 69 to 73, i.e. similar to a primitive magma from the mantle (Hart et al. 1999; Godard et al. 2009). Because of their very mafic composition and accumulation of olivine, pyroxene and plagioclase estimates of the lower crustal sections are more depleted in incompatible elements than the basalts of the upper crust, but show a comparable pattern (Fig. 4.5). Positive Sr and Eu anomalies in many lower crustal gabbroic rocks reflect the accumulation of plagioclase (Hart et al. 1999) and are complementary to more evolved lavas of the oceanic upper crust which display negative Sr and Eu anomalies.

Assimilation of crustal material is also observed. Especially, the assimilation of hydrothermally altered rocks has been studied to some detail (Coogan et al. 2003). The ascent of mafic magmas in the crust may lead to re-melting of hydrothermally altered crustal rocks at the margins of magma lenses (Gillis and Coogan 2002; Koepke et al. 2005). Assimilation-fractional crystallization (AFC) processes which lead to high Cl-contents in the lavas and also increased Sr-isotope ratios due to seawater assimilation have been observed in relatively evolved lavas (Michael and Schilling 1989; Michael and Cornell 1998; Haase et al. 2005). It is not clear whether the silicic magmas occurring on spreading axes (Fig. 4.4) form by partial melting of hydrous lower or middle crustal rocks (Koepke et al. 2007) or by AFC processes from mafic magmas (Haase et al. 2005).

MANTLE SOURCES OF THE OCEANIC CRUST

Mantle heterogeneity and mixing processes observed in MORB

The magmas forming the oceanic crust are generated by relatively high degrees of partial melting (10–20%) of rocks of the upper mantle (Langmuir et al. 1992) and thus MORB largely reflect the composition of the uppermost mantle. Early studies suggested that most MORB are depleted in incompatible elements like K and the light REE (Fig. 4.5), and consequently, depleted MORB with chondrite-normalized La/Sm ratios [$(La/Sm)_N$] less than 1 were called normal (N-) MORB (Schilling et al. 1983). However, as the along-axis sampling of oceanic spreading axes improved, the original notion of normal rocks became less clear. For example, large parts of the Mid-Atlantic Ridge erupt relatively incompatible element-enriched basalts (Fig. 4.2). In many regions of the global spreading axis system, bathymetric anomalies and voluminous off-axis volcanism coincide with these geochemical anomalies, i.e. mid-ocean-ridge segments of relatively shallow water depth erupt MORB with relatively enriched incompatible-element compositions (Fig. 4.5), so-called enriched (E-) MORB (Schilling 1973; 1985). These E-MORB frequently lie on isotopic binary mixing lines between enriched basalts from the intraplate areas and depleted MORB. Thus, enriched and relatively hot mantle from deep mantle plumes flows into the spreading axis, causing the

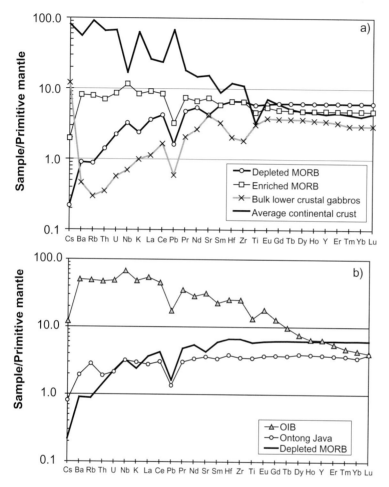

Fig. 4.5 (a) Variation of incompatible elements in depleted and enriched MORB (Sun and McDonough 1989) as well as a bulk lower oceanic crust estimate (Godard et al. 2009) compared to average continental crust (Rudnick and Fountain 1995). (b) Average ocean island basalt, depleted MORB (both from Sun and McDonough 1989) and a primitive Ontong Java Plateau basalt glass (Tejada et al. 2004). The data are normalized to the estimated concentrations of primitive mantle (Sun and McDonough 1989). See Plate 4.5 for a colour version of these images.

geochemical anomaly and increased magma production. Mixing occurs most likely between melts from both enriched and depleted parts of the mantle (Braun and Sohn 2003). An input from deep undegassed mantle sources is implied by the relatively high $^3He/^4He$ ratios (Fig. 4.6b) in many of the geochemical anomalies along the spreading axis (Graham 2002).

The scale of heterogeneity of MORB

The first studies of along-axis chemical and isotopic variation of MORB revealed a few hundred-kilometres wide anomalies close to voluminous

off-axis volcanoes like the Azores, Easter Island, or Galapagos (Fig. 4.6). The off-axis volcanism in many cases forms age-progressive volcanic chains, with the youngest volcanism closest to the spreading axis, explained by volcanism above a mantle plume rising from the lower mantle, possibly the core-mantle boundary (Duncan and Richards 1991). The relatively shallow bathymetry at plume-influenced spreading axes is at least in part caused by a thickened oceanic crust, owing to higher magma production in these regions reflecting higher temperatures in the mantle, owing to inflow of material from the lower mantle. Interactions between spreading axes and

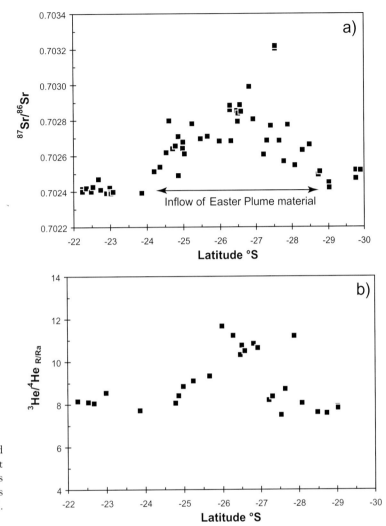

Fig. 4.6 The inflow of enriched material from the Easter Hotspot into the adjacent East Pacific Rise is shown in Sr and He isotopes (Schilling et al. 1985; Poreda et al. 1993; Haase 2002).

mantle plumes exist over distances ranging from a few tens of kilometres at Iceland to about 1000 km in the case of Réunion. Many geochemical anomalies along the spreading system are not associated with time-progressive volcanic chains or bathymetric anomalies like, for example, the 10°S anomaly on the Mid-Atlantic Ridge or the Equatorial Atlantic (Fig. 4.2), respectively (Bonatti 1990). In many of these cases the basalts are heterogeneous on a scale of a few kilometres or even within one dredge that sampled several hundred

metres along the spreading axis. The fact that the upper mantle is heterogeneous on a small scale (less than five kilometres), and yields depleted, as well as incompatible element-enriched magmas, is supported by the presence of numerous off-axis seamounts close to the spreading axes, frequently consisting of enriched tholeiites and alkali basalts having higher Sr and Pb isotope ratios than MORB from neighbouring ridge segments (Zindler et al. 1984; Castillo and Batiza 1989; Cousens 1996). Furthermore, data from lava flows along the

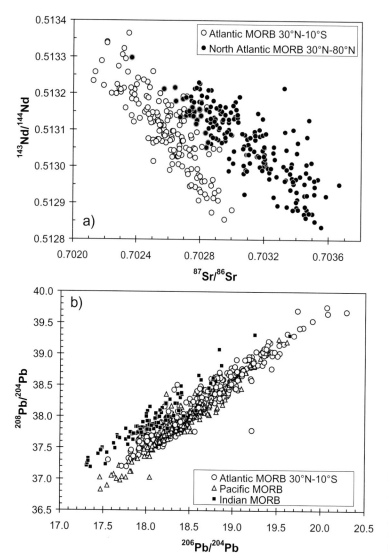

Fig. 4.7 Large-scale variations in radiogenic isotopes on a large scale between different parts of the oceanic basins: (a) variation of Sr and Nd isotopes in Atlantic MORB between 80°N and 10°S, showing a significant difference between lavas form two regions with a boundary at about 30°N; (b) comparison of the $^{208}Pb/^{204}Pb$ versus $^{206}Pb/^{204}Pb$ for Atlantic, Indian and Pacific MORB showing that Indian Ocean MORB differ from the basalts of the other two oceans in having higher $^{208}Pb/^{204}Pb$. Data from the PetDB.

spreading system frequently indicate considerable chemical and isotopic heterogeneity (Rubin et al. 2001).

Different MORB source compositions have been found on a large scale between oceans or parts of ocean basins. For example, two distinct regions exist within the North Atlantic Ocean where MORB between 30 and 80°N have higher Sr-isotope for a given Nd-isotope ratio (Fig. 4.7a).

Indian Ocean MORB is generally distinct in having higher $^{87}Sr/^{86}Sr$ and $^{208}Pb/^{204}Pb$ (Fig. 4.7b) compared to MORB from the Atlantic and Pacific Oceans (Dupré and Allègre 1983) and these variations have been used to define lateral asthenospheric mantle flow at the boundaries of the ocean basins (Klein et al. 1988; Pearce et al. 2001). In some cases the boundaries coincide with bathymetric deeps, possibly marking sites

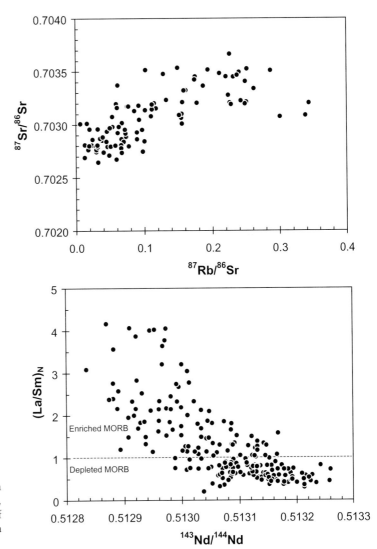

Fig. 4.8 Variation of Sr isotopes with Rb/Sr and (La/Sm)N with Nd isotopes, indicating significant variations of isotope ratios for depleted lavas. Data from the PetDB.

of asthenospheric counterflow (Klein et al. 1988). The origin of the distinct isotopic compositions in different parts of the spreading system are debated and models suggest either the presence of old subducted material in the upper mantle (Dupré and Allègre 1983) or the influence of lower-mantle material into the upper mantle (Phipps Morgan et al. 1995; Vlastélic et al. 1999). Incompatible-element ratios are correlated with radiogenic-isotope ratios suggesting that at least part of the incompatible-element enrichment reflects different mantle sources (Fig. 4.8). The observed variations in MORB compositions can either represent different binary mixing trends or different isochrons, in which case they provide age constraints on the formation of the MORB mantle source (Schilling et al. 1994; Dosso et al. 1999).

The depleted end-member source of MORB

The isotope variations of global MORB have been interpreted to represent mixing between a depleted asthenospheric end-member, and several more-enriched mantle sources introduced into the asthenosphere by deep mantle plumes (Hart et al. 1973; White et al. 1976; Phipps Morgan et al. 1995). On a global scale MORB compositions with the highest Nd and lowest Sr isotope ratios represent a mantle end-member called Depleted MORB Mantle (DMM) (Zindler and Hart 1986). This finding of a depleted reservoir in the upper mantle had important implications for understanding the chemical evolution of the Earth, and it appears that the incompatible-element-depleted upper mantle is complementary to the incompatible-element-enriched continental crust (Fig. 4.5a). However, the Sr-Nd-Pb-Hf-isotope composition of this depleted end-member is quite variable, whereas it appears to be much more homogeneous in terms of incompatible-element ratios. For example, highly depleted Atlantic MORB have relatively constant $^{87}Rb/^{86}Sr$ of 0.01 but variable $^{87}Sr/^{86}Sr$ between 0.7026 and 0.7030 (Fig. 4.8). The most depleted Atlantic MORB have $(La/Sm)_N$ of about 0.35 but $^{143}Nd/^{144}Nd$ range from 0.5130 to 0.51325 (Fig. 4.8). This probably reflects the fact that the mantle has to be relatively fertile to produce MORB magmas, i.e. there is a lower boundary of depletion below which no melt can be produced. The isotope variation possibly reflects the age of the depletion event, i.e. the time after which the mother/daughter isotope ratios became too low to produce significant radiogenic growth. Model ages of local MORB arrays in Rb-Sr, Sm-Nd and U-Pb isotope systems yield ages of a few hundred million years, which have been interpreted to represent times of fractionation of the different mantle sources (Dosso et al. 1999). Whereas there appears to be a significant heterogeneity of the depleted MORB mantle in terms of radiogenic isotopes, most stable-isotope ratios are believed to be relatively homogeneous in depleted MORB. For example, all depleted MORB have $\delta^{18}O$ of 5.5±0.5 similar to upper-mantle olivines, whereas enriched MORB may have higher values (Eiler et al. 2000). Assimilation of hydrothermally altered material also changes the $\delta^{18}O$. Furthermore, the $^3He/^4He_{R/Ra}$ (the He-isotope ratio R of the rock relative to the atmospheric He-isotope ratio Ra) of depleted MORB is relatively constant between 7 and 9, which has been interpreted as a reflection of the degassed noble-gas composition of the upper mantle owing to partial melting processes (Graham 2002).

The enriched end-member

The enriched mantle sources supplying the magmas of the oceanic crust are much more variable in chemical and isotopic compositions than the depleted mantle sources. Based on global Sr-Nd-Pb-He-isotope data, some five different enriched mantle sources have been defined (Zindler and Hart 1986; Hart et al. 1992; Hofmann 2003). The increased incompatible element concentrations and variable incompatible element ratios in enriched MORB are partly due to partial melting processes and partly due to enriched composition of the mantle source. The fact that in most cases incompatible-element enrichment correlates with radiogenic isotope ratios, implies that magma enrichment reflects distinct mantle sources and sources had variable incompatible element ratios in order to develop different isotope compositions. The most common end-member has relatively high Sr, and particularly Pb, isotope ratios and low Nd-isotope ratios compared to depleted MORB. This end-member occurs in enriched MORB but is also present in enriched off-axis seamount lavas. Generally, this enriched end-member resembles the so-called FOZO component (Hauri et al. 1994). Interestingly, near-ridge hotspots like Iceland, Galapagos and Easter Island seem to erupt magmas resembling FOZO (Hofmann 1997) whereas more extreme mantle end-members are found in hotspots distant from spreading axes on old lithosphere. In cases of plume-ridge interaction like close to Iceland, the Galapagos or Easter Island enriched MORB also have high $^3He/^4He_{R/Ra}$ (Fig. 4.6) implying that the enriched material is from the deep undegassed mantle.

THE CHEMICAL EVOLUTION OF THE OCEANIC CRUST

Evidence for variation of MORB composition with time?

Whereas the present mid-ocean ridge system and the young oceanic crust are relatively well-studied, very little is known about the chemical evolution of the oceanic crust during the history of the Earth. The oldest crust in the oceanic basins is about 200 Ma old, but very few data exist because older crust becomes inaccessible due to sedimentation. Close to the spreading axis several studies have shown that the incompatible element and isotope composition of lavas forming at one spreading segment vary with time (Perfit et al. 1994). Residual peridotites along an Atlantic fracture zone revealed significant changes in partial melting over a period of time of several million years, which is in agreement with the variation of gravimetrically determined crustal thickness (Bonatti et al. 2003). It was shown that MORB with ages older than 80 Ma apparently erupted shallower and had lower $Na_{8.0}$ and Zr/Y but higher $Fe_{8.0}$, and thus the mantle temperature may have been 50°C higher at that time (Humler et al. 1999).

Depleted mantle must have formed very early after the accretion of the Earth, and mafic magmatic rocks with depleted incompatible element compositions and high Nd-isotope ratios occur since the Early Archaean (McCulloch and Bennett 1994). However, it is debated when the first oceanic crust formed and plate tectonics began in its present form. The oldest ophiolites resembling the Phanerozoic examples are up to 2.5 Ga (Peltonen et al. 1996; Kusky et al. 2001) which is similar to the age of the oldest eclogites, indicating subduction of oceanic crust (Möller et al. 1995). Eclogite xenoliths of Archaean age from several cratons are interpreted as remnants of subducted oceanic crust (Jacob et al. 1994). Some Archaean greenstone belt sequences are interpreted as ophiolites, but differ from Phanerozoic oceanic crust in terms of composition and structure. For example, a three-kilometre-thick mafic-to-ultramafic rock sequence containing komatiites to tholeiitic basalts of the 3.5 Ga Jamestown Ophiolite Complex has been interpreted as representing Archaean oceanic crust (de Wit et al. 1987). If these rock assemblages represent oceanic crust, then the melting and formation processes must have been very different from the Phanerozoic oceanic crust.

The chemical alteration of the oceanic crust during cooling

Immediately after its formation at the mid-oceanic ridges the oceanic crust evolves chemically owing to alteration processes and exchange with seawater driven by the cooling of the crust (Alt et al. 1986; Alt 2004). The alteration process of the oceanic crust has important implications for the evolution of the Earth because the uptake of water by the crust and its release during subduction is a unique process in the Solar System forming the continental crust (Campbell and Taylor 1983). The chemical exchange of the oceanic crust with the seawater also buffers the composition of the seawater, which has been remarkably constant in pH and salinity, which, in turn, has been important for the evolution of organisms on Earth (Elderfield and Schultz 1996). The presence of hot magmas in the shallow crust at the spreading axis leads to the formation of hydrothermal circulation, the most efficient process of cooling. Thus, most of the heat of the oceanic crust and lithosphere is removed within the first 10 million years so that the thickness of the lithosphere (defined by an isotherm) increases rapidly, and the oceanic plates subside regularly with increasing age and distance from the spreading axis.

The chemical alteration of the oceanic due to hydrothermal overprinting has been studied relatively well both in samples from drillcores and from ophiolites (Staudigel et al. 1996). The uppermost 1 km of the oceanic crust consisting of volcanics and the transition to the sheeted dikes has a relatively high porosity and, as a consequence, is thoroughly altered. Metamorphic facies increases with depth from greenschist to amphibolite facies, reflecting the increasing

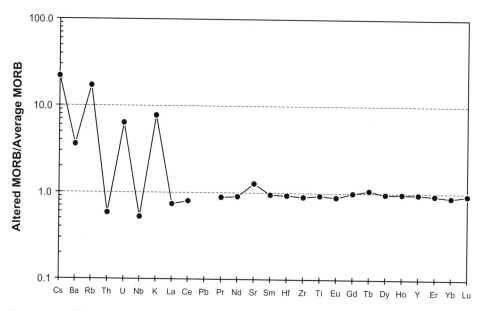

Fig. 4.9   Comparison of the incompatible element composition of an estimate of altered crust (Staudigel et al. 1996) relative to fresh depletd MORB (Sun and McDonough 1989) showing the enrichment of relatively mobile elements by hydrothermal processes.

temperature. This alteration of the pillow lavas leads to increased K_2O, H_2O and CO_2 contents but also higher $\delta^{18}O$ in the uppermost crust. Importantly, the incompatible-element composition of the upper crust is also changed, which is believed to have consequences during the subduction of this material, especially the fluid-mobile elements like U, Pb, Rb and Ba (Fig. 4.9). Reactions of seawater with mafic crust leading to silicification have been observed in Archaean rocks as old as 3.5 Ga (Hofmann and Harris 2008) implying that chemical exchange has been occurring since the formation of the first oceans. Because oceanic crust is subducted into the mantle, the chemical alteration has important consequences for the evolution of radiogenic isotope compositions in the Earth's mantle (Chase 1981; Hofmann and White 1982).

The sheeted dykes and the gabbros of the oceanic crust (Fig. 4.1) show signs of high-temperature (>350°C) alteration along veins with the formation of amphibole and secondary plagioclase and clinopyroxene, but no significant change of the chemical composition of the deeper crust. Deep faults occur mainly at slow-spreading axes, and thus deep hydrothermal metamorphism of gabbros has been observed predominantly along slow-spreading ridges (Gillis 2002; Gillis and Coogan 2002). The introduction of water deep into the crust may also lead to partial melting of portions of the crust and the formation of silicic magmas, or to assimilation of hydrothermally altered material by stagnating magmas.

The addition of material to the oceanic crust by intraplate magmatism

The composition of the oceanic crust is also affected by intrusion and extrusion of intraplate magmas (Batiza 1981; Wessel and Lyons 1997) to the crust after it was generated at the spreading axis. Off-axis magmatism occurs immediately after the crust leaves the spreading axis and forms small seamounts 5 to 15 km distant from the axis (Scheirer and Macdonald 1995). Large intraplate volcanoes like the Canary Islands or the Carolines erupt through 160 Ma-old oceanic lithosphere and the enriched ocean-island basalts (OIB) (Fig. 4.5b), and intrusives significantly alter the oceanic crust. Volcanism on old plates generally forms age-progressive chains of volcanoes caused by a stationary melting anomaly of a deep mantle plume (Morgan 1971). A deep root of these oceanic intraplate volcanoes has been observed in seismic tomography (Montelli et al. 2004) and is also supported by the relatively primitive noble-gas isotope composition (e.g. $^3He/^4He$ up to 35 R/Ra) of the magmas (Graham 2002). In most cases intraplate magmas are significantly more enriched in incompatible elements (Fig. 4.5(b)) and have more distinct isotope ratios than the basaltic rocks forming the oceanic crust (Zindler and Hart 1986; Haase 1996; Hofmann 1997). Depleted as well as enriched tholeiitic basalts dominate the near-axis seamounts (Batiza and Vanko 1984), whereas more depleted basalts are rare on intra-plate volcanoes that formed distant to the spreading axis. The volume of added material varies considerably and, for example, the volcano height of the intraplate magmatism appears to be related to the age of the lithosphere (Vogt 1974). The largest intraplate volcanoes occur at Hawaii with heights up to 10 km above the surrounding sea-floor. Furthermore, significant material is also added by intrusion and underplating of gabbroic rocks in the crust and at the Moho (ten Brink and Brocher 1987; Caress et al. 1995). Thus, in the most extreme cases of strong deep mantle plumes like Hawaii, Tahiti or Réunion, chains of large volcanoes or aseismic ridges reflect significant additional material, altering the composition of pre-existing oceanic crust (White 1993). The composition of oceanic intraplate magmas ranges from tholeiitic to highly alkaline basalts and even carbonatitic rocks, whereas evolved magmas are generally scarce. Many oceanic intraplate volcanoes evolve through stages erupting different volumes and magma compositions, where the shield-stage of the volcanism usually forms from the highest degree of partial melting, and pre- and post-shield stage lavas were generated at relatively low degrees of melting (Clague and Dalrymple 1987; Woodhead 1992).

Even more extreme additions of relatively enriched magmas have occurred at oceanic plateaus like the Ontong-Java, Manihiki or Kerguelen plateaus, which probably formed owing to the excess melting of a deep mantle-plume head beneath young oceanic lithosphere (Duncan and Richards 1991; Kerr 2003). In some of these cases like at the Ontong-Java Plateau the crust reaches thicknesses up to 35 km (Gladczenko et al. 1997) compared to the average 7 km-thick crust and thus up to $6 \times 10^7 km^3$ of basaltic material were added to the oceanic crust (Coffin and Eldholm 1994). Most of these oceanic plateau lavas are low-K tholeiitic basalts with intermediate concentrations of incompatible elements and variable isotope compositions ranging from OIB-like to depleted MORB-like (Kerr 2003; Tejada et al. 2004).

ACKNOWLEDGEMENTS

I thank the convenors Russell Harmon and Andrew Parker for the invitation to contribute to the session 'Contributions of Geochemistry to the Study of the Planet' at the 33rd International Geological Congress in Oslo.

FURTHER READING

Alt, JC. (2004) Alteration of the upper oceanic crust: mineralogy, chemistry, and processes. In: EE Davis and H Elderfield (Editors), Hydrogeology of the Oceanic Lithosphere. Cambridge University Press, Cambridge, pp. 495–533

Hofmann, AW. (2003) Sampling mantle heterogeneity through oceanic basalts: Isotopes and trace elements, Treatise on Geochemistry. Elsevier, pp. 61–101.

Klein, EM. (2003) Geochemistry of the igneous oceanic crust, Treatise on Geochemistry. Elsevier, pp. 433–63.

REFERENCES

Ahern, JL and Turcotte, DL. (1979) Magma migration beneath an ocean ridge. *Earth and Planetary Science Letters* **45**: 115–22.

Allègre, CJ and Turcotte, DL. (1986) Implications of a two-component marble-cake mantle. *Nature* **323**: 123–7.

Alt, JC. (2004) Alteration of the upper oceanic crust: mineralogy, chemistry, and processes. In: E.E. Davis and H. Elderfield (Editors), Hydrogeology of the Oceanic Lithosphere. Cambridge University Press, Cambridge, pp. 495–533.

Alt, JC, Honnorez, J, Laverne, C and Emmermann, R. (1986) Hydrothermal alteration of a 1 km section through the upper oceanic crust, Deep Sea Drilling Project hole 504B: Mineralogy, chemistry, and evolution of seawater-basalt interaction. *Journal of Geophysical Research* **91**: 10309–35.

Batiza, R. (1981) Lithospheric age dependence of off-ridge volcano production in the North Pacific. *Geophysical Research Letters* **8**: 853–6.

Batiza, R. (1989) Failed rifts. In: EL Winterer, DM Hussong and RW Decker (eds.), The eastern Pacific Ocean and Hawaii. Geological Society of America, The Geology of North America, Boulder, Colorado.

Batiza, R. (1996) Magmatic segmentation of mid-ocean ridges: a review. In: CJ MacLeod, PA Tyler and CL Walker (eds.), Tectonic, Magmatic, Hydrothermal and Biological Segmentation of Mid-Ocean Ridges. Geol. Soc. Spec. Publ., pp. 103–30.

Batiza, R. and Vanko, D. (1984) Petrology of young Pacific seamounts. *Journal of Geophysical Research* **89**: 11235–60.

Beier, C, Haase, KM, Abouchami, W, Krienitz, M-S and Hauff, F. (2008) Magma genesis by rifting of oceanic lithosphere above anomalous mantle: Terceira Rift, Azores. *Geochemistry Geophysics Geosystems* 10.1029/2008GC002112.

Bodinier, J-L, Guiraud, M, Fabriès, J, Dostal, J. and Dupuy, C. (1987) Petrogenesis of layered pyroxenites from the Lherz, Freychinède and Prades ultramafic bodies (Ariège, French Pyrénées). *Geochimica et Cosmochimica Acta* **51**: 279–290.

Bonatti, E. (1990) Not so hot 'hot spots' in the oceanic mantle. *Science* **250**: 107–11.

Bonatti, E, Ligi, M, Brunelli, D, Cipriani, A, Fabretti, P, Ferrante, V, Gasperini, L. and Ottolini, L. (2003) Mantle thermal pulses below the Mid-Atlantic Ridge and temporal variations in the formation of oceanic lithosphere. *Nature* **423**: 499–505.

Braun, MG and Sohn, RA. (2003) Melt migration in plume-ridge systems. *Earth and Planetary Science Letters* **213**: 417–30.

Campbell, IH and Taylor, SR. (1983) No water, no granites – no oceans, no continents. *Geophysical Research Letters* **10**: 1061–4.

Cannat, M. (1996) How thick is the magmatic crust at slow spreading oceanic ridges? *Journal of Geophysical Research* **101**: 2847–57.

Caress, DW. McNutt, MK, Detrick, RS and Mutter, JC. (1995) Seismic imaging of hotspot-related crustal underplating beneath the Marquesas Islands. *Nature* **373**: 600–603.

Castillo, P. and Batiza, R. (1989) Strontium, neodymium and lead isotope constraints on near-ridge seamount production beneath the South Atlantic. *Nature* **342**: 262–5.

Chase, CG. (1981) Oceanic island Pb: Two-stage histories and mantle evolution. *Earth and Planetary Science Letters* **52**: 277–84.

Chen, YJ. (1992) Oceanic crustal thickness versus spreading rate. *Geophysical Research Letters* **19**: 753–6.

Christensen, NI and Smewing, JD. (1981) Geology and seismic structure of the northern section of the Oman Ophiolite. *Journal of Geophysical Research* **86**: 2545–55.

Clague, DA and Dalrymple, GB. (1987) The Hawaiian-Emperor Volcanic Chain. U.S. Geol. Surv. Prof. Paper, pp. 5–54.

Clague, DA, Weber, WS and Dixon, JE. (1991) Picritic glasses from Hawaii. *Nature* **353**: 553–6.

Coffin, MF and Eldholm, O. (1994) Large igneous provinces: crustal structure, dimensions, and external consequences. *Reviews of Geophysics* **32**: 1–36.

Condomines, M, Morand, P and Allègre, CJ. (1981) ^{230}Th-^{238}U radioactive disequilibria in tholeiites from the FAMOUS zone (Mid-Atlantic Ridge, 36°50′N): Th and Sr isotopic geochemistry. *Earth and Planetary Science Letters* **55**: 247–56.

Coogan, LA, Mitchell, NC and O'Hara, MJ. (2003) Roof assimilation at fast spreading ridges: An investigation combining geophysical, geochemical, and field evidence. *Journal of Geophysical Research* **108**: doi: 10.1029/2001JB001171.

Cousens, BL. (1996) Depleted and enriched upper mantle sources for basaltic rocks from diverse tectonic environments in the northeast Pacific Ocean: The generation of oceanic alkaline vs. tholeiitic basalts. In: A Basu and S Hart (eds.), Earth Processes: Reading the Isotopic Code. Am. Geophys. Union, Geophys. Monogr., Washington, D.C., pp. 207–31.

Crane, K. (1985) The spacing of rift axis highs: dependence upon diapiric processes in the underlying asthenosphere? *Earth and Planetary Science Letters* **72**: 405–14.

de Wit, MJ, Hart, RA and Hart, RJ. (1987) The Jamestown OPhilite Complex, Barberton mountain belt: a section through 3.5 Ga oceanic crust. *Journal of African Earth Sciences* **6**: 681–730.

Detrick, RS, Buhl, P, Vera, E, Mutter, J, Orcutt, J, Madsen, J and Brocher, T. (1987) Multi-channel seismic imaging of a crustal magma chamber along the East Pacific Rise. *Nature* **326**: 35–41.

Dick, HJB, Lin, J and Schouten, H. (2003) An ultraslow-spreading class of ocean ridge. *Nature* **426**: 405–12.

Dosso, L, Bougault, H, Langmuir, C, Bollinger, C, Bonnier, O. and Etoubleau, J. (1999) The age and distribution of mantle heterogeneity along the Mid-Atlantic Ridge (31–41°N). *Earth and Planetary Science Letters* **170**: 269–86.

Duncan, RA and Richards, MA. (1991) Hotspots, mantle plumes, flood basalts, and true polar wander. *Reviews of Geophysics* **29**: 31–50.

Dupré, B and Allègre, CJ. (1983) Pb-Sr isotope variation in Indian Ocean basalts and mixing phenomena. *Nature* **303**: 1436.

Eiler, JM, Schiano, P, Kitchen, N and Stolper, EM. (2000) Oxygen-isotope evidence for recycled crust in the sources of mid-ocean-ridge basalts. *Nature* **403**: 530–34.

Elderfield, H and Schultz, A. (1996) Mid-ocean ridge hydrothermal fluxes and the chemical composition of the ocean. *Annual Reviews of Earth and Planetary Sciences* **24**: 191–224.

Gass, IG and Smewing, JD. (1973) Intrusion, extrusion and metamorphism at constructive margins: Evidence from the Troodos Massif, Cyprus. *Nature* **242**: 26–9.

Gillis, KM. (2002) The rootzone of an ancient hydrothermal system exposed in the Troodos ophiolite, Cyprus. *Journal of Geology* **110**: 57–74.

Gillis, KM and Coogan, LA. (2002) Anatectic migmatites from the roof of an ocean ridge magma chamber. *Journal of Petrology* **43**: 2075–95.

Gladczenko, TP, Coffin, MF and Eldholm, O. (1997) Crustal structure of the Ontong Java Plateau: Modeling of new gravity and existing seismic data. *Journal of Geophysical Research* **102**: 22711–29.

Godard, M, Awaji, S, Hansen, H, Hellebrand, E, Brunelli, D, Johnson, K, Yamasaki, T, Maeda, J, Abratis, M, Christie, DM, Kato, Y, Mariet, C and Rosner, M. (2009) Geochemistry of a long in-situ section of intrusive slow-spread oceanic lithosphere: Results from IODP Site U1309 (Atlantis Massif, 30°N Mid-Atlantic-Ridge). *Earth and Planetary Science Letters* 10.1016/j.epsl.2008.12.034.

Graham, DW. (2002) Noble gas isotope geochemistry of Mid-ocean ridge and ocean island basalts: Characterization of mantle source reservoirs. In: D Porcelli, CJ Ballentine and R Wieler (eds.), Noble gases in geochemistry and cosmochemistry. Reviews in Mineralogy. Mineralogical Society of America, Washington, D.C., pp. 247–317.

Green, DH, Falloon, TJ, Eggins, SM and Yaxley, GM. (2001) Primary magmas and mantle temperatures. *European Journal of Mineralogy* **13**: 437–51.

Haase, KM. (1996) The relationship between the age of the lithosphere and the composition of oceanic magmas: Constraints on partial melting, mantle sources and the thermal structure of the plates. *Earth and Planetary Science Letters* **144**: 75–92.

Haase, KM. (2002) Geochemical constraints on magma sources and mixing processes in Easter Microplate MORB (SE Pacific): A case study of plume-ridge interaction. *Chemical Geology* **182**: 335–55.

Haase, KM, Devey, CW, Mertz, DF, Stoffers, P and Garbe-Schönberg, C-D. (1996) Geochemistry of lavas from Mohns Ridge, Norwegian-Greenland Sea: Implications for melting conditions and magma sources near Jan Mayen. *Contributions to Mineralogy and Petrology* **123**: 223–37.

Haase, KM, Stroncik, NA, Hékinian, R and Stoffers, P. (2005) Nb-depleted andesites from the Pacific-Antarctic Rise as analogues for early continental crust. *Geology* **33**: 921–4.

Hart, SR, Blusztajn, J, Dick, HJB, Meyer, PS and Muehlenbachs, K. (1999) The fingerprint of seawater circulation in a 500-meter section of ocean gabbros. *Geochimica et Cosmochimica Acta* **63**: 4059–80.

Hart, SR, Hauri, EH, Oschmann, LA and Whitehead, JA. (1992) Mantle plumes and entrainment: Isotopic evidence. *Science* **256**: 517–20.

Hart, SR, Schilling, J-G and Powell, JL. (1973) Basalts from Iceland and along the Reykjanes Ridge: Sr isotope geochemistry. *Nature Physical Sciences* **246**: 104–7.

Hauri, EH, Whitehead, JA and Hart, SR. (1994) Fluid dynamic and geochemical aspects of entrainment in mantle plumes. *Journal of Geophysical Research* **99**: 24275–300.

Hékinian, R, Bideau, D, Hébert, R and Niu, Y. (1995) Magmatism in the Garrett transform fault (East Pacific Rise near 13°27′S). *Journal of Geophysical Research* **100**: 10163–85.

Henstock, TJ, Woods, AW and White, RS. (1993) The accretion of oceanic crust by episodic sill intrusion. *Journal of Geophysical Research* **98**: 4143–61.

Hess, PC. (1992) Phase equilibria constraints on the origin of ocean floor basalts. In: J Phipps Morgan, DK Blackman and JM Sinton (eds.), Mantle flow and melt generation at mid-ocean ridges. *Am. Geophys. Union, Geophys. Mon.,* pp. 67–102.

Hill, MN. (1957) Recent geophysical exploration of the ocean floor. *Physics and Chemistry of the Earth* **2**: 129–63.

Hirschmann, MM. (2000) Mantle solidus: Experimental constraints and the effects of peridotite composition. *Geochemistry Geophysics Geosystems* **1**: 10.1029/2000GC000070.

Hirschmann, MM, Baker, MB and Stolper, EM. (1995) Partial melting of mantle pyroxenite. *Eos Transactions of the American Geophysical Union* **76**: 696.

Hirschmann, MM and Stolper, EM. (1996) A possible role for garnet pyroxenite in the origin of the "garnet signature" in MORB. *Contributions to Mineralogy and Petrology* **124**: 185–208.

Hofmann, A and Harris, C. (2008) Silica alteration zones in the Barberton greenstone belt: A window into sub-seafloor processes 3.5–3.3 Ga ago. *Chemical Geology* **257**: 221–39.

Hofmann, AW. 1997. Mantle geochemistry: the message from oceanic volcanism. *Nature* **385**: 219–29.

Hofmann, AW. (2003) Sampling mantle heterogeneity through oceanic basalts: Isotopes and trace elements, Treatise on Geochemistry. Elsevier, pp. 61–101.

Hofmann, AW and White, WM. 1982. Mantle plumes from ancient oceanic crust. *Earth and Planetary Science Letters* **57**: 421–36.

Hooft, EE, Detrick, RS and Kent, GM. (1997) Seismic structure and indicators of magma budget along the southern East Pacific Rise. *Journal of Geophysical Research* **102**: 27319–40.

Humler, E, Langmuir, C and Daux, V. (1999) Depth versus age: new perspectives from the chemical compositions of ancient crust. *Earth and Planetary Science Letters* **173**: 7–23.

Iwamori, H, McKenzie, D and Takahashi, E. (1995) Melt generation by isentropic mantle upwelling. *Earth and Planetary Science Letters* **134**: 253–66.

Jacob, D, Jagoutz, E, Lowry, D, Mattey, D and Kudrjavtseva, G. (1994) Diamondiferous eclogites from Siberia: Remnants of Archean oceanic crust. *Geochimica et Cosmochimica Acta* **58**: 5191–207.

Jaques, AL and Green, DH. (1980) Anhydrous melting of peridotite at 0–15 Kb pressure and the genesis of tholeiitic basalts. *Contributions to Mineralogy and Petrology* **73**: 287–310.

Karson, JA. (2002) Geologic structure of the uppermost oceanic crust created at fast-to intermediate-rate spreading centres. *Annual Reviews of Earth and Planetary Sciences* **30**: 347–84.

Kay, R, Hubbard, NJ and Gast, PW. (1970) Chemical characteristics and origin of oceanic ridge volcanic rocks. *Journal of Geophysical Research* **75**: 1585–613.

Kerr, AC. (2003) Oceanic Plateaus, Treatise on Geochemistry. Elsevier, pp. 537–65.

Kinzler, RJ and Grove, TL. 1992. Primary magmas of Mid-Ocean Ridge basalts, 2. Applications. *Journal of Geophysical Research* **97**: 6907–26.

Kinzler, RJ and Grove, TL. (1993) Corrections and further discussion of the primary magmas of Mid-ocean Ridge basalts, 1 and 2. *Journal of Geophysical Research*, **98**: 22339–48,

Klein, EM. (2003) Geochemistry of the igneous oceanic crust, Treatise on Geochemistry. Elsevier, pp. 433–63.

Klein, EM and Langmuir, CH. (1987) Global correlations of oceanic ridge basalt chemistry with axial depth and crustal thickness. *Journal of Geophysical Research* **92**: 8089–8115.

Klein, EM, Langmuir, CH, Zindler, A, Staudigel, H and Hamelin, B. (1988) Isotope evidence of a mantle convection boundary at the Australian – Antarctic Discordance. *Nature* **333**: 623–9.

Koepke, J, Berndt, J, Feig, ST and Holtz, F. (2007) The formation of SiO_2-rich melts within the deep oceanic crust by hydrous partial melting of gabbros. *Contributions to Mineralogy and Petrology* **153**: 67–84.

Koepke, J, Feig, ST and Snow, J. (2005) Hydrous partial melting within the lower oceanic crust. *Terra Nova* **17**: 286–91.

Kohlstedt, DL. (1992) Structure, rheology and permeability of partially molten rocks at low melt fractions. In: J Phipps Morgan, DK Blackman and JM Sinton (eds.), Mantle flow and melt generation at Mid-Ocean Ridges. American Geophysical Union Monograph, Washington, D.C., pp. 103–21.

Kushiro, I. (2001) Partial melting experiments on peridotite and origin of Mid-ocean Ridge Basalt. *Annual Reviews of Earth and Planetary Sciences* **29**: 71–107.

Kusky, TM, Li, J-H and Tucker, RD. (2001) The Archean Dongwanzi ophiolite complex, North China craton: 2.505-billion-year-old oceanic crust and mantle. *Science* **292**: 1142–5.

Langmuir, CH and Hanson, GN. (1980) An evaluation of major element heterogeneity in the mantle sources of basalts. *Philosophical Transactions of the Royal Society of London*, **A297**: 383–407.

Langmuir, CH, Klein, EM and Plank, T. (1992) Petrological systematics of mid-ocean ridge basalts: constraints on melt generation beneath ocean ridges.

In: J. Phipps Morgan, D.K. Blackman and J.M. Sinton (Editors), Mantle flow and melt generation at mid-ocean ridges. Am. Geophys. Union Mem., Washington DC, pp. 183–280.

Macdonald, KC. (1982) Mid-Ocean Ridges: Fine scale tectonic, volcanic and hydrothermal processes within the plate boundary zone. *Annual Reviews of Earth and Planetary Sciences* **10**: 155–90.

Macdonald, KC. (2001) Mid-ocean ridge tectonics, volcanism and geomorphology. In: J. Steele, S. Thorpe and K. Turekian (eds.), Encyclopedia of Ocean Sciences. Academic Press, pp. 1798–813.

Macdonald, KC, Scheirer, DS and Carbotte, SM. (1991) Mid-Ocean Ridges: Discontinuities, segments and giant cracks. *Science* **253**: 986–94.

Magde, LS, Barclay, AH, Toomey, DR, Detrick, RS and Collins, JA. (2000) Crustal magma plumbing within a segment of the Mid-Atlantic Ridge, 35°N. *Earth and Planetary Science Letters* **175**: 55–67.

McCulloch, MT and Bennett, VC. (1994) Progressive growth of the Earth's continental crust and depleted mantle: Geochemical constraints. *Geochimica et Cosmochimica Acta* **58**: 4717–38.

McKenzie, D. (1985) The extraction of magma from the crust and mantle. *Earth and Planetary Science Letters* **74**: 81–91.

McKenzie, D and Bickle, MJ. (1988) The volume and composition of melt generated by extension of the lithosphere. *Journal of Petrology* **29**: 625–79.

Michael, PJ and Cornell, WC. (1998) Influence of spreading rate and magma supply on crystallization and assimilation beneath mid-ocean ridges: Evidence from chlorine and major element chemistry of mid-ocean ridge basalts. *Journal of Geophysical Research* **103**: 18325–56.

Michael, PJ and Schilling, J-G. (1989) Chlorine in mid-ocean ridge magmas: Evidence for assimilation of seawater-influenced components. *Geochimica et Cosmochimica Acta* **53**: 3131–43.

Möller, A, Appel, P, Mezger, K and Schenk, V. (1995) Evidence for a 2 Ga subduction zone: Eclogites in the Usagaran belt of Tanzania. *Geology* **23**: 1067–70.

Montelli, R, Nolet, G, Dahlen, FA, Masters, G, Engdahl, ER and Hung, S-H. (2004) Finite-frequency tomography reveals a variety of plumes in the mantle. *Science* **303**: 338–43.

Morgan, WJ. (1971) Convection plumes in the lower mantle. *Nature* **230**: 42–3.

Nicolas, A, Reuber, I and Benn, K. (1988) A new magma chamber model based on structural studies in the Oman ophiolite. *Tectonophysics* **151**: 87–105.

Niu, Y and Batiza, R. (1991) An empirical method for calculating melt compositions produced beneath mid-ocean ridges: application for axis and off-axis (seamounts) melting. *Journal of Geophysical Research* **96**: 21753–77.

Niu, Y and O'Hara, MJ. (2008) Global correlations of ocean ridge basalt chemistry with axial depth: a new perspective. *Journal of Petrology* **49**: 633–64.

O'Hara, MJ. (1968) Are ocean floor basalts primary magmas? *Nature* **220**: 683–6.

Pearce, JA, Leat, PT, Barker, PF and Millar, IL. (2001) Geochemical tracing of Pacific-to-Atlantic upper-mantle flow through the Drake passage. *Nature* **410**: 457–61.

Peltonen, P, Kontinen, A and Huhma, H. (1996) Petrology and geochemistry of metabasalts from the 1.95 Ga Jormua ophiolite, northeastern Finland. *Journal of Petrology* **37**: 1359–83.

Perfit, MR and Chadwick, WW. (1998) Magmatism at Mid-ocean ridges: Constraints from volcanological and geochemical investigations. In: WR Buck, PT Delaney, JA Karson and Y Lagabrielle (eds.), Faulting and Magmatism at Mid-ocean Ridges. Merican Geophysical Union, Washington, D.C., pp. 59–115.

Perfit, MR, Fornari, DJ, Malahoff, A and Embley, RW. (1983) Geochemical studies of abyssal lavas recovered by DSRV Alvin from eastern Galapagos Rift, Inca Transform, and Ecuador Rift 3. Trace element abundances and petrogenesis. *Journal of Geophysical Research* **88**: 10551–72.

Perfit, MR, Fornari, DJ, Ridley, WI, Kirk, PD, Casey, J, Kastens, KA. Reynolds, JR, Edwards, M, Desonie, D, Shuster, R and Paradis, S. (1996) Recent volcanism in the Siqueiros transform fault: picritic basalts and implications for MORB magma genesis. *Earth and Planetary Science Letters* **141**: 91–108.

Perfit, MR, Fornari, DJ, Smith, MC, Bender, JF, Langmuir, CH and Haymon, RM. (1994) Small-scale spatial and temporal variations in mid-ocean ridge crest magmatic processes. *Geology* **22**: 375–9.

Perfit, MR, Ridley, WI and Jonasson, IR. (1999) Geologic, petrologic, and geochemical relationships betwen magmatism and massive sulfide mineralization along the Eastern Galapagos Spreading Center. In: CT Barrie and MD Hannington (eds.), Volcanic-associated massive sulfide deposits: Processes and examples in modern and ancient settings. Reviews of Economic Geology. Society of Economic Geologists, Ottawa, pp. 75–100.

Pertermann, M and Hirschmann, MM. (2003) Partial melting experiments on a MORB-like pyroxenite between 2 and 3 GPa: Constraints on the presence of pyroxenite in basalt source regions from solidus location and melting rate. *Journal of Geophysical Research* **108**: doi: 10.1029/2000JB000118.

Phipps Morgan, J, Morgan, WJ, Zhang, Y-S and Smith, WHF. (1995) Observational hints for a plume-fed, suboceanic asthenosphere and its role in mantle convection. *Journal of Geophysical Research*, **100**: 12753–67.

Poreda, RJ, Schilling, J-G and Craig, H. (1993) Helium isotope ratios in Easter microplate basalts. *Earth and Planetary Science Letters* **119**: 319–29.

Presnall, DC, Dixon, JR, O'Donnell, TH and Dixon, SA. (1979) Generation of mid-ocean ridge tholeiites. *Journal of Petrology* **20**: 3–35.

Presnall, DC and Hoover, JD. (1984) Composition and depth of origin of primary mid-ocean ridge basalts. *Contributions to Mineralogy and Petrology* **87**: 170–8.

Presnall, DC and Hoover, JD. (1987) High pressure phase equilibrium constraints on the origin of mid-ocean ridge basalts. In: BO Mysen (ed.), Magmatic Processes: Physicochemical Principles. The Geochemical Society, pp. 75–89.

Puchelt, H and Emmermann, R. (1983) Petrogenetic implications of tholeiitic basalt glasses from the East Pacific Rise and the Galapagos Spreading Center. *Chemical Geology* **38**: 39–56.

Purdy, GM, Kong, LSL, Christeson, GL and Solomon, SC. (1992) Relationship between spreading rate and the seismic structure of mid-ocean ridges. *Nature* **355**: 815–817.

Ringwood, AE. (1975) Composition and petrology of the Earth's mantle. McGraw-Hill, New York, 618 pp.

Robinson, JAC and Wood, BJ. (1998) The depth of the spinel to garnet transition at the peridotite solidus. *Earth and Planetary Science Letters* **164**: 277–84.

Roeder, PL and Emslie, RF. (1970) Olivine-liquid equilibrium. *Contributions to Mineralogy and Petrology* **29**: 275–89.

Rubin, KH, Smith, MC, Bergmanis, EC, Perfit, MR, Sinton, JM and Batiza, R. (2001) Geochemical heterogeneity within mid-ocean ridge lava flows: insights into eruption, emplacement and global variations in magma generation. *Earth and Planetary Science Letters* **188**: 349–67.

Rudnick, RL and Fountain, DM. (1995) Nature and composition of the continental crust: a lower crustal perspective. *Reviews of Geophysics* **33**: 267–309.

Ryan, MP. (1987) Neutral buoyancy and the mechanical evolution of magmatic systems. In: BO Mysen (ed.),

Magmatic Processes: Physicochemical Principles. The Geochemical Society, pp. 259–87.

Ryan, MP. (1993) Neutral buoyancy and the structure of Mid-ocean Ridge magma chambers. *Journal of Geophysical Research* **98**: 22321–38.

Saal, AE, Hauri, EH, Langmuir, CH and Perfit, MR. (2002) Vapour undersaturation in primitive mid-ocean-ridge basalt and the volatile content of Earth's upper mantle. *Nature* **419**: 451–5.

Salters, VJM and Hart, SR. (1989) The hafnium paradox and the role of garnet in the source of mid-ocean ridge basalts. *Nature* **342**: 420–2.

Scheirer, DS and Macdonald, KC. (1993) Variation in cross-sectional area of the axial ridge along the East Pacific Rise: Evidence for the magmatic budget of a fast spreading center. *Journal of Geophysical Research* **98**: 7871–85.

Scheirer, DS and Macdonald, KC. (1995) Near-axis seamounts on the flanks of the East Pacific Rise, 8°N to 17°N. *Journal of Geophysical Research* **100**: 2239–59.

Schilling, J-G. (1973) Iceland mantle plume: geochemical study of Reykjanes Ridge. *Nature* **242**: 565–71.

Schilling, J-G. (1985) Upper mantle heterogeneities and dynamics. *Nature*, **314**: 62–7.

Schilling, J-G, Hanan, BB, McCully, B, Kingsley, RH and Fontignie, D. (1994) Influence of the Sierra Leone mantle plume on the equatorial Mid-Atlantic Ridge: A Nd-Sr-Pb isotopic study. *Journal of Geophysical Research* **99**: 12005–28.

Schilling, J-G, Kingsley, RH and Devine, JD. (1982) Galapagos hot spot-spreading center system 1. Spatial petrological and geochemical variations (83°W–101°W). *Journal of Geophysical Research* **87**: 5593–610.

Schilling, J-G, Sigurdsson, H, Davis, AN and Hey, RN. (1985) Easter microplate evolution. *Nature* **317**: 325–331.

Schilling, J-G, Zajac, M, Evans, R, Johnston, T, White, W, Devine, JD and Kingsley, R. (1983) Petrologic and geochemical variations along the Mid-Atlantic Ridge from 29°N to 73°N. *American Journal of Science* **283**: 510–86.

Sigurdsson, H. (1981) First-order major element variation in basalt glasses from the Mid-Atlantic Ridge: 29°N to 73°N. *Journal of Geophysical Research* **86**: 9483–502.

Sinha, MC, Navin, DA, MacGregor, LM, Constable, S, Peirce, C, White, A, Heinson, G and Inglis, MA. (1997) Evidence for accumulated melt beneath the slow-spreading Mid-Atlantic Ridge. *Philosophical*

Transactions of the Royal Society of London **355**: 233–53.

Sinton, JM and Detrick, RS. (1992) Mid-ocean ridge magma chambers. *Journal of Geophysical Research* **97**: 197–216.

Staudigel, H, Plank, T, White, W and Schmincke, H-U. (1996) Geochemical fluxes during seafloor altera- tion of the basaltic upper oceanic crust: DSDP sites 417 and 418. In: GE Bebout, DW Scholl, SH Kirby and JP Platt (eds.), Subduction top to bottom. Am. Geophys. Un. Geophys. Monogr., Washington, DC, pp. 19–38.

Staudigel, H, Tauxe, L, Gee, JS, Bogaard, P, Haspels, J, Kale, G, Leenders, A, Meijer, P, Swaak, B, Tuin, M, MC Van Soest, Verdurmen, EAT and Zevenhuizen, A. (1999) Geochemistry and intrusive directions in sheeted dikes in the Troodos Ophiolite: Implications for Mid-Ocean Ridge Spreading Centers. *Geochemistry Geophysics Geosystems* **1**: 1999GC000001.

Stewart, MA, Klein, EM and Karson, JA. (2002) Geochemistry of dikes and lavas from the north wall of the Hess Deep Rift: Insights into the four- dimensional character of crustal construction at fast spreading mid-ocean ridges. *Journal of Geophysical Research* **107**: 10.1029/2001JB000545.

Stolper, E and Walker, D. (1980) Melt density and the average composition of basalt. *Contributions to Mineralogy and Petrology* **74**: 7–12.

Suen, CJ and Frey, FA. (1987) Origins of the mafic and ultramafic rocks in the Ronda peridotite. *Earth and Planetary Science Letters* **85**: 183–202.

Sun, S-s and McDonough, WF. (1989) Chemical and isotopic systematics of oceanic basalts: implications for mantle composition and processes. In: A.D. Saunders and M.J. Norry (Editors), Magmatism in the ocean basins. Geol. Soc. Spec. Publ., London, pp. 313–45.

Takahashi, E and Kushiro, I. (1983) Melting of a dry peridotite at high pressures and basalt magma genesis. *American Mineralogist* **68**: 859–79.

Tejada, MLG, Mahoney, JJ, Castillo, PR, Ingle, SP, Sheth, HC and Weis, D. (2004) Pin-pricking the elephant: evidence on the origin of the Ontong Java Plateau from Pb-Sr-Hf-Nd isotopic characteris- tics of ODP Leg 192 basalts. In: JG Fitton, JJ Mahoney, PJ Wallace and AD Saunders (eds.), Origin and evolution of the Ontong Java Plateau. Geological Society of London, Special Publications, London, pp. 133–50.

ten Brink, US and Brocher, TM. (1987) Multichannel seismic evidence for a subcrustal intrusive complex under Oahu and a model for Hawaiian volcanism. *Journal of Geophysical Research* **92**: 13687–707.

Vlastélic, I, Aslanian, D, Dosso, L, Bougault, H, Olivet, JL and Géli, L. (1999) Large-scale chemical and thermal division of the Pacific mantle. *Nature* **399**: 345–50.

Vogt, PR. 1974. Volcano spacing, fractures, and thick- ness of the lithosphere. *Earth and Planetary Science Letters* **21**: 235–52.

Vogt, PR and Jung, WY. (2004) The Terceira Rift as hyper- slow, hotspot-dominated oblique spreading axis: A comparison with other slow-spreading plate bounda- ries. *Earth and Planetary Science Letters* **218**: 77–90.

Wendt, JI, Regelous, M, Niu, Y, Hékinian, R and Collerson, KD. (1999) Geochemistry of lavas from the Garrett Transform Fault: insights into mantle hetero- geneity beneath the eastern Pacific. *Earth and Planetary Science Letters* **173**: 271–84.

Wessel, P and Lyons, S. (1997) Distribution of large Pacific seamounts from Geosat/ERS-1: Implications for the history of intraplate volcanism. *Journal of Geophysical Research* **102**: 22459–75.

White, RS. (1984) Atlantic oceanic crust: seismic struc- ture of a slow-spreading ridge. In: IG Gass, SJ Lipard and AW Shelton (eds.), Ophiolites and Oceanic Lithosphere. Geol. Soc. Spec. Publ., pp. 101–11.

White, RS. (1993) Melt production rates in mantle plumes. *Philosophical Transactions of the Royal Society of London*, **A342**: 137–53.

White, WM, Schilling, J-G and Hart, SR. (1976) Evidence for the Azores mantle plume from strontium isotope geochemistry of the Central North Atlantic. *Nature* **263**: 659–63.

Woodhead, JD. (1992) Temporal geochemical evolution in oceanic intra-plate volcanics: a case study from the Marquesas (French Polynesia) and comparison with other hotspots. *Contributions to Mineralogy and Petrology* **111**: 458–67.

Zindler, A and Hart, SR. (1986) Chemical geodynamics. *Annual Reviews of Earth and Planetary Sciences* **14**: 493–571.

Zindler, A, Staudigel, H and Batiza, R. (1984) Isotope and trace element geochemistry of young Pacific seamounts: implications for the scale of upper mantle heterogeneity. *Earth and Planetary Science Letters* **70**: 175–95.

5 Silicate Rock Weathering and the Global Carbon Cycle

SIGURDUR R. GISLASON[1] AND ERIC H. OELKERS[2]

[1]Institute of Earth Sciences, University of Iceland, Reykjavik, Iceland
[2]LMTG, UMR CNRS 5563, Université Paul-Sabatier, Toulouse, France

ABSTRACT

Silicate weathering and soil formation is critical to global-scale processes since silicates constitute about 83% of the rocks exposed at Earth's land surface. On geological time scales, atmospheric carbon dioxide content has been balanced by the weathering of Ca-Mg-silicate rocks, the burial and weathering of sedimentary organic matter and degassing of volcanoes. A large number of laboratory and field-based studies have been performed to better understand the link between silicate weathering, climate and the long-term carbon cycle. In this chapter we summarize some of this past work and suggest directions for future research.

INTRODUCTION

The link between increasing atmospheric CO_2 concentration and increasing global temperature has become one of the greatest concerns of the past decade (e.g. Gunter et al. 1996; Lackner 2003; Broecker 2005; Oelkers and Schott 2005; Oelkers and Cole 2008). Over the short term, the atmospheric carbon concentration is buffered by its interaction with vegetation, soils and the ocean (Houghton 2007), whereas over longer geological timescales atmospheric carbon dioxide content has been balanced by fluxes in and out of rocks; the weathering of Ca-Mg-silicate rocks, the burial and weathering of sedimentary organic matter; and degassing of volcanoes (e.g. Ebelmen 1845, 1847; Urey 1952; Garrels and Perry 1974; Holland 1978; Walker et al. 1981; Berner et al. 1983; Wallmann 2001; Berner 2004).

Jacques Joseph Ebelmen (1814–52), a French chemist, first published the idea that basalt weathering and associated processes control the levels of CO_2 and O_2 in the atmosphere in 1845. This connection between weathering and atmospheric composition was not independently deduced until more than 100 years later (Berner and Maasch, 1996). The origin of the link between wreathing and climate stem from reactions of divalent metal cations. As silicates weather, they release cations such as Ca and Mg, which can combine with dissolved CO_2 to form carbonate minerals. The negative feedback between atmospheric CO_2 and weathering stems from the observation that the weathering rates of silicate rocks

Frontiers in Geochemistry: Contribution of Geochemistry to the Study of the Earth, First edition. Edited by Russell S. Harmon and Andrew Parker.
© 2011 Blackwell Publishing Ltd. Published 2011 by Blackwell Publishing Ltd.

increase with temperature. Rising atmospheric CO_2 content increases average global temperature, which as a consequence accelerates silicate weathering rates and in turn increases CO_2 mineralization as carbonate rocks. A large number of laboratory and field-based studies have been performed in the recent past to understand better this link between silicate weathering and long-term global climate (e.g. Gislason and Eugster 1987; Brady 1991; Brady and Carroll 1994; Bluth and Kump 1994; Gibbs and Kump 1994; White and Blum 1995; Gislason et al. 1996, 2006, 2009; Brady and Gislason 1997; Gaillardet et al. 1999a; White et al. 1999a; Kump et al. 2000; Berg and Banwart 2000; Dessert et al. 2003; Olivia et al. 2003; Dupré et al. 2003; Ribe et al. 2004; Pokrovsky et al. 2005; Golubev et al. 2005; Wolff-Boenisch et al. 2006; Anderson 2007; Louvat et al. 2008). The goal of this chapter is to summarize some of this past work and to suggest directions for future research.

WEATHERING: DEFINITIONS AND PROCESSES

In the sense used here, 'weathering' is the decomposition of terrestrial rocks, soils and their constituent minerals through direct and indirect contact with the atmosphere. Weathering occurs *in situ* and is often coupled to 'erosion', which involves the movement of dissolved constituents by water and of rocks and minerals via water, ice, wind and gravity. Weathering of rocks can take place in the ocean, in suspension, or in sediments and sea-floor lavas can also be weathered.

Weathering is a consequence of two processes: mechanical breakdown and chemical dissolution. Mechanical weathering, involves the breakdown of materials by physical processes. By contrast, chemical weathering involves the direct attack of rocks by atmospheric or biologically produced chemicals and is commonly water-mediated (Berner and Berner 1996; Garrels and Mackenzie 1971). This chemical attack leads to rock dissolution which can be accelerated by acids derived primarily from the atmosphere, plants, bacteria

and pollution. Together, mechanical and chemical weathering result in the breakdown of rock, the formation of soil, and the production of dissolved constituents in solutions (Berner and Berner 1996). The products of mechanical and chemical weathering may be transported away from their site of origin or may remain in place as residual materials (Garrels and Mackenzie 1971). Taken together, weathering and erosion leads to 'denudation', the removal of material from the Earth's surface, leading to elevation reduction and relief in landforms. At steady state, soil profiles become constant, and weathering, erosion and denudation rates are equivalent. As such, much of what is currently known about weathering rates stems from the study of denudation rates and *vice versa*.

The weathering of silicates is critical to global-scale processes, since silicates constitute about 83% of the rocks exposed at Earth's land surface, whereas carbonates cover about 16% and evaporates little over 1% (Meybeck 1987). Of the silicate rocks, shales are the most abundant, comprising 40% of all silicates exposed at the surface, followed by sandstone at 19%, metamorphic rocks at 18%, plutonic igneous rocks at 13% and volcanic igneous rocks at 9.5% (Meybeck 1987). Together, metamorphic and igneous rocks are often referred to as 'crystalline rocks' (Blatt and Jones 1975; Nesbitt and Young 1984). It is the weathering and erosion of 'crystalline silicate rocks' that is the subject of this chapter.

Weathering begins by the breakdown of crystalline rocks, and thus a good starting point for studying weathering is the composition of crystalline rocks. Crystalline rocks range from Si-rich granites and rhyolites to Si-poor gabbros, basalts and ultramafic rocks. A summary of major silicate materials comprising the crystalline crust exposed at the surface is presented in Table 5.1. The major minerals/solid phases are plagioclase feldspar, quartz, volcanic glass, K-feldspar, biotite and muscovite. These minerals are not homogeneously distributed among rocks. Minerals containing the divalent metal cations required to remove CO_2 from the atmosphere – predominantly Ca, Mg and Fe – are concentrated in the

Table 5.1 Average mineralogical composition (vol.%) of the exposed 'cystalline' continental crust (Nesbitt and Young 1984).

Plagioclase	34.9
Quartz	20.3
Volcanic glass	12.5
Orthoclase	11.3
Biotite	7.6
Muscovite	4.4
Chlorite	1.9
Amphiboles	1.8
Pyroxenes	1.2
Olivines	0.2
Oxides	1.4
Other	2.6

Si-poor rocks, whereas quartz, muscovite and K-feldspar are concentrated in the Si-rich silicates. Chemical weathering of 'crystalline rocks' and soil formation are dominated by the alteration of plagioclase and volcanic glasses (Nesbitt and Young 1984), which accounts for close to half of the crystalline rock exposed at the surface. Quartz comprises roughly 20% of the remaining crystalline rock, but this mineral is relatively inert (Table 5.1; Nesbitt and Young 1984; Taylor and McLennan 1985).

The weathering of crystalline rocks produces both authigenic and clastic sediments and, ultimately, sedimentary rocks. The coarsest sediments derived from weathering of crystalline rocks are abundant in unstable minerals such as plagioclase and K-feldspar. Finer-grained sands tend to be dominated by relatively inert quartz. Still finer-grained sediments comprise shales, which are rich in clay minerals formed from the weathering of crystalline rocks. Weathering of mafic rocks produces shales rich in smectites, i.e. a family of clays with high cation-exchange capacity that swell when immersed in water, whereas weathering of granite results in shales rich in kaolinite and illite clays with low cation-exchange and shrink-swell capacities (Cox and Lowe 1995). The average shale is composed of 40 to 60% clay, 20 to 30% quartz, 5 to 10% feldspar, and minor amounts of iron oxide, carbonate, organic matter

and other components (Veizer and Mackenzie 2003 and references therein).

WEATHERING, EROSION AND ATMOSPHERIC CO_2

The weathering of silicates and removal of CO_2 from the atmosphere as carbonates usually occurs as a non-stoichiometric reaction, involving dissolution of the primary crystalline or glassy silicate, the precipitation of secondary clay minerals, and a release of Ca, and/or Mg and sometimes Fe to solution. For example, the weathering of Ca-rich plagioclase ($CaAl_2Si_2O_8$) is commonly coupled to the precipitation of the clay allophane, $Al_2SiO_5 \cdot 2.53H_2O$ (Wada et al. 1992) by the reaction:

$$\mathbf{CaAl_2Si_2O_8} + 2\ CO_2 + 5.53H_2O \rightarrow \quad (1a)$$

$$\mathbf{Al_2SiO_5\ 2.53H_2O} + Ca^{++} + H_4SiO_4^o + 2\ HCO_3^- \quad (1b)$$

$$Ca^{++} + H_4SiO_4^o + 2\ HCO_3^- \rightarrow \mathbf{CaCO_3} + $$
$$H_4SiO_4^o + CO_2 + H_2O \quad (1c)$$

Note that the CO_2 in this reaction serves as a proton donor driving the first part of this reaction forward (1a → 1b). The second part of this reaction (1c) brings together dissolved calcium and bicarbonate ions to form calcite ($CaCO_3$) and release part of the CO_2 originally consumed by the plagioclase dissolution. The weathering and erosion of Ca-plagioclase in this overall reaction causes net removal of one mole of CO_2 for each mole of plagioclase consumed. The first part of the reaction (1a → 1b) can occur either on the Earth's surface or in the ocean, by the dissolution of sediments in suspension or on the ocean floor. In contrast, carbonate precipitation generally occurs in the oceans, as most surficial waters associated with silicate terrains remain undersaturated with respect to calcite. Carbon-dioxide drawdown described by Eqn. 1 stems from the combined effect of weathering on the continents, and transport of weathered material by erosion to the oceans, where subsequently carbonate minerals can form in the ocean.

MINERAL DISSOLUTION AND PRECIPITATION RATES

Mineral dissolution

Much of what is known about chemical weathering rates derives from laboratory dissolution and precipitation experiments. An exhaustive number of experimental dissolution-rate studies have been performed over the past 30 years (e.g. Schott et al. 2009). These studies have shown that rates depend strongly on the type of mineral, temperature of dissolution and the composition of the aqueous solution. At constant temperature, the variation of rates with solution composition stems from two factors: (i) the dissolution mechanism, and (ii) the distance of the solution from equilibrium such that (Aagaard and Helgeson 1982; Oelkers 2001a)

$$r = r_+ s (1 - exp(-A/\sigma RT)) \qquad (2)$$

where r refers the dissolution rate, r_+ designates the far-from-equilibrium dissolution rate, s is the mineral-fluid surface area, A represents the chemical affinity (e.g. the driving force) of the dissolution reaction, which is 0 at equilibrium, R is the gas constant and T denotes absolute temperature, σ signifies a stoichiometric coefficient equal to the number of moles of activated complexes created during the dissolution of one mole of the dissolving mineral or glass, which was observed equal to a value of 1 for quartz (Berger et al. 1994) and basaltic glass (Daux et al. 1997) when the hydrated basaltic glass layer is normalized to one Si atom and 3 for alkali-feldspars (Gautier et al. 1994). Equation 2 indicates that the dissolution rate decreases systematically to zero as equilibrium is approached.

The variation of the far-from-equilibrium dissolution rate, r_+, with solution composition depends on the dissolution mechanism. Basaltic glass and the alkali-feldspars dissolve via H-rich, Al-poor precursors and, thus, $r_+ \propto \left(a_{H^+}^3 / a_{Al^{3+}} \right)^n$, Mg pyroxenes dissolve via a H-rich, Mg-poor precursor and, thus, $r_+ \propto \left(a_{H^+}^2 / a_{Mg^{2+}} \right)^n$, and quartz, anorthite and olivine dissolve via protonated or de-protonated surface species and thus $r_+ \propto \left(a_{H^+}^n \right)$ and/or $r_+ \propto \left(a_{OH^-}^n \right)$, where a_i refers to the activity of the subscripted aqueous species and n refers to a constant (Oelkers et al. 1994; Pokrovsky and Schott 2000; Oelkers 2001b; Oelkers and Schott 1995, 2001; Oelkers and Gislason 2001). These reaction mechanisms lead to a commonly described variation of far-from-equilibrium dissolution rates with pH. For alkali-feldspars and basaltic glass, r_+ decreases with increasing pH at acid conditions, minimizes at near-to-neutral pH, and increases with pH at basic conditions. For pyroxene and olivine, r_+ decreases continuously with increasing pH to at least pH = 10.

Examples of the variation of the far-from-equilibrium dissolution rates at 25°C for a number of rock-forming minerals and glasses are shown as a function of pH in Fig. 5.1. Rates shown in this figure are for the stoichiometric dissolution of the mineral, where no secondary minerals are formed, and the release of elements is in the same proportion as the dissolving mineral. Near-to-neutral far-from-equilibrium dissolution rates range by over 15 orders of magnitude. The fastest dissolving minerals are salts like halite and gypsum. The dissolution rates of these minerals are so rapid that they are commonly measured using rotating disk techniques; the sample mineral is rotated at more than 100 rpm and rates extrapolated to infinite rotation speed to obtain the far-from-equilibrium dissolution rates. The common carbonate minerals, calcite and dolomite, have intermediate far-from-equilibrium dissolution rates, and the rates for silicate minerals are relatively slow. The sluggish dissolution rates of many silicate minerals results from the fact that strong Si-O bonds need to be broken for their dissolution (c.f. Oelkers 2001a). Nevertheless, the far-from-equilibrium dissolution rates of the silicate minerals themselves vary considerably; the near-to-neutrality dissolution rate of the Mg-rich olivine, forsterite, is close to five orders of magnitude faster than that of the K-Al phyllosilicate mineral, illite.

The release of elements from a crystalline silicate rock undergoing dissolution depends on its mineralogical composition and will not be stoichiometric, even in the absence of secondary

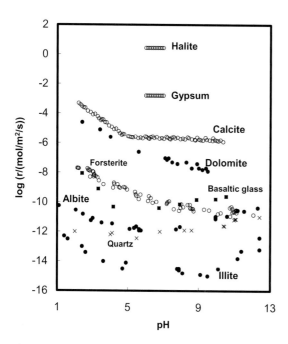

Fig. 5.1 Dissolution rate of salts, minerals and basaltic glass, normalized to BET surface area at 25–30°C as a function of pH. The data illustrated are from the following sources: halite (Alkattan et al. 1997); gypsum (Raines and Dewars 1997); calcite (Busenberg and Plummer 1986); dolomite (Chou et al. 1989); forsterite (Pokrovsky and Schott 2000); basaltic glass (Gislason and Oelkers 2003); albite (Chou and Wollast 1985); quartz (Brady and Walther 1990); illite (Kohler et al. 2003).

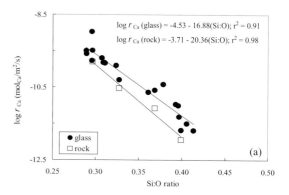

Fig. 5.2 Ca release rate from glass and crystalline rocks as a function of their Si:O, ratio ranging from gabbroic composition at the far left to granitic composition at the right. The filled circles and open squares represent results reported by Wolff-Boenisch et al. (2006), and the lines correspond to a least-squares linear regression fit of these data. The equations and the coefficients of correlation (r^2) of these fits are provided in the figure.

mineral precipitation. This is due to the fact that each mineral in the rock will dissolve at a different rate, and each mineral has a distinct chemical composition. An estimate of the rate at which each constituent element is released by the far-from-equilibrium dissolution of a rock can be made by summing up the contribution to the flux of the element of interest for each of its constituent minerals. An example of the result of one such calculation is shown in Fig. 5.2, which shows the far-from-equilibrium release rate of Ca from crystalline and glassy rocks as a function of the Si-to-O ratio of the rock at 25°C and pH = 4. Granitic rocks, which are crystalline and Si-rich, are represented by the open square in the lower

right part of the figure, whereas mafic, Si-poor crystalline rocks like gabbro are represented by the open square in the upper left portion of Fig. 5.2. Calculations suggest that Ca is released two orders of magnitude faster from gabbro than from granite, and that glassy rocks release Ca to solution by a factor of two to three faster than crystalline rocks of corresponding composition. Thus the fastest-dissolving silicates are glassy mafic and ultramafic rocks.

Comparing laboratory dissolution rates to the field

The direct comparison of laboratory-derived estimates of silicate weathering rates to those in the field is problematical. Laboratory, generated dissolution rates, such as shown in Figs. 5.1 and 5.2, are commonly based on experiments performed at far-from-equilibrium conditions, and in the absence of precipitation of secondary phases. Under natural conditions in the field, fluids can remain in contact with a mineral surface for far longer than in most laboratory experiments, causing the fluid to approach equilibrium with

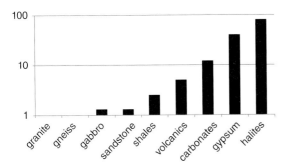

Fig. 5.3   Relative chemical denudation rates of various major rock types normalized to granite. Results shown in this figure were obtained from studies of monolthologic streams examined by Meybeck (1987) and Louvat (1997).

the dissolving minerals. As quantified theoretically through Eqn. 2, dissolution rates slow to zero at equilibrium, so just the approach to equilibrium can lower rates by orders of magnitude compared to their laboratory measured far-from-equilibrium counterparts. Fluids that remain in contact with minerals for an extended time in the field can also become supersaturated with respect to secondary mineral phases. Some of these supersaturated phases may precipitate, thus limiting the release of certain elements during weathering.

Some of the differences between field and laboratory dissolution rates can be assessed by comparison of Figs. 5.1 and 5.2 with Fig. 5.3. Figure 5.3 illustrates the relative chemical denudation rates of various rocks relative to that of granite. The chemical denudation rates in this figure were determined from the dissolved chemical fluxes in rivers draining monolithic catchments; as the dissolved constituents in the rivers are the product of chemical weathering of the single catchment lithology, it seems likely that they should reflect the dissolution rates of these rocks. The relative order of chemical denudation rates shown in Fig. 5.3 is similar to that deduced from laboratory rates: halites and gypsum weather fastest, carbonates exhibit intermediate weathering rates, and silicates weather the slowest. Yet the relative difference between field denudation rates and laboratory-based dissolution rates are dramatically different. Whereas halite and gypsum far-from-equilibrium dissolution rates at neutral conditions are more than approximately 10 orders of magnitude faster than those of silicates, their field chemical rates are no more than two orders of magnitude faster. A major reason for this smaller difference in the field is that the rates shown in Figs. 5.1 and 5.2 are based on far-from-equilibrium conditions. The faster a mineral dissolves, the more rapidly it will approach equilibrium in nature. As equilibrium is approached, dissolution rates slow down, thus attenuating the differences in the dissolution rates between faster- and slower-dissolving minerals. Natural surface waters are also commonly supersaturated with respect to secondary minerals, which additionally constrain element-release from river catchments containing primary minerals that dissolve rapidly. Furthermore, river catchments are never truly monolithologic. For example, small quantities of calcite in the fractures or joints of granite can increase dramatically the calcium flux out of granitic catchments (e. g. White et al. 1999a and 1999b).

Secondary mineral precipitation

The quantification of silicate mineral precipitation rates has been problematical, both in the laboratory and in the field. Whereas all minerals dissolve when exposed to an undersaturated fluid, many minerals do not appear to precipitate in supersaturated fluids. Moreover, there lacks an accurate geochemical model for predicting at which conditions a supersaturated fluid will begin to precipitate a particular secondary phase. For example, many surface waters are strongly supersaturated with respect to quartz; an example is shown in Fig. 5.4. Quartz, however, precipitates rarely if at all at Earth-surface conditions (Dove and Rimstidt 1994; Gislason et al. 1997). By contrast, quartz readily precipitates in sedimentary basins at temperatures in excess of approximately 80°C (Bjorlykke and Egeberg 1993; Oelkers et al. 1996).

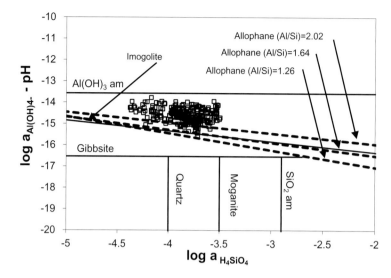

Fig. 5.4 Logarithmic activity diagram showing phase relations among Al-Si bearing minerals in northeastern Iceland river waters at 5°C. The symbols correspond to measured northeastern Iceland river water compositions and the lines represent equilibrium between water and the indicated mineral. Figure modified after Eiriksdottir, Gislason and Oelkers (2010).

The secondary silicate phases that precipitate during weathering are dominated by clay minerals, particularly kaolinite, illite, allophane, smectites and their amorphous precursors (e. g. Banfield et al. 1991; Wada et al. 1992; Cox and Lowe 1995; Veizer and Mackenzie 2003). However, relatively few studies of clay mineral precipitation rates are available. As such, clay mineral precipitation rates remain one of the biggest obstacles to the application of geochemical modelling codes to quantifying chemical weathering in nature. There are several reasons why so few studies of clay mineral precipitation rates exist, including: (i) difficulties in quantifying the precipitating clay mineral phase and its composition; (ii) poor understanding of the thermodynamic properties of clay minerals, which depend on both the composition of the precipitating phase and its crystal size; and (iii) the small particle size of clays, which makes it difficult to keep the clay minerals in an experimental reactor. Among the few experimental studies performed on clay mineral precipitation, most have been performed on kaolinite (c.f. Nagy et al. 1991; Devidal et al. 1997; Yang and Steefel 2008). These studies suggest that kaolinite can readily precipitate at Earth-surface conditions, though the rates depend strongly on the surface chemistry and structure of the clay sur-

faces. Field evidence also attests to the rapid formation of clay minerals at the Earth's surface. For example, many surface waters are close to equilibrium with the clay minerals present in their aqueous system. For example, the compositions of river waters of northeastern Iceland are plotted on the aqueous activity diagram shown in Fig. 5.4. River waters appear to be supersaturated or close to equilibrium with allophane, the Al-silicate clay commonly found in these river catchments (Arnalds 2005).

MINERAL SURFACE AREA

Another critical parameter in applying laboratory measured rates and geochemical models to quantifying chemical weathering and chemical denudation rates that remains poorly understood is the mineral-fluid interfacial surface area. Mineral dissolution and precipitation rates are surface-controlled reactions and their rates are proportional to the surface area of the mineral or glass in contact with the reactive fluid (see Eqn. 2). Mineral surface areas increase dramatically during weathering. The surface area of a smooth 10 cm basalt cube is 0.02 cm^2/g. By contrast, the surface area of the basaltic suspended material collected from the rivers of northeastern Iceland, measured by a Brunauer, Emmett and Teller (BET) gas adsorption technique, range from 100,000 to 800,000 cm²/g (Oelkers et al. 2004). If all other factors remain the same, just the mechanical weathering and erosion of silicate rocks from the protolith to form river-suspended material can increase dissolution rates by over six orders of magnitude!

FIELD MEASUREMENTS OF SILICATE WEATHERING AND DENUDATION RATES

A number of different approaches have been taken in an attempt to determine silicate weathering rates in the field (c.f. White 2005). One method that has been successful in generating field-based weathering rates of individual minerals uses a mass-balance approach. This method compares the concentration differences of a mineral, element, or isotopic species in a weathered rock, referred to as 'regolith', versus that in its original protolith. Effects of external factors such as compaction or erosion can be overcome by comparing relative mobility of various elements in the system to one or more elements assumed to be immobile (see e.g. Craig and Loughnan 1964; Brimhall and Dietrich 1987). This approach has been applied successfully by White and co-workers (White et al. 2001, 2002), who converted elemental mobilities to changes in mineralogy during weathering, using a series of linear equations describing mineral stoichiometries. In this manner, weathering rates of individual minerals can be deduced. Another direct approach to defining weathering rates of individual minerals is by comparing mineral abundances in soil chronosequences (e.g. White et al. 1996). Soil chronosequences are defined as a group of soils that differ in age, but have similar parent materials and formed under similar climatic and geomorphic conditions (Jenny 1941). Such studies provide weathering rates that are the average of the tens to millions of years period of rock alteration. This method for generating weathering rates of individual minerals can be directly compared with dissolution rates of primary minerals measured in the laboratory. Weathering rates of K-feldspars and plagioclases generated from field studies of this kind range from $10^{-16.5}$ to 10^{-13} mol m^{-2} s^{-1}, which are approximately two to five orders of magnitude slower than corresponding far-from-equilibrium rates measured during laboratory studies (White and Brantley 2003, and Fig. 5.1). Reasons for these differences include: (i) the slowing of rates as equilibrium is approached, as it is likely that the soil water responsible for weathering in soils is close to equilibrium with the dissolving minerals; and (ii) the age of the rocks, as weathering rates appear to slow with the duration of weathering (Gislason et al. 1996, White and Brantley 2003).

Insight into silicate weathering rates on a larger scale can be obtained through measurements of denudation rates based on river fluxes (e.g. Meybeck 1987; Bluth and Kump 1994; Gislason et al. 1996; Gaillardet et al. 1999a; Dessert et al. 2003; Pokrovsky et al. 2005; Vigier et al. 2006; Zakharova et al. 2005; Louvat et al. 2008). The advantage of such studies is that they provide an integrated average of denudation rates over a large area and take account of the large number of coupled processes that contribute to denudation. By contrast to mass balance and soil chronosequence studies, the river-flux approach provides a near-instantaneous look at weathering and denudation processes. In river studies, the flux of dissolved material being transported from the continents to the oceans has been attributed to chemical denudation, and that of suspended material has been attributed to mechanical denudation.

FACTORS AFFECTING SILICATE WEATHERING AND DENUDATION RATES

Runoff

Both laboratory and field studies suggest a number of factors that would influence chemical

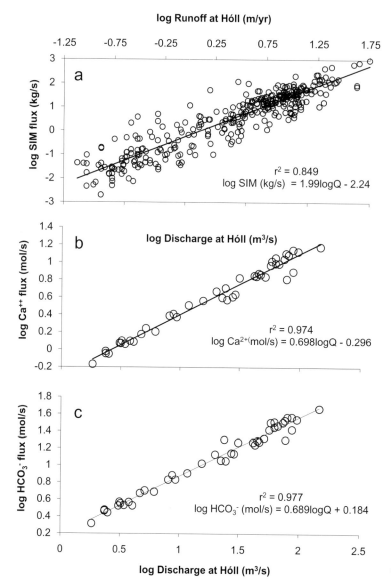

Fig. 5.5 Variation of the river suspended inorganic material (SIM) fluxes and the dissolved Ca^{2+} and HCO_3^- fluxes with discharge in the Jokulsa River at Hóll in northeastern Iceland. Data cover the period 1963–2004 in (a) but 1998–2004 in (b) and (c). The symbols represent measured fluxes reported by Gislason et al. (2009), and the lines correspond to a least-squares linear regression fit of these data. The equations and the coefficients of correlation (r^2) of these fits are provided in the figure.

weathering rates measured in the field. First and foremost is runoff. In the extreme case, where sufficient mineral dissolution occurs so that minerals are essentially at equilibrium with their co-existing fluids, chemical denudation rates are proportional to runoff. If mineral dissolution rates are extremely slow, increasing runoff will just lead to dilution and chemical denudation rates

being independent of runoff (Gislason et al. 2009). An example of how the variation of chemical and mechanical denudation rates vary with runoff for a catchment in northeastern Iceland is shown in Fig. 5.5. Chemical denudation rates (r_w) of Ca^{2+} and HCO_3^- in this figure are consistent with

$$r_w = aQ^b \qquad (3)$$

where Q is the measured runoff and a and b are constants. Note that the diagrams in Fig. 5.5 show the logarithm of the denudation rates and runoff, therefore the constants and a and b appear as intercept and slope of the linear regressions. The constant b for the linear regression 'best-fit' lines shown in Fig. 5.5 is approximately 0.7, suggesting that the chemical denudation rate is neither sufficiently fast to maintain equilibrium with the dissolving minerals ($b = 1$) nor sufficiently slow so that dissolution is negligible ($b = 0$). In such cases, increasing runoff leads to increasing weathering rate due to: (i) an increase in the degree of undersaturation of primary minerals, thus enhancing dissolution rates since the chemical affinity (A in Eqn. 2 increases with undersaturation; (ii) a decrease in the degree of supersaturation of secondary phases, thus slowing or preventing their precipitation and even resulting in their dissolution if they get undersaturated; (iii) a change in the solution composition that affects the dissolution rate at high degrees of undersaturation according to the respective reaction mechanisms; and (iv) an increase in runoff increases the reactive surface area available for fluid-rock interaction as the fluid can reach surfaces at high-flow stage not available at low runoff. Also shown in Fig. 5.5 is the variation of suspended river inorganic material (SIM) flux out of the river catchment as a function of runoff. The SIM flux, and thus mechanical denudation, increases far faster with increasing runoff than chemical denudation. The constant b in the linear regression 'best-fit' shown in this figure is approximately 2. This strong dependence of mechanical denudation on runoff arises, in part, from the augmented ability of rivers to transport particles as its flow rate increases (Gislason et al. 2009).

The strong dependence of chemical and mechanical denudation on runoff leads to dramatic seasonal variations. An example of the chemical and mechanical denudation fluxes of basaltic rocks out of the Brú river catchment in Northeast Iceland during the year 1976 is shown in Fig. 5.6(a). Daily chemical and mechanical denudation fluxes vary spectacularly over the year. Maximum fluxes are associated with high

runoff, which occurs during flood events and during the summer melt season in this glacial river, and minimum fluxes are observed during the winter months. Such seasonal variations will depend on the local climate and the tendency of a catchment to experience catastrophic flooding events. For example, it is likely that non-glaciated catchments at high elevation and/or latitude would experience maximum runoff during the spring snow melt rather than during the mid- and late-summer, as is the case with the glaciated river example shown in Fig. 5.6(a). The observation that mechanical and chemical denudation fluxes are so variable throughout the year suggests that many river catchment studies performed to assess denudation rates that are based on analysis of only one or several samples may provide very misleading and/or unrepresentative results.

A longer time-series, estimating the chemical and mechanical denudation rates in this river over 30 years, is shown in Figs. 5.6(b) and (c), and further illustrates the annual chemical and mechanical weathering rate cycles. Chemical and mechanical denudation rates vary dramatically during each annual cycle. The delivery of Ca via chemical and mechanical denudation varies by more than one and three orders of magnitude, respectively, each year, in the example shown in Fig. 5.6.

Temperature

The link between runoff and chemical and mechanical denudation fluxes provides a direct connection between climate and weathering/denudation rates. Moreover, runoff is linked to temperature as increasing temperature leads to: (i) increased melting of glaciers at high altitudes and latitudes; (ii) increased precipitation of rain and snow due to increasing humidity; and (iii) more intense catastrophic flooding events (Groisman et al. 2005). A direct correlation between runoff and temperature is shown in Fig. 5.7. Annual runoff from the basaltic Hóll catchment in northeastern Iceland increased by a factor of two in response to a 4°C increase in the average annual temperate. As

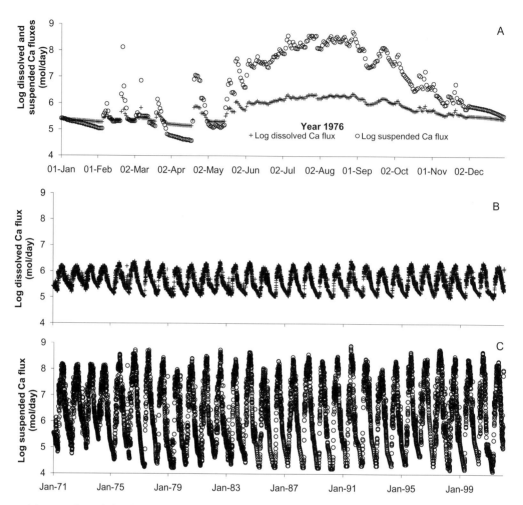

Fig. 5.6 (a) Logarithm of the daily suspended and dissolved Ca flux in the Jokulsa River at Holl in northeastern Iceland during 1976. The crosses and circles represent dissolved and suspended fluxes, respectively. (b) Daily dissolved Ca flux in the Jokulsa a Dal River at Bru in northeastern Iceland from 1971 to 2004. (c) Daily suspended Ca flux in Jokulsa a Dal River at Bru from 1971 to 2004. Data from Gislason et al. (2006).

chemical and mechanical denudation rates are directly linked to runoff, this increase leads to an increase of more than 50% in the chemical denudation flux and an increase of more than 300% in the mechanical denudation flux to the ocean.

This link between silicate weathering/ denudation rates and temperature is less clear for granites. Olivia et al. (2003) showed that the positive influence of temperature on chemical denu-

dation of granites is only apparent when runoff values are higher than 1000 mm/yr. High runoffs are often associated with mountainous areas where soils are constantly eroded. In the lowlands, the formation of thick soils protects the bedrock from chemical weathering. In the extreme, thick tropical soils contain no primary minerals (Stallard and Edmond 1983; Dupré et al. 2003).

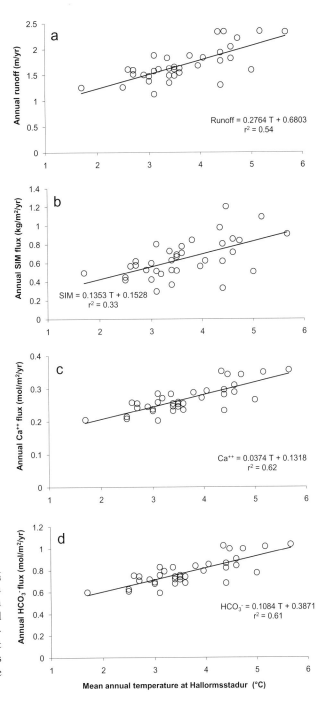

Fig. 5.7 Annual runoff and SIM, Ca^{2+}, and HCO_3^- fluxes in the Jokulsa River at Holl in northeastern Iceland from 1961 to 2004 plotted as a function of temperature. The symbols refer to data reported by Gislason et al. (2009), whereas the lines correspond to the least-squares linear regression fit of these data. The equations and the coefficients of correlation (r^2) of these fits are provided in the Figure.

Table 5.2   Selected activation energies of silicate weathering/denudation derived from field studies.

Material	Activation Energy (kJ/mol)	References
Feldspar	77	Velbel (1993)
Granitoid rocks	51	White et al. (1999a)
Basaltic rocks	42.3	Dessert et al. (2001)
Silicates rocks of the Yamuna River catchment	50-80	Dalai et al. (2002)
Granitic rocks	48.7	Olivia et al., (2003)
Quartz in soils	15	Richards and Kump (2003)
Silicate rocks	74 ± 29	West et al. (2005)
Basalts	70	Navarre-Sitchler and Brantley (2007)
Basaltic catchments	26-90	Gislason et al. (2009)

The temperature variation of weathering and denudation rates can be quantified using some sort of an Arrhenius equation such as

$$r = k\,exp(-E_a/RT) \qquad (4)$$

where the rate, r, is related to temperature T via the activation energy E_a, R is the gas constant and k represents a constant. Activation energies have been determined from both experimental measurements and regression analysis of field data. A selection of activation energies reported in the literature is provided in Table 5.2. Activation energies from field studies range from 20 to 95 kJ/mol. This range is similar to that obtained from laboratory studies (c.f. Brantley 2003; Gislason and Oelkers 2003). Considering an activation energy of 62.8 kJ/mol, Berner et al. (1983) estimated that silicate weathering rates increase by an order of magnitude with each 10°C increase in temperature.

Topography

There is a strong feedback between topography and weathering. This link stems in part from the fact that steep slopes promote removal of material

away from the weathering site helping expose fresh rock surfaces to reactive aqueous solutions (Stallard and Edmond 1983; Lyons et al. 2005; NIR Goldsmith et al. 2008a, 2008b; Carey et al. 2005; Harmon et al. 2009). This link has motivated a large number of studies of the impact of orogenesis, and in particular the Himalayan uplift, on weathering and the global carbon cycle (c.f. Galy and France-Lanord 1999; Krishnaswami and Singh 2005). Raymo and Ruddiman (1992) proposed that CO_2 drawdown from the increased silicate weathering resulting from the Himalayan uplift would lead to global cooling. A positive feedback between mechanical weathering and climate would result as cooler temperatures would increase mechanical weathering rates due to the effects of frost-thaw cycles. This positive feedback, however, has been shown to be inconsistent with the global carbon and oxygen record (Derry and France-Lanord 1996; Godderis and Francois 1996). These more recent studies concluded that carbon consumption by Ca- and Mg-silicate weathering is minor within the Ganges and Brahmaputra rivers (Galy and France-Lanord 1999), but the main sink of carbon due to the Himalayan orogenesis is the burial of organic carbon within the Bengal fan (France-Lanord and Derry 1997).

THE RELATIVE IMPORTANCE OF CHEMICAL AND MECHANICAL PROCESSES

The role of chemical versus mechanical weathering/denudation on influencing global chemical cycles remains a topic of debate. Whereas chemical weathering releases material to solution by dissolution of silicate rocks on the continents, mechanical weathering coupled to erosion can release material to solution by dissolution in the ocean. The mechanical denudation of young volcanic terrains produces nearly 45% of the global suspended material transport to the oceans (Milliman and Syvitski 1992), and much of this material is of basaltic composition, rich in divalent metal cation-bearing minerals and glasses that are highly reactive in the ocean (Stefánsdóttir and Gislason 2005; Pogge von Strandmann et al. 2008). There are several reasons why volcanic and tectonically active islands contribute a relatively large quantity of suspended material to the oceans. The first is because they are located near the oceans, have high relief, and their rocks are readily affected by both mechanical and chemical denudation processes. High relief also tends to increase rainfall on young volcanic terrains, and their minerals and glasses are highly undersaturated in both river and ocean waters. By contrast, mechanical denudation of durable granite-gneiss terrains is dominated by iron oxides and clay minerals transport, owing to the increased average age of these rocks. This material tends to be relatively inert in ocean water compared to the suspended material originating from young volcanics. Gaillardet et al. (1999b) and Eiriksdottir et al. (2008) quantified the weathering intensity of river suspended material on a scale from 1 to 25, reflecting its reactivity once exposed to seawater. The weathering intensity for the volcanic material originating from Iceland is 1.1 to 3.2, that of the Azores is 1.8, that of Reunion is 1.6, and that of Java is 4.0. In contrast, the low-relief, inactive, continental catchments, which contribute negligibly, compared to their extent, to suspended material transport to the ocean such as the Orinoco, Niger and Amazon rivers, have weathering intensities of 20 or greater.

The total flux of silicate mechanical denudation exceeds that of chemical denudation on a global scale. Berner and Berner (1996) estimated that 90 Mt/yr of dissolved Ca originating from silicate chemical weathering on the continents is transported to the oceans. In comparison, the total mass of suspended particulate Ca transported to the oceans by rivers was estimated to be 333 Mt/yr (Martin and Meybeck 1979; Berner and Berner 1996). The percentage of this mass released in dissolved form to ocean water is difficult to assess, in part, because of sediment burial. Nevertheless, evidence suggests that much of this Ca is added to the oceans before its deep burial. For example, the concentration of Ca in deep-sea clay is only half of that of the global river-suspended material (Martin and Whitfield

1983). The ultimate fate of Ca, after being added to the sedimentary record, is the formation of carbonate minerals; in excess of 90% of Ca in sedimentary rocks is situated in carbonate minerals (Ronov and Yaroshevsky 1969; Garrels and Mackenzie 1971). Much of this transformation occurs during diagenesis.

Taking account of the sum of three contributions: (i) the exchange of Ca for Na on clay mineral surfaces in estuaries; (ii) diffusive flux from ocean sediments; and (iii) the dissolution of Ca-bearing silicate mineral and glass suspended material before its deep burial, Gislason et al. (2006) concluded that mechanical weathering could be as important as chemical weathering to the global carbon cycle. This conclusion was later confirmed by (Wallman et al. 2008) by studying the weathering of anoxic marine sediment cores. In addition, Gislason et al. (2006) showed that the effect of climate on mechanical denudation is stronger than that of chemical denudation so that mechanical weathering and denudation has an important role in CO_2 buffering and climate moderation.

WEATHERING AND DENUDATION AT A GLOBAL SCALE

A major motivation for studying silicate weathering rates is in an attempt to quantify the global carbon cycle. Over geological timescales, atmospheric CO_2 is strongly influenced by the release of CO_2 to the atmosphere by Earth degassing, including volcanic and metamorphic activity and the consumption of CO_2 by carbonate mineral formation driven by silicate weathering (Eq. 1). These conflicting processes have been incorporated into global-scale carbon cycle models, most notably GEOCARB (e.g. Berner et al. 1983; Berner and Kothavala 2001; Berner 2004) which have provided a description of atmospheric CO_2 content over the past 550 million years. Estimates of current global-scale silicate weathering rates rely heavily on field studies of the world's main rivers (e.g. Milliman and Meade 1983; Milliman and Syvisky 1992; Maybeck 1979; Martin and Meybeck 1979; Martin and Whitfield 1983;

Zakharova et al. 2005). From such studies, Berner and Berner (1996) estimated that the average chemical denudation rate of the continents, including silicates, carbonates and evaporates, to be 26 tons/km^2/yr, and mechanical denudation rate to amounts to be 226 tons/km^2/yr. Assuming an average rock density of 2.7 g/cm^3, this denudation rate equals 9.4 cm of continental denudation per 1000 years. Gaillardet et al. (1999a) calculated a similar chemical denudation rate for the continents of 24 tons/km^2/yr, of which only 26% was due to silicate-weathering. Taking account of global silicate weathering rates on the continents and the volcanic islands, Dessert et al. (2003) concluded that 30 to 35% of CO_2 consumed by the global silicate-weathering is due to basalt weathering, despite the fact that basalt covers less than 5 per cent of the continental surface. Gaillardet et al. (1999a) calculated that 11.7×10^{12} moles of CO_2, or 0.14 gigatons of carbon per year, is consumed by silicate weathering on land (steps 1a and 1b in eqn. 1). This CO_2 consumption rate can be compared to the current atmospheric carbon content of 800 gigatons carbon, and the flux of CO_2 into and out of the ocean and the terrestrial biosphere, both estimated to be approximately 100 Gt C/year (Houghton 2007). At first glance, this comparison suggests that silicate weathering plays only a negligible role in the global carbon cycle. However, carbonate mineral formation stores CO_2 over far longer timescales than either the ocean or the biosphere; the residence time of carbon in the biosphere is approximately 20 years, in the oceans it is approximately 400 years, but the residence time of carbon in carbonate rocks is in excess of millions of years (Houghton 2007).

CONCLUSIONS

The weathering and denudation of silicate rocks remains an important and challenging area of active research. The challenge is related to the complex nature of weathering, which is affected by temperature, runoff, biogenetic activity, the chemical and mineralogical composition of the parent rock, and tectonic activity. As such, it is not possible to describe completely this phenomenon in detail in this short chapter. Excellent sources of further information on silicate weathering can be found in the books, and special issues of *Chemical Geology* and *Elements* referred to in the 'Supplemental reading'.

ACKNOWLEDGEMENTS

We thank S. Callahan, O. Pokrovsky, J. Schott, K. Burton, D. Wolff-Boenisch, E. S. Eiriksdottir, B. Sigfússon, A. Stefánsson, G. Gisladottir, E. Gunnlaugsson and W. Broecker for insightful discussion and encouragement throughout this study. This work was supported by the Centre National de la Recherche Scientifique and the European Commission Marie Curie Grants MIR and MIN-GRO (MEST-2005-012210 and MRTN-2006-31482).

FURTHER READING

Berner, EK and Berner, RA. (1996) *Global Environment: Water, Air, and Geochemical Cycles: New Jersey,* Prentice Hall, 376 pp.

Gislason, SR, Oelkers, EH and Bruno, J. (eds.) (2002) Geochemistry of crustal fluids – Fluids in the crust and fluxes at the Earth's surface. *Chem. Geo.* **190**: 1–431.

Drever, JI, (ed.) (2003) *Surface and Ground Water, Weathering, and Soils Vol. 5 Treatise on Geochemistry* (eds. HD Holland and KK Turekian), Elsevier – Pergamon, Oxford 626 pp.

Anderson, SP and Blum, WE. (eds.) (2003) Controls on chemical weathering. *Chem. Geo.* **202**: 191–507.

Berner, RA. (2004) *The Phanerozoic Carbon Cycle.* Oxford University Press, Oxford **150** pp.

Brantley, SL, White, TS and Ragnarsdottir, KV. (eds.) (2007) The critical zone: where rocks meets life. *Elements* **3**: 307–38.

Oelkers, EH and Schott, J. (eds.) (2009) Thermodynamics and Kinetics of Water-Rock Interactions. *Review in Mineralogy & Geochemistry* Volume **70**: 207–58. The Geochemical Society and the Mineralogical Society of America, Chantilly, Virginia, 569 pp.

REFERENCES

Aagaard, P and Helgeson, HC. (1982) Thermodynamic and kinetic constraints on reaction rates among minerals and aqueous solutions: I. Theoretical considerations. *Amer. J. Sci.* **282**: 237–85.

Alkattan, M, Oelkers, EH, Dandurand, J-L and Schott, J. (1997) Experimental studies of halite dissolution kinetics. I. The effect of saturation state and the presence of trace metals. *Chem. Geo.* **137**: 201–21.

Anderson, SP. (2007) Biogeochemistry of glacial landscape systems. *Ann. Rev. Earth Planet. Sci.* **35**: 375–99.

Arnalds, Ó. (2005) Icelandic soils. In: C Caseldine, A Russell, J Harð ardóttir and Ó Knudsen (eds.) *Iceland – Modern Processes and Past Environments*, Elsevier, Amsterdam, 309–18.

Banfield, JF, Jones, BF and Veblen DR. (1991) An AEM-TEM study of weathering and diagenesis, Aber Lake, Oregon: I. Weathering reaction in the volcanic. *Geochim. Cosmochim. Acta* **55**: 2781–93.

Berg, A and Banwart, SA (2000) Carbon dioxide mediated dissolution of Ca-feldspar: Implications for silicate weathering. *Chem. Geol.* **163**: 25–42.

Berner, EK and Berner, RA. (1996) *Global Environment: Water, Air, and Geochemical Cycles: New Jersey*, Prentice Hall, 376 pp.

Berner, RA. (2004) *The Phanerozoic Carbon Cycle.* Oxford University Press, Oxford.

Berner, RA and Maasch KA. (1996) Chemical weathering and controls on atmospheric O_2 and CO_2: Fundamental principles were enunciated by JJ Ebelmen in 1845. *Geochim. Cosmochim. Acta* **60**: 1633–7.

Berner, RA and Kothavala, Z. (2001) GEOCARB III. A revised model of atmospheric CO_2 over Phanerozoic time. *Am. J. Sci.* **301**: 182–204.

Berner, RA, Lasaga, AC and Garrels, RM (1983) The carbonate-silicate geochemical cycle and its effect on atmospheric carbon dioxide over the past 100 million years: *American Journal of Science* **283**: 641–83.

Berger, G, Cadoré, E, Schott, J and Dove, P (1994) Dissolution rate of quartz in Pb and Na electrolyte solutions. Effect of the nature of surface complexes and reaction affinity. *Geochim. Cosmochim. Acta* **58**: 541–51.

Bjorlykke, K and Egeberg, PK. (1993) Quartz cementation in sedimentary basins. *AAPG Bull.* **77**: 1538–46.

Blatt, H and Jones, RL. (1975) Proportions of exposed igneous rocks, metamorphic, and sedimentary rocks. *Geol. Soc. Amer. Bull.* **86**:1085–8.

Bluth, GJS and Kump, LR. (1994) Lithologic and climatologic controls of river chemistry. *Geochim. Cosmochim. Acta* **58**: 2341–59.

Brady, PV. (1991) The effect of silicate weathering on global temperature and atmospheric CO_2. *Journal of Geophys. Res.* **96** (B11): 18101–6.

Brady, PV and Carroll, SA. (1994) Direct effects of CO_2 and temperature on silicate weathering: Possible implications for climate control. *Geochim. Cosmochim. Acta* **58**: 1853–6.

Brady, PV and Gislason, SR. (1997) Seafloor weathering controls on atmospheric CO2 and global climate. *Geochim. Cosmochim. Acta* **61**: 965–97.

Brady, PV and Walther, JV. (1990) Kinetics of quartz dissolution at low temperatures. *Chem. Geo.* **82**: 253–64.

Brantley, SL. (2003) Reaction kinetics of primary rock-forming minerals under ambient conditions, pp. 73–117. In: JI Drever (ed.) *Surface and Ground Water, Weathering, and Soils, Vol. 5 Treatise on Geochemistry* (eds. HD Holland and KK Turekian), Elsevier – Pergamon, Oxford.

Brimhall, GH and Dietrich, WE. (1987) Constitutive mass balance relations between chemical composition, volume, density. Porosity, and strain in metasomatic hydrochemical systems: results on weathering and pedogenesis. *Geochim. Cosmochim. Acta* **51**: 567–87.

Broecker, WS (2005) Global warming: Take action or wait? *Jökull* **55**: 1–16.

Busenberg, E and Plummer, LN. (1986) A comparative study of the dissolution and crystal growth kinetics of calcite and aragonite. In: FA Mumpton (ed.) *Studies in diagenesis, U.S.G.S. bull.* **1578**: 139–66.

Carey, AE, Gardner, CB, Goldsmith, ST, Lyons, WB and Hicks, DM. (2005) Organic carbon yields from small, mountainous rivers, New Zealand, *Geophysical Research Letters*, **32**, 15, L15404.

Chou, L, Garrels, M and Wollast, R. (1989) Comparative Study of the kinetics and mechanisms of dissolution of carbonate minerals. *Chem. Geol.* **78**: 269–82.

Chou, L and Wollast, R. (1985) Steady state kinetics and dissolution mechanisms of albite. *Am. J. Sci.* **285**: 963–93.

Cox, R and Lowe, DR. (1995) A conceptual review of regional-scale controls on the composition of clastic sediment and the co-evolution of continental blocks and their sedimentary cover. *J. Sed. Res.* **A65**: 1–12.

Craig, DC and Loughnan, FC. (1964) Chemical and mineralogical transformation accompanying the weathering of basic volcanic rocks from New South Wales. *Aust. J. Soil Res.* **2**: 218–34

Dalai, TK, Krishnaswarmi, S and Sarin, MM. (2002) Major ion chemistry in the headwaters of the Yamuna river system: Chemical weathering, its temperature dependence and CO_2 consumption in the Himalaya. *Geochim. Cosmochim. Acta* **66**: 3397–416.

Daux, V, Guy C, Advocat, T, Crovisier, J-L and Stille, P. (1997) Kinetic Aspects of basaltic glass dissolution at 90°C: Role of silicon and aluminum. *Chem. Geol.* **142**: 109–28.

Derry, LA and France-Lanord, C. (1996) Neogene growth of the sedimentary organic carbon reservoir. *Paleoceanography* **11**: 267–75.

Dessert, C, Dupré, B, Francois, LM, Schott, J, Gaillardet, J, Chakrapani, G and Bajpai, S. (2001) Erosion of Deccan Traps determined by river geochemistry: Impact on the global climate and the Sr-87/Sr-86 ratio of seawater. *Earth Planet. Sci. Lett.* **188**: 459–74.

Dessert, C, Dupré, B, Gaillardet, J, Francois, LM and Allégre, CJ. 2003. Basalt weathering laws and the impact of basalt weathering on the global carbon cycle. *Chem. Geol.* **202**, 257–73.

Devidal, JL, Schott, J and Dandurand, JL. (1997) An experimental study of kaolinite dissolution and precipitation kinetics as a function of chemical affinity and solution composition at 150°C, 40 bars, and pH 2, 6.8 and 7.8. *Geochim. Cosmochim. Acta* **61**: 5165–86.

Dove, PM and Rimstidt, DJ. (1994) Silica – Water Interactions. *Rev. Min.* **29**: 259–308.

Dupré, B, Dessert, CD, Olivia, P, Goddéris, Y, Viers, J, Francois, L, Millot, R and Gaillardet, J. (2003) Rivers, chemical weathering and Earth's climate. *C. R. Geosci.* **335**: 1141–60.

Ebelmen, JJ, (1845) Sur les produits de la décomposition des especes minérales de la famille des silicates. *Annales des Mines, 4ᵉ Series* **7**: 3–66.

Ebelmen, JJ. (1847) Sur la décomposition des roches. *Annales des Mines, 4ᵉ Series* **12**: 627–54.

Eiriksdottir, ES, Louvat, P, Gislason, SR, Óskarsson, N and Hardardóttir, J. (2008) Temporal variation of chemical and mechanical weathering in NE Iceland: Evaluation of a steady state model of erosion. *Earth Planet. Sci. Lett.* **272**: 78–88.

Eiriksdottir, ES, Gislason, SR and Oelkers, EH. (2010) Climatic effects on chemical weathering of basalts in NE-Iceland: temperature, solution composition saturation state and dissolution rates. *Geochim. Cosmochim. Acta* (submitted).

France-Lanord, C and Derry, LA. (1997) Organic carbon burial forcing of the carbon cycle from Himalaya erosion. *Nature* **390**: 65–7.

Galy, A and France-Lanord, C. (1999) Weathering processes in the Ganges -Brahmaputra basin and the riverine alkalinity budget. *Chem. Geo.* **159**: 31–60.

Gaillardet, J, Dupré, B, Louvat, P and Allègre, CJ. (1999a) Global silicate weathering and CO_2 consumption rates deduced from the chemistry of large rivers. *Chem. Geol.* **159**: 3–30.

Gaillardet, J, Dupré, B and Allègre, CJ. (1999b) Geochemistry of large river suspended sediments: Silicate weathering or recycling tracers? *Geochim. Cosmochim. Acta* **63**: 4037–51.

Garrels, RM and Mackenzie, FT. (1971) *Evolution of Sedimentary Rocks: New York*, W. W. Norton and Company, Inc., 397 pp.

Garrels, RM and Perry, EA. (1974) Cycling of carbon, sulfur, and oxygen through geologic time. In: ED Goldberg (ed.) *The Sea*, Vol. 5, pp. 303–16. Wiley.

Gautier, J-M, Oelkers, EH and Schott, J. (1994) Experimental study of K-feldspar dissolution rates as a function of chemical affinity at 150°C and pH 9. *Geochim. Cosmochim. Acta* **58**: 4549–60.

Gibbs, MT, Kump, LR. (1994) Global chemical erosion during the last glacial maximum and the present: Sensitivity to changes in lithology and hydrology. *Paleoceanography* **9**: 529–43.

Gislason, SR, Arnórsson, S and Ármannsson, H. (1996) Chemical weathering of basalt in Southwest Iceland: Effects of runoff, age of rocks and vegetative/glacial cover. *Am. J. Sci.* **296**: 837–907.

Gislason, SR and Eugster, HP. (1987) Meteoric water-basalt interactions: I. A laboratory study. *Geochim. Cosmochim. Acta* **51**: 2827–40.

Gislason, SR, Heaney, PJ, Oelkers, EH and Schott, J. (1997) Kinetic and thermodynamic properties of moganite, a novel silica polymorph. *Geochimica Cosmochimica Acta* **61**: 1193–204.

Gislason, SR and Oelkers, EH. (2003) The mechanism, rates, and consequences of basaltic glass dissolution: II. An experimental study of the dissolution rates of basaltic glass as a function of pH at temperatures from 6°C to 150°C. *Geochim. Cosmochim. Acta* **67**: 3817–32.

Gislason, SR, Oelkers, EH, Snorrason, Á. (2006) The role of river suspended material in the global carbon cycle. *Geology* **34**: 49–52.

Gislason, SR, Oelkers, EH, Eiriksdottir, ES, Kardjilov, MI, Gisladottir, G, Sigfusson, B, Snorrason, A, Elefsen, SO, Hardardottir, J, Torssander P and Oskarsson, N. (2009) Direct evidence of the feedback between climate and weathering. *Earth Planet. Sci. Let.* **277**: 213–22.

Godderis, Y and Francois, LM. (1996) Balancing the Cenozoic carbon and alkalinity cycles: constrints from isotopic records. *Geophys. Res. Let.* **23**: 3743–6.

Goldsmith, ST, Kao, S-J and Carey, AE. (2008a) Geochemical fluxes from the ChoShui River during Typhoon Mindulle, July 2004. *Geology* **36**: 483–6.

Goldsmith, ST, Carey, AE, Lyons, WB and Hicks, M. (2008b) Geochemical fluxes and weathering on high standing islands: Taranaki and Manawatu-Wanganui Regions, New Zealand. *Geochim. Cosmochim. Acta* **72**: 2248–67.

Golubev, SV, Pokrovsky, OS and Schott, J. (2005) Experimental determination of the effect of dissolved CO_2 and the dissolution rates of Mg and Ca silicates at 25°C. *Chem. Geol.* **217**: 227–38.

Groisman, PY, Knight, RW, Easterling, DR, Karl, TR, Hegerl, CG and Razuvaev, VAN. (2005) Trends in intense precipitation in the climate record. *J. Clim.* **18**: 1326–50.

Gunter, WD, Bachu, S, Law, DHS, Marwaha, V, Drysdale, DL, MacDonald, DE, McCann, TJ. (1996) Technical and economic feasibility of CO_2 disposal in aquifers within the Alberta sedimentary basin, Canada. *Energy Conv. Man.* **37**: 1135–42.

Harmon, RS, Lyons, WB, Long, DT, Mitasova, H, Gardner, CB, Welch, KA and Witherow, RA. (2009) Geochemistry of Four Tropical Montane Watersheds, Central Panama. *Applied Geochemistry* **24**: 624–40.

Holland, HD. (1978) *The Chemistry of the Atmosphere and Oceans*. Wiley.

Houghton, RA. (2007) Balancing the global carbon budget. *Ann. Rev. Earth Planet. Sci.* **35**: 313–47.

Jenny, H. (1941) *Factors of Soil Formation*, McGraw-Hill.

Krishnaswami, S and Singh, SK. (2005) Chemical weathering in the river basins of the Himalaya, India. *Current Sci.* **89**: 841–9.

Köhler, SJ, Dufaud, F and Oelkers, EH. (2003) An experimental study of illite dissolution kinetics as a function of pH and temperatures from 5 to 50°C. *Geochim. Cosmochim. Acta* **67**: 3583–94.

Kump, LR, Brantley, SL and Authur, MA. (2000) Chemical weathering, atmospheric CO_2 and climate. *Ann. Rev. Earth Planet. Sci.* **28**: 611–67.

Lackner, KS. (2003) A guide to CO_2 sequestration. *Science* **300**: 1677–8.

Louvat, P. (1997) *Étude géochimique de l'érosion fluviale d'îles volcaniques à l'aide des bilans d'éléments majeurs et traces*. Ph D thesis, University of Paris, Paris, 322 pp.

Louvat, P, Gislason, SR and Allégre, CJ. (2008) Chemical and mechanical erosion rates in Iceland as deduced from river dissolved and solid material. *Ame. J. Sci.* **308**: 679–726.

Lyons, WB, Carey, AE, Hicks, DM and Nezat, C. (2005) Chemical weathering in high-sediment-yielding watersheds, New Zealand, *J. Geophys. Res.* **100**: F01008, doi: 10.1029/2003JF000088.

Martin, JM and Meybeck, M. (1979) Elemental mass-balance of material carried by major world rivers. *Mar. Chem.* **7**: 173–206.

Martin, J-M and Whitfield, M. (1983) The significance of the river input of chemical elements to the ocean. In CS Wong, EA Boyle, KW Bruland, JD Burton and ES Goldberg (eds.) Trace Metals in Sea Water: New York, Plenum, 265–96.

Meybeck, M. (1979) Concentrations des eaux fluviales en éléments majeurs et apports en solution aux océans, *Revue de Géologie Dynamique et Géographie Physique* **21**: 215–46.

Meybeck, M. (1987) Golbal chemical weathering of surficial rocks estimated from river dissolved loads. *Am. J. Sci.* **287**: 401–28

Milliman, JD and Meade, RH. (1983) World-wide delivery of river sediments to the ocean. *J. Geo.* **91**: 1–21.

Milliman, JD and Syvitski, JPM. (1992) Geomorphic/ tectonic control of sediment discharge to the ocean: The importance of small mountainous rivers. *J. Geol.* **100**: 525–44.

Nagy, KL, Blum, AE and Lasaga, AC. (1991) The dissolution and precipitation kinetics of kaolinite at 80°C and pH 3- the dependence on solution saturation state. *Am. J. Sci.* **291**: 649–86.

Navarra-Sitchler, A and Brantley, S. (2007) Basalt weathering across scales. *Earth Planet. Sci. Lett.* **261**: 321–34.

Nesbitt, HW and Young, GM. (1984) Prediction of some weathering trends of plutonic and volcanic-rocks based on thermodynamic and kinetic considerations *Geochim. Cosmochim. Acta* **48**: 1523–34.

Oelkers, EH. (2001a) General kinetic description of multioxide silicate mineral and glass dissolution. *Geochim. Cosmochim. Acta* **65**: 3703–19.

Oelkers, EH. (2001b) An experimental study of forsterite dissolution rates as a function of aqueous Mg and Si concentrations. *Chem. Geo.* **175**: 485–94.

Oelkers, EH, Bjorkum, PA and Murphy, WM. (1996) A petrographic and computational investigation of quartz cementation and porosity reduction in North Sea sandstones. *Am. J. Sci.* **296**: 420–52.

Oelkers, EH and Cole, DR. (2008) Carbon dioxide sequestration: A solution to a global problem. *Elements* **4**: 305–10.

Oelkers, EH and Gislason, SR. (2001) The mechanism, rates, and consequences of basaltic glass dissolution: I. An experimental study of the dissolution rates of basaltic glass as a function of aqueous Al, Si, and oxalic acid concentration at 25°C and pH = 3 and 11. *Geochim. Cosmochim. Acta* **65**: 3671–81.

Oelkers, EH and Gislason, SR, Eiriksdottir, ES, Elefsen, SO and Hardardotti, J. (2004) The role of suspended material in the chemical transport of eastern Icelantic rivers. In: RB Wanty and RR Seal II (eds.) *Water-Rock Interaction*, Taylor and Francis Group, London, UK, pp. 865–8.

Oelkers, EH and Schott, J. (1995) Experimental study of anorthite dissolution and the relative mechanism of feldspar hydrolysis. *Geochim. Cosmochim. Acta.* **59**: 5039–53.

Oelkers, EH and Schott, J. (2001) An experimental study of enstatite dissolution rates as a function of pH, temperature, and aqueous Mg and Si concentration, and the mechanism of pyroxene/pyroxenoid dissolution. *Geochim. Cosmochim. Acta* **65**: 1219–31.

Oelkers, EH, Schott, J. (2005) Geochemical Aspects of CO_2 Sequestration. *Chem. Geo.* **217**: 183–186.

Oelkers, EH, Schott, J and Devidal, J-L. (1994) The effect of aluminium, pH, and chemical affinity on the rates of aluminosilicate dissolution reactions. *Geochim. Cosmochim. Acta* **58**: 2011–24.

Olivia, P, Viers, J and Dupré, B. (2003) Chemical weathering in granitic environments. *Chem. Geol.* **202**: 225–56.

Pogge von Strandmann, PAF, James, RH, Van Calsteren, P, Gislason, SR, Burton, KW. (2008) Lithium, magnesium and uranium isotope behaviour in the estuarine environment of basaltic islands. *Earth and Planetary Science Letters* **274**: 462–71.

Pokrovsky, OS and Schott, J. (2000) Kinetics and mechanism of forsterite dissolution at 25°C and pH from 1 to 12. *Geochim Cosmochim. Acta* **64**: 3313–25.

Pokrovsky, OS, Schott, J, Kudryavtzev, DI and Dupré, B. (2005) Basalt weathering in Central Siberia under permafrost conditions. *Geochim. Cosmochim. Acta* **69**: 5659–80.

Raines, M and Dewers, T. (1997) Mixed transport reactions control of gypsum dissolution kinetics in aqueous solutions and initiation of gypsum karst. *Chem. Geo.* **140**: 29–48.

Raymo, ME and Ruddiman, WF. (1992) Tectonic forcing of late Cenozoic climate. *Nature* **357**: 117–22.

Ribe, CS, Kirchner, JW and Finkel, RC. (2004) Erosion and climatic effect on long-term chemical weathering rates in granitic landscapes spanning diverse climate regimes. *Earth Planet. Sci. Let.* **224**: 547–62.

Richards, PL and Kump, LR. (2003) Soil pore-water dristributiosn and the temperature feedback between weathering in soils. *Geochim. Cosmochim. Acta* **67**: 3803–15.

Ronov, AB and Yaroshevsky, AA. (1969) Earth's crust geochemistry. In: RW Fairbridge, (ed.) *Encyclopedia of Geochemistry and Environmental Sciences: New York*, Van Nostrand, pp. 243–86.

Schott, J, Pokrovsky, OS and Oelkers, EH. (2009) The link between mineral dissolution/precipitation kinetics and solution chemistry. In: EH Oelkers and J Schott (eds.) Thermodinamic and Kinetics of Water-rock Interactions, *Review in Mineralogy & Geochemistry* **70**: 207–58. The Geochemical Society and the Mineralogical Society of America, Chantilly, Virginia, USA.

Stallard, RF and Edmond, JM. (1983) Geochemistry of the Amazon 2. The influence of geology and weathering environment on the dissolved load. *J. Geophys. Res.* **88**: 9671–88.

Stefánsdóttir, MB, Gislason, SR. (2005) The source of suspended matter and suspended matter/seawater interaction following the 1996 outburst flood from the Vatnajökull Glacier, Iceland. *Earth Planet. Sci. Lett.* **237**: 433–52.

Taylor, SR and McLennan, SM. (1985) *The Continental Crust: Its Composition and Evolution.* Blackwell, Oxford, UK.

Urey, HC. (1952) The Planets: Their Origin and Development. Yale Univ. Press.Vebel M.A., 1993. Temperature dependence of silicate weathering in nature – How strong a negative feedback on long-term accumulation of atmospheric CO_2 and global greenhouse warming. *Geology* **21**: 1059–62.

Velbel MA. (1993). Temperature dependence of silicate weathering in nature – How strong a negative feedback on long-term accumulation of atmospheric CO_2 and global greenhouse warming. *Geology* **21**: 1059–62.

Veizer, J and Mackenzie, FT. (2003) Evolution of Sedimentary rocks. *Tretis.* **7**: 369–407.

Vigier, N, Burton, KW Gislason, SR, Rogers, NW, Duchene, S, Thomas, L, Hodge, E, and Schaefer, B. (2006) The relationship between riverine U-series disequilibria and erosion rates in a basaltic terrain, *Earth Planet. Sci. Let.* **249**: 258–73.

Wada, K, Arnalds, O, Kakuto, Y, Wilding, LP and Hallmark, CT. (1992) Clay minerals of four soils formed in eolian and tehpra materials in Iceland. *Geoderma* **52**: 351–65.

Walker, JCG, Hays, PB, Kasting, JF. (1981) A negative feedback mechanism for the long-term stabilization of Earth's surface temperature. *J. Geophys. Res.* **86**: 9776–82.

Wallmann, K. (2001) Controls on the Cretaceous and Cenozoic evolution of seawater composition, atmospheric CO_2 and climate. *Geochim. Cosmochim. Acta* **65**: 3005–25.

Wallman, K, Aloisi, G, Haeckel, M, Tishchenko, P, Pavlova, G, Greinert, J, Kutterolf, S, Eisenhauer, A. (2008) Silicate weathering in anoxic marine sediments, *Geochim. Cosmochim. Acta.* **72**: 2895–918.

West, AJ, Galy, A and Bickle, M. (2005) Tectonic and climate control on silicate weathering. *Earth Planet. Sci. Lett.* **235**: 211–28.

White, AF. (2005) Natural weathering rates of silicate minerals. *Treatise on Geochemistry* **5**: 133–68.

White, AF and Brantley, SB, 2003. The effect of time on the weathering of silicate minerals: Why do weathering rates differ from the laboratory and the field. *Chem. Geo.* **202**, 479–506.

White, AF, Blum, AE. (1995) Effects of climate on chemical weathering in watersheds. *Geochim. Cosmochim. Acta* **59**: 1729–47.

White, AF, Blum, AE, Schultz, MS, Bullen, TD, Harden, JW and Peterson, ML. (1996) Chemical weathering of a soil chronosequence on grantic alluvium. 1. Reacton rates based on changes in soli mineralogy. *Geochim. Cosmochim. Acta.* **60**: 2533–50.

White, A, Blum, AE, Bullen, TD, Davison, B, Vivit, DV, Schulz, M and Fitzpatrick, J. (1999a) The effect of temperature on experimental and natural chemical weathering rates of granitoid rocks. *Geochim. Cosmochim. Acta* **63**: 3277–91.

White, A, Bullen, TD, Davison, B, Schulz, M and Clow, DW. (1999b) The role of disseminated calcite in the chemical weathering of granitoid rocks. *Geochimica et Cosmochimica Acta* **63**: 1939–53.

White, AF, Blum, AE, Stonestrom, DA, Bullen, TD, Schulz, MS, Huntington, TG and Peters, NE. (2001) Differential rates of feldspar weathering in granitic regoliths. *Geochim. Cosmochim Acta* **65**: 847–69.

White, AF, Blum, AE, Schultz, MS, Huntington, TG, Peters, NE and Stonestrom, DA. (2002) Chemical weathering of the Pamola granite: solute and regolith element fluxes and the dissolution rate of biotite. In: *Water-rock interaction, Ore Deposits, and Environmental Geochemistry: A tribute to David A Crerar*, R Hellmann and SA Wood (eds.) The Geochemical Society, St. Louis, pp. 37–59.

Wolff-Boenisch, D, Gislason, SR and Oelkers, EH. (2006) The effect of crystallinity on dissolution rates and CO2 consumption capacity of silicates. *Geochim. Cosmochim. Acta* **70**: 858–70.

Yang, L and Steefel, CI. (2008) Kaolinite dissolution and precipitation kinetics at 22°C and pH 4. *Geochim. Cosmochim. Acta* **72**: 99–116.

Zakharova, EA, Pokrovsky, OS, Dupre, B and Zaslavskaya, MB. (2005) Chemical weathering of silicate rocks in Aldan Shield and Baikal Uplift: Insights from long-term seasonal measurements of solute fluxes in rivers. *Chem. Geo.* **214**: 223–48.

6 Geochemistry of Secular Evolution of Groundwater

TOMAS PACES

Czech Geological Survey, Czech Republic

ABSTRACT

Magmatic, metamorphic, fossil, meteoric and oceanic water are interconnected by slow geologic and fast hydrologic cycles. Advances in isotopic geochemistry bring a deeper insight into the evolution of the genetic types. Tracers with a wide dating range enable us to follow secular chemical changes in sedimentary basins. Common existence of fossil brines in oil fields, fluid inclusions in minerals and in fractures in old crystalline platforms suggest that the geologically old stagnant waters are a result of secular development of the Earth's crust. Changes in contents of specific elements and isotopic ratios indicate an intimate relationship between deep tectonic processes and composition of groundwater. A simple diagram is used to visualize the changes and relationships.

INTRODUCTION

The secular evolution of natural water is determined by a global geological cycle upon which is superimposed a fast hydrological cycle. Water-rock interaction during both cycles changes chemical and isotopic composition of groundwater while at the same time influencing the composition of rocks. The duration of interactions ranges from hours to days during the hydrological cycle. Duration of the geological cycle is in the order of 10^8 years according to the rate of sea-floor spreading. The mass flow of free-gravitation water in the continental zone of active water exchange is estimated to be 1.05×10^{19} g/yr, whereas the mass flow in deeper zones of restricted circulation in continental and subcontinental blocks is 6.6×10^{17} g/yr. The flow of physically and chemically bound water in sediments and sedimentary rocks of the continental and subcontinental blocks is estimated to be about 6×10^{15} g/yr and in granitic and basaltic rocks of continents it is 4×10^{13} g/yr (Zverev 2009). Hence, a very large timespan needs to be considered when the secular evolution of groundwater is concerned. The involvement of water during the development of the Earth's crust and during hydrological cycles is illustrated in Fig. 6.1.

Chemical and isotopic compositions of natural waters result from physical-chemical processes between water, rock and atmosphere and the consequent mixing of different groundwater and surface water bodies. Frequently, this interaction is influenced by biota and increasingly by anthropogenic impacts. Palaeo-ocean water, generated as a product of early Earth degassing, mantle-crust fractionation and volcanism throughout Earth history has been a source of all water in the geo-

Frontiers in Geochemistry: Contribution of Geochemistry to the Study of the Earth, First edition. Edited by Russell S. Harmon and Andrew Parker. © 2011 Blackwell Publishing Ltd. Published 2011 by Blackwell Publishing Ltd.

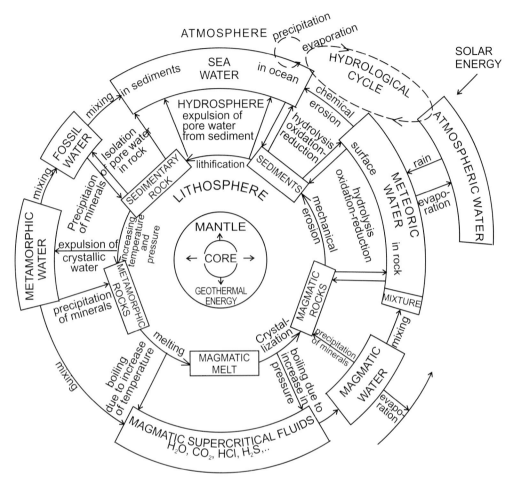

Fig. 6.1 Geological and hydrological cycles; the rectangles are reservoirs of matter; arrows represent fluxes of matter and energy.

logical and hydrological cycles. Evaporation of ocean water transferred water molecules through the atmosphere to continents.

HISTORICAL CONCEPTS

Water can be viewed as a continuum of different genetic types. D.E. White applied geochemical data in combination with hydrogeological and geophysical information in an attempt to define the genetic water types (White 1957, 1969, 1974;

White et al. 1973). Another important contribution of White's work was a notion of interrelationships between groundwater, oceanic water, water in oil fields and ore-forming fluids. After White's pioneering work, relationships between different water types have been supported by numerous studies of other authors (e.g. Ohmoto and Rye 1968; Rye and Haffty 1969; Ohmoto 1986; Knauth, and Beeunas, 1986; Rye 1993, Nesbitt, and Muehlenbachs, 1995, Magenheim et al., 1995, Mojzsis, Harrison, and Mckeegan, 2001; Naden et al. 2003; Yardley 2005).

With respect to their origin, natural waters can be classified as follows:

Atmospheric water – vapour, water, and ice in atmosphere;

Meteoric water – atmospheric water infiltrated to regolith and rocks after passing through the hydrological cycle;

Ocean water – water contained in oceans and ocean water infiltrated to rocks;

Fossil water – water in rocks closed off from the hydrological cycle for geologically long periods of time;

Metamorphic water – water released from rocks during their thermal and pressure metamorphism;

Magmatic water – water originally dissolved in magma and released from it during crystallization of minerals;

Juvenile water – magmatic water released from the Earth's mantle.

There are many other terms commonly used to describe natural waters, such as oil-filed water, formation water, connate water, glacial water, volcanic water, geothermal water, pore water, soil moisture, mineral fluid inclusion water, etc. However, these terms relate water to environments in which it was sampled for investigation and not to its ultimate origin.

The genetic classification of water is useful because it relates water not only to different Earth reservoirs, but also to geological processes and to residence time of water bodies in various geological environments.

Despite numerous data on various types of water, there are no simple geochemical criteria that permit unequivocal definition of individual genetic types (White et al. 1973; White 1974). The most useful criterion is the isotopic composition of oxygen and hydrogen in water molecules. Following isotopic analyses of meteoric water by Epstein and Mayeda (1953) and Friedman (1953), Craig (1961) published the now classic 'meteoric water line' (MWL) on a δD versus $\delta^{18}O$ diagram. The stable-isotope distribution and fractionation in natural waters is discussed in more detail by Hoefs in Chapter 7, this volume. A summary of stable-isotope data on fluid inclusions and other

types of water by many authors indicates a possible range of H- and O- isotope compositions of magmatic water (White et al. 1973; White 1974; Yardley 2009). This diagram (Fig. 6.2) is frequently used as a basis for distinguishing genetic water types and tracing the evolution of water subject to isotopic fractionation during physical processes such as evaporation, isotopic exchange of oxygen and hydrogen between water and rock, dehydration and hydration of minerals and ultra-filtration through clays. The interpretation of H and O isotopic data by many authors since the 1950s to 1960s has shown that the data should be interpreted in connection with geological and hydrological information about each groundwater system before their evolutionary significance is proved.

NEW ADVANCES

Tracing the secular evolution of ground water in the context of the geological and hydrological cycles is now improved owing to new advances in isotopic geochemistry (de Groot 2004). Traditional radioactive and stable isotopes of light elements such as H, O, C, S and N have been supplemented with new data on isotopes such as ^7Li, ^{11}B, ^{37}Cl, ^{81}Br, ^{44}Ca, $^3He/^4He$, ^{53}Cr, ^{65}Cu, ^{30}Si and data on concentrations of rare-earth elements, and noble gases (e.g. Baum 1996; Moeller 2000; Loosli and Purtschert; 2005 Torgersen et al. 1991). Groundwater tracers with a wide age range (^3H, $^3H/^3He$, ^{85}Kr, ^{39}Ar, ^{14}C and ^4He, e.g. Beyerle et al. 1999; Lehmann et al. 2003; Loosli et al. 2001; Purtschert et al. 2001) allow secular chemical changes of water in sedimentary basins to be investigated and better understood (Mazor 1997; Edmunds and Shand 2008). Changes in contents of Rn, F, CO_2 and an increase in the ratio $^3He/^4He$ of dissolved helium in water during earthquakes indicate a relationship between deep tectonic processes and composition of groundwater (Weise et al. 2001). The common existence of fossil Na-Ca-Cl brines in oil fields, fluid inclusions in minerals and in fractures in old crystalline platforms suggests that the geologically-old stagnant waters

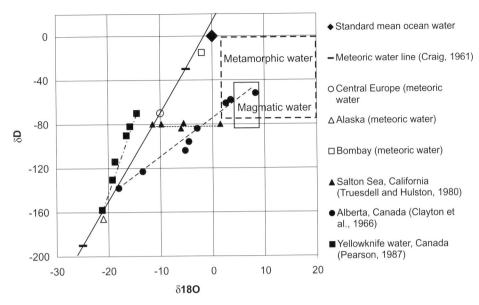

Fig. 6.2 Diagram of δD-δ18O indicates fractionation trends in water and different genetic water types. The solid line is the Meteoric Water Line of Craig (1961); the dash-and-dot line is the isotopic shift observed in the Yellowknife mining district of Canada by Pearson (1987); the dashed line defines the fractionation trend due to evaporation recorded for an oil-field basin in Alberta, Canada (Clayton et al. 1966); the dotted line describes the path of isotopic exchange between rock and geothermal water in the Salton Sea basin of California (Craig 1966); and rectangles denote data on magmatic and metamorphic waters derived from fluid inclusions (Yardley 2009).

are a result of secular development of the Earth's crust (White 1974; Frape and Fritz 1987; Mullis, et al., 1994; Scambelluri and Philippot 2001). An occurrence of a Na-SO$_4$ brine (147 g/l) in crystalline basement of a Tertiary basin indicates that it is a result of an evolution of Tertiary volcanic volatiles dissolved in a playa lake (Paces and Smejkal 2004). The formation of such brines requires a subsequent oxidation and evaporation, crystallization of fossil salts in fractures of rocks and final leaching of the salts by present-day meteoric water charged with magmatic CO$_2$. Such a set of complex steps is needed to explain the evolution of brines in the Eger rift fractures of the Bohemian Massif of central Europe (Paces 1987; Paces and Smejkal 2004; Frape et al. 2007). This case indicates that evolution of deep groundwater can often involve several processes and mixing to yield the chemical composition of water, which we sample for analysis today.

Direct evidence of geochemical evolution of water and related volatiles such as CO$_2$, CH$_4$, N$_2$ and noble gases comes from investigations in the shallow zone of continental and ocean crust reached by drillings and from studies of fluid inclusions in minerals (e.g. Samson et al. 2003; Andersen et al. 2001; Shen et al. 2008; Dolnicek et al. 2009; Kovalevych et al. 2009). Further conclusions can be reached by investigating rocks of deep origin exposed today at the surface displaying evidence of past fluid activity (Scambelluri and Philippot 2001; Yardley 2009).

The concept of secular evolution needs to be specified within a particular timescale. There are three groups of temporal problems: (i) recent evolution of waters during human interference with natural conditions; (ii) secular evolution during the last 30,000 years; and (iii) secular evolution during Earth's history. This separation results from the different methods that can be

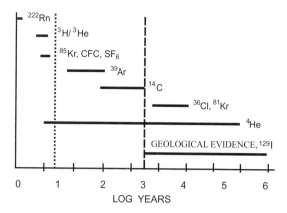

Fig. 6.3 Indicators of the timescales of secular evolution of water. The doted line separates recent evolution from secular evolution during last 30,000 years; the dash line separatcs cvolution during 30,000 years from that of evolution during geological timescale of the Earth's hydrosphere (adapted from Purtschert 2008, with data in Fehn et al. 1994 and Moran et al. 1995).

used to date groundwater, as illustrated in Fig. 6.3. Radioactive isotopes and some man-made chemicals introduced to the natural environment (e.g. freons) are used to estimate the age of very young water, while geological evidence has to be used to define evolution of old waters and waters involved in magmatic and metamorphic processes.

The chemical composition of young groundwaters is vulnerable to anthropogenic impacts. Such groundwater often exhibits either sudden changes or systematic trends in their chemistry. Chemical evolution of groundwater during the last 30,000 years is controlled by meteoric water-rock interaction, the rate of percolation, and mixing with other genetic types of water. Despite the long residence times of some cold ground waters, chemical equilibrium is reached in few partial systems only. Equilibrium is often reached in carbonate systems, in water-ferric hydroxide systems, in reactions of groundwater with secondary amorphous aluminosilicates and ion exchange reactions with clays and mineral surfaces (Siegel and Andersholm 1994; Barnes and Back 1964; Beaucaire et al. 2008). Equilibria between cold groundwater and primary alumino-

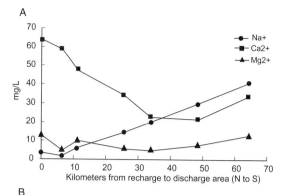

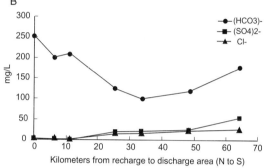

Fig. 6.4 Evolution of Palaeocene ground water in Cenomanian aquifer of the Cretaceous basin of the Bohemian Massif; A – Trends in major cations, B – trends in major anions (Paces et al. 2008).

silicates and quartz are not reached (Paces 1972; Drever 1988; Ohse et al. 1984). The water-mineral interaction is irreversible in this case. However, the interaction becomes often reversible at elevated temperatures in geothermal systems (Giggenbach 1988). In such a case, equilibria are reached between geothermal aqueous solutions and aluminosilicates and also sulphide minerals. The chemical composition of low-temperature groundwaters is maintained steady not by thermodynamic equilibrium but by a hydrodynamic and chemical steady state defined by flow rate and kinetics of the irreversible dissolution reactions (Paces 1974). Figure 6.4 shows a very slow evolution of chemical composition of water in Cenomanian aquifer in the Bohemian Massif (Central Europe) that was dated along the flow trajectory (Paces et al. 2008). Data on ^{85}Kr, ^{39}Ar and ^{14}C indicate that the water is more than 1000 years old. The activities of ^{14}C decrease from 54 to 6 pmC. A model of the residence time with NETPATH code (Plummer et al. 1991) indicates that the evolution of water spans from a few hundred years to about 17,000 years (Corcho Alvarado et al. 2005; Purtschert 2008; Paces et al. 2008). The gradual increase in sodium from recharge area to discharge of the aquifer, while chlorides are low and steady, indicates a very slow dissolution of residual sodium feldspars in the sandstones. Considering the long residence time of water (17,000 yr) and the increase in the concentration of sodium by 40 mg/L, the rate of dissolution R is $3.25.10^{-18}$ mole cm^{-3}s^{-1}. A rate constant for dissolution of oligoclase, k, measured in the laboratory under 1 atm CO_2 pressure, is $1.66.10^{-16}$ mole cm^{-2}s^{-1} (Busenberg and Clemency 1976). Since $R = (s/V)k$, where s is the reactive area of feldspar in sandstone, V is the volume of water, the observed rate of the sodium increase corresponds to a reactive surface of oligoclase s/V 0.02 cm^{2}cm^{-3}. This is a realistic value with respect to a rare occurrence of feldspars in the sandstone. During this irreversible process, groundwater stays in chemical equilibrium with precipitating calcite. Therefore, Ca^{2+} and HCO_3^- decrease during the percolation of water as pH increases due to hydrolysis of feldspars. The concentrations of Ca^{2+} and HCO_3^- start to increase near the discharge point because the water mixes with artesian water from lower part of the aquifer.

Diffusion, convective movement, dissolution of salts and membrane filtration through clay layers can concentrate brines during their geological history (Clayton et al. 1966; Desaulniers et al. 1986; Hanor and McIntosh 2006, 2007). Figure 6.5 illustrates a possibility of diffusion of chloride ions during the last 65 million years from the subjacent and poorly permeable Permo-Carboniferous rocks containing fossil brine to the fresh groundwater Cenomanian aquifer of the Bohemian Cretaceous basin (Jetel 1970; Paces 1983). The diffusion propagates chloride ions upwards through the Permo-Carboniferous sediments. The diffusion equation was fitted to

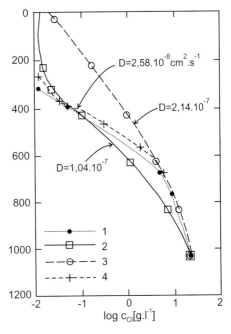

Fig. 6.5 Diffusion model for the concentration gradient of Cl- in a Permo-Carboniferous aquifer under Cenomanian sandstones of the Bohemian Cretaceous basin (Paces 1983). 1 – observed gradient in a drill hole (Jetel 1970); 2 – gradient calculated using data of the drill hole section from 319 to 1032 m depth; 3 – gradient calculated using data of the section from 766 to 1032 m depth; 4 – gradient calculated using data for the section from 393 to 668 m depth.

observed gradients to yield the diffusion coefficient D:

$$\frac{c(z,t)-c_{z=0}}{c_{t=0}-c_{z=0}} = erfc\left(\frac{z}{2\sqrt{Dt}}\right), \qquad (1)$$

where chloride concentration c (z, t) changes with depth z and time t from original low concentration at 65 Ma, $c_{t=0}$, owing to diffusion from original reservoir of brine at a basement of the Permo-Carboniferous semi-permeable aquifer, $c_{z=0}$.

The best fit to data in the Cenomanian aquifer in the section at 250 to 650 m depth yields an unrealistically low diffusion coefficient D of $2.6 \times 10^{-8}\,cm^2\,s^{-1}$. In the deeper Permo-Carboniferous section at 650 to 1032 m depth, the coefficient is $2.1 \times 10^{-7}\,cm^2\,s^{-1}$. A single value of D $(1.04 \times 10^{-7}\,cm^2\,s^{-1})$ does not explain the observed gradient for the entire depth profile. The coefficient $2.1 \times 10^{-7}\,cm^2\,s^{-1}$ is almost two orders of magnitude lower than molecular diffusion in free aqueous solution. Either the diffusion started later than 65 Ma, or a tortuosity of the semi-permeable Permo-Carboniferous layer is high and retards diffusion. This model indicates that diffusion is a feasible process to operate during evolution of stagnant groundwater bodies in sediments with low permeability. Under such conditions, diffusion can increase concentration of Cl^- in aquifers above stagnant water bodies.

While diffusion is probably able to increase concentrations in layers above stagnant brine bodies, membrane filtration through clay layers increases concentrations of anions and large-diameter cations, such as Cl^-, HCO_3^- and Ca^{2+}, in the subjacent stagnant brines (White 1965; Berry 1969; Zhou X 1992). This process is probably the reason for the evolution of Ca-Cl-HCO_3 brines from ancient seawater and Na-Cl brines. There are two types of fossil brines with respect to concentration of sulphate anions: brines having very small concentration of sulphates and those with high concentration of sulphate ions. The low-sulphate brines originally contained bacteria, which consumed sulphates and transferred them into volatile H_2S (Rees 1973; Jørgensen 1982). By contrast, brines with high concentrations of sulphates are typically those derived from volcanic playa lakes, where sulphur emitted from volcanoes dissolved in volcanic lakes. The resulting oxidized brines penetrated the bottom of the lakes to form brines rich in sulphates and chloride (Arad 1969; Smejkal 1981; Paces and Smejkal 2004).

Geochemical processes produce characteristic couples of chemical components during secular evolution of groundwaters. The characteristic couples are: (i) Ca+C– water in which the sum of calcium and inorganic carbon species originate from hydrolysis of aluminosilicates and carbonates, with the presence of biogenic CO_2 in a surface zone of fast hydrological circulation; (ii) Ca+S– water in which the sum of calcium and sulphate results from oxidation of rocks containing sulphide minerals and carbonates, with the presence of O_2 in a surface zone of fast hydrological circulation; (iii) Na+C– water in which the sum of sodium and carbonate species is released during intensive hydrolysis of sodium aluminosilicates owing to an input of endogenous CO_2 in a deeper zone of slow hydrological circulation; (iv) Na+Cl– water in which the sum of sodium and chloride is preserved in connate sedimentary waters or dissolved from salt strata; (v) Na+S– water in which the sum of sodium and sulphur is a component, which occurs in waters related to post-volcanic activities in continental rifts.

Total dissolved solids (TDS) distinguish between groundwater circulating fast in the surface zone and groundwater in stagnant deeper-water bodies. Considering the genetic significance of the chemical couples, the chemical evolution of groundwater can be illustrated in a 3D space with coordinates $x = Log\ S/(Na+C)$, $y = Log\ (Ca+C)/(Na+Cl)$, and $z = Log\ TDS$. The plot of chemical composition of groundwaters of various origin and history yields a 'genetic' diagram shown in Fig. 6.6 (Paces 1987).

An example of the plot with data on waters at different stages of their evolution is illustrated in Fig. 6.7(a) and (b). The genetic meaning of the diagram is not strict. There are processes leading to chemical compositions of water that lie outside the fields defined in the diagram (for example dissolution of sediments with gypsum, etc.). However, many samples of waters of different chemical composition seem to follow the suggested trends.

The diagram in Fig. 6.6 separates groundwaters in an exogenous zone, whose chemical composition is result of hydrolysis and oxidation of rock during relatively fast circulation, from waters from a deeper zones of restricted circulation, where fossil water and residual salts are stored for prolonged geological time. The diagram separates

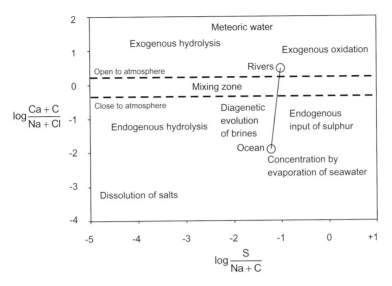

Fig. 6.6 Genetic diagram with fields of groundwaters at different stages of their evolution.

fossil waters evolved in sedimentary oilfield basins from fossil waters evolved in crystalline continental shields. Most waters and their salts have been derived ultimately from the palaeo-oceans. However, there are surface waters and groundwaters enriched in volcanic sulphur and chlorine of magmatic or metamorphic origin. Such waters sometimes occur in continental rifts. (Paces and Smejkal 2004).

CONCLUSIONS

The chemical composition of groundwater evolves owing to water-rock interaction, dissolution of fossil salts, diffusion, membrane filtration and mixing of waters of different origins. The evolution depends on the residence time of water in various Earth reservoirs. The residence time ranges from few hours to approximately 10^8 years. Some radioactive isotopes and man-made chemicals released into the environment enable the quantitative estimation of the residence time of water from a few years up to 30,000. Beyond this

time, the evolution of chemical composition of groundwaters can be related to geological and hydrogeological processes. Despite early papers on dating of very old waters with the use of [36]Cl and [129]I (Fehn et al. 1994; Moran et al. 1995), exact dating of very old fossil waters is still uncertain.

Evolutions of brines in sedimentary basins and crystalline shield are different, as indicated by different shifts in ^{2}H and [18]O isotopes and different trends in chemical composition. The different chemical trends are illustrated on the 'genetic' diagram Fig. 6.7 (a), (b). A characteristic evolution is observed in groundwaters related to ancient volcanic activities in continental rifts. While fossil brines in sedimentary basins and crystalline shields are low in sulphur, brines evolved in continental rifts contain increased amounts of sulphur of volcanic origin.

ACKNOWLEDGEMENT

Preparation of this paper was financially supported by grant 1M0554 of the Ministry of

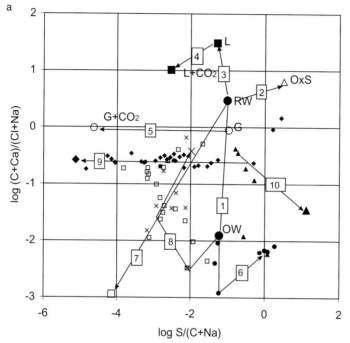

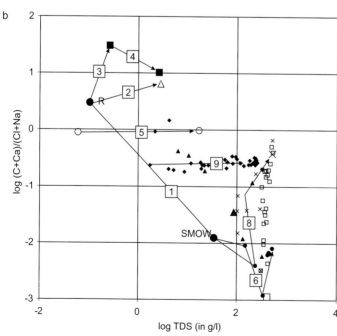

Fig. 6.7 (a) Genetic diagram with chemical data on groundwaters of different origin and evolution. [1] Tie-line between mean river water and mean oceanic water; [2] oxidation of sulphides; [3] hydrolysis of limestone; [4] hydrolysis of limestone accelerated by presence of CO_2; [5] hydrolysis of granitic rocks accelerated with presence of CO_2; [6] evaporation residuum of sea water; [7] dissolution of halite strata; [8] oil-field brines and brines from salt domes and sedimentary basins; [9] fossil water from crystalline rocks of the Canadian Shield; [10] sulphur-rich volcanic playa lakes and saline waters from continental rifts. (b) Genetic diagram related to total dissolved solids (TDS) in groundwaters of different origin and evolution.

Education, Youth and Sport of the Czech Republic. The author thanks Russell Harmon for his careful editing of the paper.

FURTHER READING

Clark, ID and Fritz, P. (1997) Environmental Isotopes in Hydrogeology. Lewis, Boca Raton, Fl.

Edmunds, WM and Shand, P. (eds.) (2008) Natural Ground Water Quality. Blackwell Publishing.

Fritz, P and Frape, SK. (eds.) (1987). Saline Water and Gases in Crystalline Rocks. GAC Special Paper 33. Geological Association of Canada. ISBN 0-919216-32-3.

Lerman, A. 1979. Geochemical Processes Water and Sediment Environments. John Wiley & Sons.

White, DE. (1957) magmatic, connate, and metamorphic waters. *Bull. Geol. Soc. Amer.* **68**: 1659–82.

Yardley, BWD. (2009) The role of water in the evolution of the continental crust. *J. Geol. Soc.* **166**: 585–600.

REFERENCES

Andersen, T, Frezotti, M-L and Burke, EAJ. (eds.) (2001) Fluid inclusions: Phase Relationships-Methods-Applications. *Lithos* **55**, Special issue, Amsterdam.

Arad, A. (1969) Mineral springs and saline lakes of the Western Rift Valley, Uganda. *Gechim. Cosmochim Acta* **33**: 1169–81.

Barnes, I and Back, W. (1964) Geochemistry of iron-rich ground water of Southern Maryland. *Jour. Geol.* **72**: 435–47.

Baum, M. (1996) Controls on the fractionation of isovalent trace elements in magmatic and aqueous systems: Evidence from Y/Ho and Zr/Hf, and lanthanide tetrad effect. *Contrib. Mineral. Petrol.* **123**: 323–33.

Beaucaire, C, Michelot, J-L, Savoye, S et al. (2008) Groundwater characterisation and modelling of water-rock interaction in an argillaceous formation (Tournemire, France). *Applied Geochemistry* **23**: 2182–97.

Berry, FAF. (1969) Relative factors influencing membrane filtration effects in geological environments. *Chem. Geol.* **4**: 295–301, Amsterdam.

Busenberg, E and Clemency, CV. (1976) The dissolution kinetics of feldspars at 25°C and 1 atm CO_2 partial pressure. *Geochim. Cosmochim. Acta* **40**: 41–9.

Beyerle, U, Aschbach Hertg, W, Hofer, M et al. (1999) Infiltration of river water to a shallow aquifer investigated wit 3H/3He, noble gas and CFCs. *Journal of Hydrology* **220**: 169–85.

Corcho, JA, Purtschert, R, Hinsby K et al. (2005) 36Cl in modern groundwater dated by a multi tracer approach: A case study of Quaternary sand aquifers in the Odense Pilot River Basin, Denmark. *Appl. Geochemistry* **20**: 599–609.

Craig, H. (1961) isotopic variations in meteoric waters. *Science* **133**: 1702–3.

Craig, H. (1966) isotopic composition and origin of the Red Sea and Salton Sea brines. *Science* **154**: 1544–7.

Clayton, RN, Friedman, I, Graff, DL et al. (1966) The origin of saline formation waters, 1. Isotopic composition. *J. Geophys. Res.* **71**: 3869–82.

de Groot, PA (ed.) (2004) *Handbook of stable isotope analytical techniques.* Vol. **1**, Elsevier, Amsterdam.

Desaulniers, DE, Kaufmann, RS, Cherry, JA et al. (1986) 37Cl-35Cl variations in a diffusion-controlled groundwater systems. *Geochim. Cosmochim. Acta* **50**: 1757–64.

Dolnicek, Z, Fojt, B, Prochaska, W et al. (2009) Origin of the Zalesi U-Ni-Co-As-Ag/Bi deposit, Bohemian Massif, Czech Republic: Fluid inclusions and stable isotope constraints. *Mineralium Deposita* **44**: 81–97.

Drever, JI (1988) *The Geochemistry of Natural Waters.* Prentice Hall, Englewood Cliffs, New Jersey.

Edmunds, WM and Shand, P (eds.) (2008) *Natural Water Quality.* Blackwell Publishing, Oxford.

Epstein S. and Mayeda T. (1953) Variation of O-18 content of waters from natural sources. *Geochimica et Cosmochimica Acta* **4**: 213–224.

Friedman I. (1953) Deuterium content of natural waters and other substances. *Geochimica et Cosmochimica Acta* **4**: 89–103.

Fehn, U, Moran, JE, Teng, RTD et al. (1994) Dating and tracing of fluids using I-129 and Cl-36 – results from geothermal fluids, oil-field brines and formation waters. *Nuclear Instruments & Methods in Physics Research Section B-Beam Interactions with Materials and Atoms* **92**: 380–4.

Frape, SK and Fritz, P. (1987) Geochemical trends for ground waters from the Canadian Shield. In: SK Frape, and P Fritz (eds.) *Saline Water and Gases in Crystalline Rocks. Geol. Asoc. Canada Spec. Pap.* **33**: 19–38.

Frape, SK, Shouakar-Stash, O, Pačes, T et al. (2007) Geochemical and isotopical characteristics of the waters from crystalline and sedimentary structures of the Bohemian Massif. In: TD Bullen and Y Wang (eds.) *Water – Rock Interaction*, 727–33, Balkema, Taylor and Francis Group, London, ISBN 978-0-415-45136-9.

Giggenbach, WF (1988) Geothermal solute equilibria. Derivation of Na-K-Mg-Ca geoindicators. *Geochim. Cosmochim. Acta* **52**: 2749–65.

Friedman I. (1953) Deuterium content of natural waters and other substances. *Geochimica et Cosmochimica Acta* **4**: 89–103.

Hanor, JS and McIntosh, JC. (2006) Are secular variations in seawater chemistry reflected in the compositions of basinal brines? *Journal of Geochemical Exploration* **89**: Special Issue, 153–6.

Hanor, JS and McIntosh, JC. (2007) Diverse origins and timing of formation of basinal brines in the Gulf of Mexico sedimentary basin. *Geofluids* **7**: 227–37.

Jetel, J. (1970) Hydrogeology of the Permocarboniferous and Cretaceous in the profile line Melnik-Jested. *Sbornik Geologickych Ved, HIG* **7**: 7–42, Academia, Praha.

Jørgensen, BB. (1982) Ecology of the bacteria of the sulphur cycle with special reference to anoxic-oxic interface environments. *Phil. Trans. R. Soc. Lond.* **B** **298**: 543–561.

Knauth, LP and Beeunas, MA. (1986) Isotope geochemistry of fluid inclusions in Permian halite with implications for the isotopic history of ocean water and the origin of saline formation waters. *Geochim. Cosmochim. Acta* **50**: 419–33.

Kovalevych, V, Paul, J and Peryt, TM. (2009) Fluid inclusions in halite from the rot (lower Triassic) salt deposit in central Germany: Evidence for seawater chemistry and conditions of salt deposition and recrystallization. *Carbonates and Evaporites* **24**: 45–57.

Lehmann, BE, Purtschert, R, Loosli, HH et al. (2003) A comparison of groundwater dating with 81Kr, 36Cl and 4He in 4 wells of the Great Artesian Basin, Australia. *Earth and Planetary Science Letters* **211**: 237–50.

Loosli, HH, Blaser, P, Darling, G et al. (2001) Isotopic methods and their hydrogeochemical context in the investigation of palaeowaters. In: WM Edmunds and CJ Milne (eds.) Palaeowaters in Coastal Europe: *Evolution of Groundwater since the Late Pleistocene.* Spec. Publ. 189, Geol. Soc. London, 193–212.

Loosli, HH and Purtschert R. (2005) Rare gases. In: P Aggarwal, JR Gat and K Froehlich (eds.) *Isotopes in Water Cycle: Past, Present and Future of a Developing Science.* IAEA, Vienna, 91–5.

Magenheim, AJ, Spivack, AJ, Michael, PJ et al. (1995) Chlorine stable isotope composition of the oceanic crust: implications for Earth's distribution of chlorine. Earth *Planet. Sci. Lett.* **131**: 427–32.

Mazor, E. (1997) *Chemical and Isotopic Groundwater Hydrology. The Applied Approach.* (2nd edition) Marcel Dekker, Inc., New York. ISBN: 0-8247-9803-1.

Moeller, P. (2000) Rare earth elements and yttrium as geochemical indicators of the source of mineral and thermal waters. In: I Stober and K Bucher (eds.) Hydrogeology of Crystalline Rocks. Kluwer Academic Publ. 227–46.

Mojzsis, SJ, Harrison, TM and Mckeegan, KD. (2001) Oxygen isotope evidence from ancient zircons for liquid water at Earth's surface 4300 Myr ago. *Nature* **409**: 178–181.

Moran, JE, Fehn, U and Hanor, JS. (1995) Determination of source ages and migration patterns of brines from the US Gulf Coast basin using I-129. *Geochim. Cosmochim. Acta* **59**: 5055–69.

Mullis, J, Dubessy, J, Poty, B et al. (1994) Fluid regimes during late stages of a continental collision: physical, chemical and stable isotope measurements of fluid inclusions in fissure quartz from a geotreverse through the Central Alps, Switzerland. *Geochim. Cosmochim. Acta* **58**: 2239–67.

Naden, J, Kilias, SP, Cheliotis, I et al. (2003) Do fluid inclusions preserve $\delta^{18}O$ values of hydrothermal fluids in epithermal systems over geological time? Evidence from palao- and modern geothermal systems, Milos Island, Eegean Sea. *Chem. Geol.* **197**: 143–59.

Nesbitt, BE and Muehlenbachs, K. (1995) Geochemical studies of the origins and effects of synorogenic crustal fluids in the southern Omineca Belt of British Columbia, Canada. *Geol. Soc. Am. Bull.* **107**: 1033–55.

Ohmoto, H. (1986) Stable isotope geochemistry of ore-deposits. *Reviews in Mineralogy* **16**: 491–559.

Ohmoto, H and Rye RO. (1968) Ores of Bluebell mine British Columbia – A product of meteoric water. *Econ. Geol.* **63**: 699.

Ohse, W, Matthess, G and Pekdeger, A. (1984) Equilibrium and disequilibrium between pore waters and minerals in the weathering environment. In: JI Drever (ed.) The Chemistry of Weathering. NATO ASI *Series C: Mathematical and Physical Sciences* **149**: 211–29, D. Reidel Publ. Comp., Dordrecht.

Paces, T. (1972) Chemical characteristics and equilibration in natural water-felsic rock-CO_2 System. *Geochim. Cosmochim. Acta*, **36**: 217 240, Oxford.

Pačes, T. (1983) Principles of Geochemistry of Waters. Academia, Praha (in Czech).

Paces, T. (1987) Hydrochemical evolution of saline waters from crystalline rocks of the Bohemian Massif (Czechoslovakia). In: P Fritz and SK Frape (eds.) Saline Water and Gases in Crystalline Rocks, Geol. Assoc. Canada, Special Paper 33: 145–56.

Paces, T, Corcho Alvarado, JA., Herrmann, Z et al. (2008): The Cenomanian and Turonian Aquifers of the Bohemian Cretaceous Basin, Czech Republic. In: WM Edmunds and P Shand (eds.) *Natural Water Quality*, 372–390, Blackwell Publishing, Oxford.

Paces, T and Smejkal, V. (2004) Magmatic and fossil components of thermal and mineral waters in the Eger River continental rift (Bohemian massif, central Europe). In: RB Wanty and RR Seal II (eds.) *Water-Rock Interaction*, 167–72, Taylor and Francis Group, London, AA. Balkema Publishers, ISBN 90 5809641 6.

Pearson, FJ Jr. (1987) Models of mineral controls on the composition of saline ground waters of the Canadian Shield. In: P Fritz and SK Frape (eds.) *Saline Water and Gases in Crystalline Rocks, Geol.* Assoc. Canada, Special Paper 33: 39–51.

Plummer, LN, Prestemon, EC and Parkhurst, DL. (1991) NETPATH: An interactive code for interpreting NET geochemical reactions from chemical and isotopic data along a flow PATH. In: YK Kharaka and AS Maest (eds.) *Water-Rock Interaction, 239–42*, A.A. Balkema, Rotterdam.

Purtschert, R, Lehmann, BE and Loosli, HH. (2001) Grouindwater dating and subsurface processes inves-tigated by noble gas isotopes (37Ar, 39AR, 85Kr, 222Rn, 4He). In: R Cidu (ed.) *Water-Rock Interaction*, WRI-10, 1569-1573, AA. Balkema Publ., Lisse.

Purtschert, R. (2008) Timescales and tracers. In: WM Edmunds and P Shand (eds.) *Natural Water Quality*, 91–108, Blackwell Publishing.

Rees, CE. (1973) A steady-state model for sulphur isotope fractionation in bacterial reduction processes. *Geochim. Cosmochim. Acta* **37**: 1141–62.

Rye, RO. (1993) The evolution of magmatic fluids in the epithermal environment. *Econ. Geol.* **63**: 232–8.

Rye, RO and Haffty, J. (1969) Chemical composition of hydrothermal fluids responsible for lead-zinc deposits at Providencia, Zacatecas, Mexico. *Econ. Geol.* **64**: 629–43.

Samson, I, Anderson, A and Marshall, D. (eds.) (2003) Fluid Inclusionds: Analysis and interpretation. Mineralogical Association of Canada, Short Course, Spec. Vol. 32, Vancouver.

Scambelluri. M and Philippot, P. (2001) Deep fluids in subduction zones. *Lithos* **55**: 213–27.

Shen, K, Zhang, Z, Yan, L et al. (2008) Composition and evolution of fluids in the continental orogen: A study of fluid inclusions in high-pressure granulites from the Namche Barwa area, Tibet of southwest China. *Acta Petrologica Sinica* **24**: 1488–500.

Siegel, MD and Andersholm, S. (1994) Geochemical evolution of groundwater in the culebra dolomite near the waste isolation pilot-plant, southeastern New-Mexico, USA. *Geochim. Cosmochim. Acta* **58**: 2299–323.

Smejkal, V. (1981) Stable isotopes pf Carbon, Oxygen and Sulfur and fossil origin of Na-SO4-Cl in mineral waters of the 'Karlovy Vary' type. *Acta Univ. Carol., Geol.* **1**: 71–84, Praha.

Torgersen, T, Habermehl, MA, Phillips, FM et al. (1991) Chlorine 36 dating of very old groundwater 3. Further studies in the Great Arthesian bain. *Water Res. Res.* **27**: 3201–13.

Weise, SM, Bräuer,K, Kämpf, H et al. (2001) Transport of mantle volatiles through the crust straced by seis-mically released fluids: a natural experiment in the earthquake swarm area Vogtland/NW Bohemia, central Europe. *Tectonophysics* **336**: 137–50.

White, DE (1957) magmatic, connate, and metamorphic waters. *Bull. Geol. Soc. Amer.* **68**: 1659–82.

White, DE. (1965) Saline waters of sedimentary rocks. In: Subsurface Environments, *A Symposium Amer. Assoc. Petrol. Geol. Mem.* **4**: 342–66, Tulsa.

White, DE. (1969) Thermal and mineral water of the United States – Brief review of possible origins. XXII

Int. Geol. Congress, vol. 19, *Mineral and Thermal Waters of the World*, B- Oversea Countries, 269–86, Prague.

White, DE. (1974) Diverse origins of hydrothermal ore fluids. *Econ. Geol.* **69**: 954–73.

White, DE, Barnes, I and O'Neil JR. (1973) Thermal and mineral waters of nonmeteoric origin, California Coast Ranges. *Geol. Soc. Amer. Bull.* **84**: 547–60.

Yardley, BWD. (2005) Metal concentrations in crustal fluids and their relationship to ore formation. *Econ. Geol.* **100**: 613–32.

Yardley, BWD. (2009) The role of water in the evolution of the continental crust. *J. Geol. Soc.* **166**: 585–600.

Zhou X, LC. (1992) Hydrogeochemistry of deep formation brines in the central Sichuan Basin, China. *Jour. Hydrology* **138**: 1–15.

Zverev, VP. (2009) Subsurface Waters and Evolution of the Earth. Nauchnyi Mir, Moscow, ISBN 978-5-91522-077-4 (in Russian, Engl. abstract).

7 Stable Isotope Geochemistry: Some Perspectives

JOCHEN HOEFS

Geowissenschaftliches Zentrum, University of Göttingen, Göttingen, Germany

ABSTRACT

Over the last 50 years, stable isotope geochemistry has grown continuously, and today is an integral part of many different areas of geoscience research. This article offers a brief overview of stable isotopes as geochemical fingerprints. It discusses the advances in analytical technologies and demonstrates how progress in recent years has opened new research fields by measuring, for example, more than one rare isotope.

INTRODUCTION

The year that stable-isotope geochemistry became an independent branch of geochemistry is generally regarded to be 1947, launched by two developments: the introduction of a special type of mass spectrometer developed by A Nier and co-workers (Nier et al. 1947) which enabled the parallel and precise measurement of isotope ratios of two gases; and (ii) the now 'classic' paper by Urey (1947) on the theoretical thermodynamic basis of stable isotope fractionation.

Frontiers in Geochemistry: Contribution of Geochemistry to the Study of the Earth, First edition. Edited by Russell S. Harmon and Andrew Parker. © 2011 Blackwell Publishing Ltd. Published 2011 by Blackwell Publishing Ltd.

Today, it is accepted convention to express isotope ratios measured for samples as δ-values in permil (‰) deviations from a standard reference gas. The δ-value is thus defined:

$$\delta \left(in \; \text{\textperthousand} \; deviation \right) = \left(\frac{R_{\text{Sample}}}{R_{\text{Standard}}} - 1 \right) \times 10^3 \qquad (1)$$

In a broad sense, stable-isotope studies can be applied to two different categories of earth science problems: (i) the determination of temperatures of formation of minerals and gases; and (ii) as tracers to determine the origin of rocks, fluids, gases and other geomaterials or as monitors of chemical processes within the natural environment. Major geochemical reservoirs on Earth – like the mantle, the ocean and sedimentary organic matter – have distinct stable-isotope signatures that can be used as characteristic 'fingerprints'. The remainder of this chapter will concentrate largely on the second type of problems rather than the first. For a contemporary review of stable-isotope geothermometry, the reader is referred to discussions by Valley and Cole (2001), Sharp (2007) and Hoefs (2009).

HISTORICAL PERSPECTIVES

When I started working in the field in the mid-1960s, stable-isotope geochemistry had already made important strides from its stage of infancy, but stable-isotope measurements were still restricted to a few laboratories around the world, which generally were specialized in the analysis of one particular element such as C, O or S. Topics of prime research interest at that time, which are discussed individually in the four sections that follow below, were:

(i)   isotope characteristics of different water types,
(ii) hydrothermal ore deposits,
(iii) carbonate palaeothermometry, and
(iv) biosignatures.

Isotope characteristics of different water types

Arguably the most fundamental contribution of stable-isotope geochemistry to the geosciences is the observation of a near-linear relationship between the H- and O-isotope composition (i.e. δD and $\delta^{18}O$ values) of meteoric waters (Craig 1961; Dansgaard 1964). This relationship, generally known as the 'Meteoric Water Line' (MWL), was first defined by Craig (1961) and represents the foundation for a majority of hydrology and palaeoclimatology studies using stable isotopes. Furthermore, it represents a characteristic 'fingerprint' of waters of unknown origin in the subsurface, providing an effective means for evaluating sources and flow paths of groundwaters. In general, groundwaters show no seasonal variation in δD and $\delta^{18}O$-values, and have an isotopic composition close to the mean annual precipitation, which is strong evidence for direct recharge to an aquifer (Gat 1971). In arid regions, evaporative losses before and during recharge may shift the isotopic composition of groundwater toward higher δ-values. In some arid regions, groundwater may be classified as 'palaeowater', which was recharged under meteorological conditions different from today and which may imply ages of water of several thousand years (Gat and Issar 1974).

An important implication of the MWL, to be discussed in more detail, is the question how deep meteoric waters might penetrate into the crust. Because meteoric water interacts with rocks in the subsurface, oxygen and, to a lesser degree, hydrogen isotope exchange reactions, take place through a variety of water-rock interaction processes as surface waters circulate through the crust. As a consequence, meteoric waters become enriched in ^{18}O, while the rocks become depleted in ^{18}O, the extent depending on the temperature and on the water/rock ratio. Mainly through the work of HP Taylor and co-workers, it has become well-established that many igneous intrusions have interacted with heated meteoric water to depths of three to five km on a very large scale (see e.g. Criss and Taylor 1986 and references therein). An extreme example of meteoric water-rock interaction, which can be traced even to mantle depths, is shown in Fig. 7.1. Very recently, Wang and Eiler (2008) concluded that the small oxygen isotope differences observed in Hawaiian lavas were a consequence of interactions between the magmas in shallow subsurface reservoirs and their surrounding volcanic rocks. On the east side of the Hawaiian Islands, where rainfall is high, basalts were hydrothermally altered, either subaerially by meteoric waters or in the submarine environment by seawater, so that magma-rock interaction resulted in small but consistent magma ^{18}O depletions. By contrast, basaltic country rocks were not affected by such hydrothermal alteration on the western side of Hawaii, where the annual rainfall is low. This study demonstrated the same kinds of meteoric-hydrothermal interactions for the Hawaiian Islands that are responsible for the large ^{18}O depletions of Iceland (e.g. Hattori and Muehlenbachs 1982; Harmon and Hoefs 1995 and references therein).

Unusually low $\delta^{18}O$-values have been observed in ultra-high pressure (UHP) rocks from the Dabie-Sulu area in China (e.g. Zheng et al. 1998 and Xiao et al. 2006 among others). UHP rocks are characterized by coesite and microdiamond in eclogite and other crustal rocks, which is strong evidence that a large segment of continental crust was subducted to mantle depths. The extremely low $\delta^{18}O$ values observed in Fig. 7.1 result from meteoric water interaction prior to ultra-high-pressure metamorphism. Surprisingly, these rocks have preserved their low $\delta^{18}O$ values, indi-

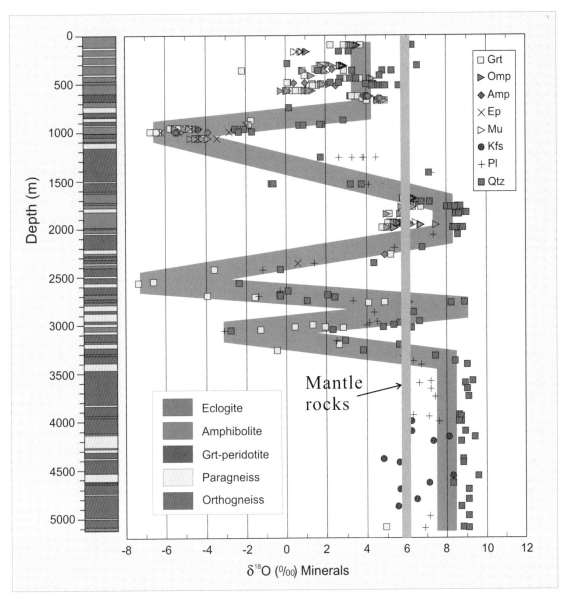

Fig. 7.1 A $\delta^{18}O$ continuous profile from the Chinese Continental Scientific Drilling (CCSD) project, drilling entirely through ultra-high-pressure rocks (Xiao et al., unpublished).

cating a short residence time at mantle depths followed by a rapid uplift.

Hydrothermal ore deposits

The determinations of light isotopes of H, C, O and S in hydrothermal ore deposits have provided information about the diverse origins of ore fluids, about temperatures of mineralization, and about physiochemical conditions of mineral deposition. In contrast to early views, which assumed that almost all hydrothermal deposits owed their genesis to magmas, stable-isotope investigations have convincingly demonstrated that ore formation has taken place within the Earth's crust by recycling processes of fluids, metals, sulphur and carbon. A huge amount of literature exists on more-or-less all economically important hydrothermal ore deposits, which has been summarized in numerous journal and book reviews (see e.g. Ohmoto and Rye 1979; Ohmoto 1986; Taylor 1987).

Inasmuch as water is the dominant constituent of ore-forming fluids, knowledge of its origin is fundamental to any theory of ore genesis. There are two ways for determining δD and $\delta^{18}O$ values of ore fluids: (i) by direct measurement of fluid inclusions contained within hydrothermal minerals, or (ii) by analysis of hydroxyl-bearing minerals and the subsequent calculation of the isotopic composition of fluids from known temperature-dependent mineral-water fractionations, assuming that minerals were precipitated from solutions under conditions of isotope equilibrium. This indirect method of deducing the isotope composition of ore deposits is more frequently used, because it is technically easier. Numerous stable-isotope studies have indicated that all types of water may become ore-forming fluids and that during the various stages of ore formation different types of water may become dominant (e.g. Ohmoto 1986; Taylor 1997). This basic analytical approach has been continuously used over the past 40 years and it has been only recently that new analytical tools have emerged.

The basic principles to be followed in the interpretation of $\delta^{34}S$ values in ore deposits were eluci-

dated by Sakai (1968), and subsequently extended by Ohmoto (1972). Of special importance for a correct deduction about the origin of sulphur in a sample, as well as of carbon, are the oxygen fugacity and the pH of the ore-forming fluid. The isotopic composition of a hydrothermal sulphide is determined by a number of factors such as: (i) isotopic composition of the hydrothermal fluid from which the mineral is deposited; (ii) temperature of deposition; (iii) chemical composition of the dissolved-element species including pH and fO_2 at the time of mineralization; and (iv) relative amount of the mineral deposited from the fluid. The first parameter is characteristic of the source of sulphur, the three others relate to the geochemical conditions of deposition. Of special importance is the oxygen fugacity of a hydrothermal fluid, because of the large isotope fractionation between sulphate and sulphide. As a consequence, small changes in fO_2 may result in large changes in the S-isotope composition of either sulphide or sulphate, which will be balanced by a significant change in the ratio of sulphate to sulphide. If the oxidation state of the fluid is below the sulphate/H_2S boundary, then $^{34}S/^{32}S$ ratios of sulphides will be insensitive to redox shifts.

Carbonate thermometer and stable isotope palaeoclimatology

In 1947, H. Urey suggested that variations in the temperature of precipitation of calcium carbonate from water should lead to measurable variations in the $^{18}O/^{16}O$ ratio of precipitated calcium carbonate. He postulated that determination of temperatures of ancient oceans should be possible by measuring the ^{18}O content of fossil-shell calcite. However, before a meaningful temperature calculation can be carried out for a fossil carbonate shell, several assumptions have to be fulfilled. Shell-secreting organisms to be used for palaeotemperature studies must have been precipitated in isotope equilibrium with ocean water. For oxygen isotopes, most organisms precipitate $CaCO_3$ close to equilibrium; if disequilibrium prevails, the isotopic difference from equilibrium is rather small.

The isotopic composition of an aragonite or calcite shell will remain unchanged until the shell material dissolves, and recrystallizes during diagenesis. In most shallow depositional systems, C- and O-isotope ratios of calcitic shells are fairly resistant to diagenetic changes, but many organisms have a hollow structure, allowing diagenetic carbonate to be added. With increasing depths of burial and time, the chances of diagenetic effects generally increase (e.g. Schrag 1999).

Most oceanic palaeoclimate studies have concentrated on foraminifera. In many cases analyses have been made both of planktonic and benthonic species. Since the first pioneering paper of Emiliani (1955), numerous cores from various sites of the Deep Sea Drilling Project (DSDP) and Ocean Drilling Program (ODP) have been analysed and, when correlated accurately, have produced a well-established oxygen isotope curve for the Pleistocene and Tertiary (see, e.g. Emiliani 1978; Shackleton 1986, 1997). Emiliani (1955) introduced the concept of 'isotopic stages', which subsequently was refined by Shackleton and Opdyke (1973), by designating stage numbers for identifiable events in the marine foraminiferal O-isotope record for the Pleistocene. Odd numbers identify interglacial or interstadial (warm) stages, whereas even numbers define ^{18}O-enriched glacial (cold) stages. O-isotope analysis of deep-sea core foraminifera over the past 30 years has demonstrated that similar δ^{18}O-variations are observed contemporaneously in all areas of the world (Shackleton 1997). These studies have provided a unique window through which to study and understand Plio-Pleistocene

climate changes and their causes. With independently-dated timescales, such as can be derived from the magnetic variations in deep-sea sediment cores, these systematic δ^{18}O variations result in synchronous isotope signals in the sedimentary record because the mixing time of the oceans is relatively short (10^3 years). These signals provide stratigraphic markers, enabling correlations between cores which may be thousands of kilometres apart, and the isotopic variations observed through time are guides to both the isotopic composition of seawater and the global volume of continental ice sheets. Figure 7.2 shows the oceanic O-isotope curve for the Pleistocene. This diagram exhibits several striking features: the most obvious one is the cyclicity; furthermore, fluctuations never go beyond a certain maximum value on either side of the range. This seems to imply that very effective feedback mechanisms are at work to arrest the cooling and warming trends at some maximum level. The 'sawtooth'-like curve in Fig. 7.2 is characterized by very steep gradients, with maximum cold periods followed immediately by maximum warm periods.

Ice cores from the polar regions (see e.g. Johnson et al. 1972; Dansgaard et al. 1993; Jouzel 2003) and tropical mountain glaciers in High Andes (Thompson et al. 1979) and Africa (Thompson et al. 2002) have yielded palaeoclimate records analogous to those observed in deep-sea sediment cores, but with a higher degree of resolution for the Holocene and latest Pleistocene because of their more rapid accumulation rates (Masson et al. 2000).

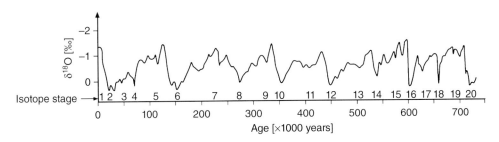

Fig. 7.2 Composite δ^{18}O fluctuations in the foraminifera species *G sacculifer* from Caribbean cores (Emiliani 1978).

Biosignatures

It has long been recognized (see e.g. Thode et al. 1949; Craig 1953) that biological processes significantly fractionate the isotopes of C and S, which leads to characteristic biosignatures in sedimentary rocks that will be preserved in the geological record. A topic of broad and continuing interest relates to isotopic indicators of life early in Earth history or on Mars. As a note of caution, however, it is still uncertain whether abiogenic processes may lead to isotope compositions of C similar to those that are indicative of biological activity. What has been discussed and debated for more than 50 years (see e.g. the debate between Craig 1954 and Rankama 1954) is the question whether or not the C-isotope ratios observed in the oldest rocks on Earth can be taken as evidence for the presence of life.

This controversy about the meaning of graphite δ^{13}C-values from early Archaean (3.8 to 3.5 Ga) rocks has continued unabated to the present day (see e.g. Bolhar et al. 2004). The validity of using carbon isotopes in the search for early life hinges on the assumption that metabolism by early organisms produced a C-isotope fractionation effect similar to that subsequently observed during the Phanerozoic. For these very ancient high-grade metamorphic rocks, the thermal history of the metasediments has to be considered. For example, early Archaean metasedimentary rocks from Greenland have δ^{13}C values that range from −50 to −22‰, which could be interpreted as being of biogenic origin (Mojzsis et al. 1996). These very negative δ^{13}C values were measured for tiny particles of reduced carbon within apatite grains, which apparently had been shielded from metamorphic alteration. However, a case also can be made that graphite from Archaean rocks may be of abiotic, hydrothermal origin. Also, more and more experimental and natural evidence has been presented in recent years that Fischer-Tropsch type reactions may produce abiogenic organic matter that exhibit δ^{13}C values similar to those of characteristic biogenic organic matter (see, e.g. Taran et al. 2007).

A similar debate has occurred about the meaning of S-isotope ratios, and especially about the question of when bacterial sulphate reduction commenced on Earth. Considering a typical difference in S isotope values of 20 to 60‰ between marine sulphate and bacteriogenic sulphide in present-day sedimentary environments, similar fractionations in ancient sedimentary rocks may be interpreted as an evidence for the activity of sulphate-reducing bacteria. Thus, the presence or absence of such fractionations in sedimentary rocks may constrain the time of emergence of sulphate-reducing bacteria. Sedimentary pyrite depleted in ^{34}S has been observed as far back in geological time as the Precambrian. There is, however, ongoing debate about the timing of the onset of bacterial reduction in the geological record.

In early Archaean sedimentary rocks, most sulphides and the rare sulphates have δ^{34}S values near 0‰ (Monster et al. 1979; Cameron 1982), which has been interpreted as indicating an absence of bacterial reduction in the Archaean. Shen and Buick (2004) argued that the large spread in δ^{34}S values of microscopic pyrite grains aligned along growth faces of former gypsum in the 3.47 Ga North Pole barite deposit in Australia represents the oldest evidence for microbial sulphate reduction. However, the question still remains whether thermochemical or hydrothermal processes can cause similar fractionations. Recent analytical progress in measuring all four stable isotopes, ^{32}S, ^{33}S, ^{34}S and ^{36}S suggests that additional evidence can be gained that allows the differentiation of hydrothermal versus biological processes (Ono et al. 2006) even when δ^{34}S-values are inconclusive. Using this approach Ueno et al. (2008) argued that quadruple sulphur isotopes in barite and pyrite from the Early Precambrian Dresser Formation, Australia indicate the existence of microbial sulphate reduction at 3.5 billion years ago.

Fe isotopes are another isotope system which has been discussed actively in the recent literature. Virtually all reduction from Fe^{3+} to Fe^{2+} at the Earth's surface is mediated by the metabolism of dissimilatory bacteria. Biological processes therefore may produce measurable Fe-isotopic fractionations because the metabolic processing of Fe involves a number of steps such as transport across membranes that may fractionate Fe isotopes. Experiments with dissimilatory Fe-reducing bacteria of the genus *Shewenella* indicate that iron-isotope fractionations are a function of Fe(III) reduction rates: at low rates the produced Fe^{2+} is isotopically depleted by 1.3‰, whereas the depletion is up to 3‰ at high rates (Johnson et al. 2008).

Fe-isotope fractionation has been also observed during bacterial iron oxidation (Croal et al. 2004). Under anaerobic conditions, Fe(II)-oxidizing phototrophs may produce a ferrihydrite precipitate that is enriched in ^{57}Fe by 1.5‰ relative to its aqueous Fe(II) source. Controversy still exists whether the Fe-isotope variations observed in microbial experiments are primarily controlled by kinetic or non-biological equilibrium factors. Since abiotic iron reduction/oxidation reactions may reveal fractionations similar in direction and magnitude to microbial reactions (Skulan et al. 2002), the presence of Fe-isotope variations is not, in itself, conclusive evidence of biotic activity. This complicates the ability to use iron isotopes to identify microbiological processing in the rock record (Balci et al. 2006).

Of special significance in this connection are biosignatures on Mars, because McKay et al. (1996) have claimed that Martian meteorite ALH 84001 – found in Antarctica – contains evidence of past Martian life. Various kinds of apparent biosignatures, with characteristic isotope compositions, have been suggested: organic matter, carbonate minerals, magnetite grains, sulphide minerals. After intensive investigations during the last decade, none of these proposed biosignatures is today considered valid.

PROGRESS IN ANALYTICAL TECHNIQUES AND DEVICES DURING THE LAST YEARS

The 1980s and 1990s

During the 1960s and 1970s, analytical devices and procedures remained more-or-less the same, with only incremental improvements in mass-spectrometry and sample extraction and processing. The standard procedure during this time for H-, C-, O- and N-isotope analysis was to prepare the sample in an extraction/preparation system physically removed from the mass spectrometer, and subsequently to introduce the purifed gas to the mass spectrometer for isotope-ratio measurement.

This situation changed in a major way during the 1980s with the introduction of a series of new analytical approaches to stable-isotope analysis. The first major change was introduction of continuous-flow isotope ratio mass spectrometry, which permitted the sample analysis of microgram quantities of sample. Table 7.1 compares conventional off-line techniques with on-line techniques. Clearly, the most important parameter in this list is the change in sample size from milligram quantities to microgram quantities that can be routinely analysed. This allows, for instance, the isotopic analysis of individual

Table 7.1 Comparison of offline dual inlet and online continuous flow methods.

Offline method (dual inlet)	vs.	Online method (continuous flow)
Offline sample preparation	↔	Online sample preparation
Offline purification of gases	↔	Purification of gases by GC column
Large sample size (mg)	↔	Small sample size (µg)
Direct inlet of sample gas	↔	Sample gas inlet via carrier gas
Pressure adjustment of both gases	↔	No adjustment of pressure linearity and stability of the system are necessary conditions
Sample/standard change (>6 times)	↔	One peak per sample
δ-value calculated from statistical mean	↔	δ-value calculated by peak integration and reference gas
Little problems with sample sample homogeneity	↔	Problems with sample homogeneity

organic compounds from sedimentary organic matter separated by gas chromatography.

Organic matter in the geosphere is a complex mixture of source organisms having variable biosynthetic pathways and detrital remains. The determination of $\delta^{13}C$-values of bulk organic matter is thus unable to distinguish between different carbon sources. Immediately after burial of the biological organic material into sediments, complex diagenetic changes occur in the organic matter. Two processes have been proposed to explain the observed changes in carbon isotope composition. The first is preferential degradation of organic compounds that have different isotope composition compared to the preserved organic compounds. Since easily degradable organic compounds like amino-acids are enriched in ^{13}C compared to the more resistant compounds like lipids, degradation causes a shift to slightly more negative values. Alternatively, isotope fractionations due to metabolism of microorganisms will produce new compounds having different isotopic compositions from the original source material. A classic example has been presented by Freeman et al. (1990), who analysed hydrocarbons from the Messel shale in Germany (see Table 7.2). While the major portion of the analysed hydrocarbons reflect the primary biological source material, some hydrocarbons present in low concentrations

are extremely ^{13}C-depleted, indicating their secondary microbial origin in a methane-rich environment. Later studies, summarized by Peckmann and Thiel (2005), have documented even larger ^{13}C depletions ($\delta^{13}C$ values as low as −120‰) in various biomarkers that have been formed from diverse methane-using taxa.

Another micro-analytical approach that has had a profound effect on stable-isotope geochemistry is laser-assisted extraction of O, C and S compounds from a variety of samples (see, e.g. Sharp 1990). This approach allows for the high-precision analysis of much smaller amounts of material than required by traditional approaches. In order to achieve precise and accurate measurements, the samples have to be evaporated completely, because steep thermal gradients during laser heating can induce large isotopic fractionations. Thermal effects can be overcome by vaporizing samples with ultraviolet (UV) KrF and ArF lasers, thus making possible the *in situ* O-isotope analysis of silicate minerals (Wiechert and Hoefs 1995). Using a laser-assisted technique, Kelley and Fallick (1990) demonstrated high-precision and spatially resolved analysis of S-isotopes in sulphide. Laser-assisted analytical methods generally reveal greater isotope heterogeneity than conventional approaches. As a rule of thumb: the smaller the scale of measurement, the larger the sample heterogeneity.

Over the past decade, the most common application of laser-assisted stable-isotope analysis is the O-isotope analysis of small grains of silicate minerals, especially olivines, garnets and zircons, which are difficult to react with the conventional fluorination technique. Mattey et al. (1994) and Ionov et al. (1994) demonstrated a very restricted range of ^{18}O variation in mantle peridotites. A series of studies by J Eiler and co-workers has shown that olivines from genetically different types of basaltic lavas show subtle but resolvable $^{18}O/^{16}O$ ratio differences owing to their tectonic setting (Eiler et al. 1996, 1997a, 2000). The largest variability in oxygen-isotope composition of olivine phenocrysts has been found in subduction-related basalts. O-isotope variations can constrain the contributions of subducted sediments

Table 7.2 Carbon isotope composition of individual branched and cyclic alkanes from the Eocene Messel Shale (Freeman et al. 1990).

Peak	$\delta^{13}C$	compound
1	−22.7	norpristane
2	−30.2	C19 acyclic isoprenoid
3	−25.4	pristane
4	−31.8	phytane
5	−29.1	C23 acyclic isoprenoid
8	−73.4	C32 acyclic isoprenoid
9	−24.2	isoprenoid alkane
10	−49.9	22,29,30-trisnorhopane
11	−60.4	isoprenoid alkane
15	−65.3	30-norhopane
19	−20.9	lycopane

and fluids (Dorendorf et al. 2000; Eiler et al. 2000; Bindeman et al. 2005).

Stable-isotope analysis at even greater spatial resolution and at smaller sample sizes can be achieved by secondary-ion mass spectrometry (SIMS), which allows a spatial resolution that is an order of magnitude better than laser techniques. In SIMS, the surface of the specimen is sputtered with a focused primary ion beam, and then ejected secondary ions are collected and analysed with a mass spectrometer to determine isotopic composition (Giletti and Shimizu 1989). The main problem with this technique is that measurements must be corrected for instrumental mass fractionation and matrix effects (Eiler et al. 1997b). The latest version of the ion microprobe technology has overcome these difficulties and can achieve precise analysis of isotope ratios at the ±0.1‰ level. Studies by Valley and Graham (1993), Eiler et al. (1995) and Valley et al. (1998) illustrated the potential of spatially resolved stable-isotope analysis to a broad spectrum of geochemical problems.

2000–2010

With the beginning of the new millennium, the introduction of the multiple-collector inductively coupled mass spectrometry (MC-ICP-MS) has revolutionized stable-isotope geochemistry, because the technique enables research on natural isotope variations for a wide range of elements across the periodic table, particularly the transition and heavy metal elements, which previously could not have been measured with adequate precision. This new technique is also of relevance for the light elements because it allows more combinations of isotope systems to be considered in addressing a specific geochemical problem. For the light isotopes, combined uses of isotope systems have been undertaken for carbonate, nitrate, sulphate and quartz, using the traditional analytical methodologies, but with the introduction of MC-ICP-MS combinations of many more are possible; such studies are starting to be conducted and many more will be done in the near future. This opens new possibilities for investiga-

tion of the natural environment using isotope variations of metal elements such as Fe, Mn, Cu and Zn. For example, such metal isotope studies can provide new insights into ore deposit deposition (Markl et al. 2006a, 2006b), into processes associated with marine calcification (Eisenhauer et al. 2009), and the evolution of the ocean and atmosphere (Johnson et al. 2008). By combining published ages curves for carbon, sulphur and iron isotopes, Johnson et al. (2008) demonstrated that major changes in oxidation/reduction conditions occurred in the age period from 3.1 to 2.4 Ga.

As previously described, reconstruction of Holocene-Pleistocene palaeoclimate has centred on the excellent marine deep-sea sediment and ice core records. However, there are significant gaps in present understanding of past climate change in terrestrial environments that have begun to be addressed as a consequence of the technological advances through studies of speleothems, the secondary $CaCO_3$ precipitates that form in caves. Building on the early work of Hendy and Wilson (1968), recent studies utilizing SIMS analysis for O-isotope analysis at a spatial resolution of 20 to 30 μm along a stalagmite growth axis are able to provide precisely-dated records of both long-term, astronomically forced climate change as well as short-duration climate changes such as Heinrich Events, Dansgaard-Oeschger Cycles and events related to changes in ocean circulation such as monsoonal and El Nino-Southern Oscillation phenomena (Kolodny et al. 2003; Fairchild et al. 2006; Treble et al. 2005, 2007).

Mass-dependent and mass-independent isotope effects

It has been a common belief that isotope effects for most natural reactions arise solely because of isotopic mass differences. This means that for an element with more than two isotopes, such as oxygen and sulphur, the enrichment of ^{18}O relative to ^{16}O or ^{34}S relative to ^{32}S is expected to be approximately twice as large as the enrichment of ^{17}O relative to ^{16}O, or as the enrichment of ^{33}S relative to ^{32}S. Thus, it was thought that $\delta^{17}O$, $\delta^{33}S$

and $\delta^{36}S$ carry no additional information about material formation, because isotope fractionations are mass-dependent, so that:

$\delta^{17}O \approx 0.5\ \delta^{18}O$, $\delta^{33}S \approx 0.5\ \delta^{34}S$, and $\delta^{36}S \approx 2.0\ \delta^{34}S$

Deviations from the mass-dependent fractionation pattern were first identified for oxygen isotopes from meteoritic silicate and oxide minerals (Clayton et al. 1973). Later, work on atmospheric ozone (Thiemens and Heidenreich 1983; Mauersberger 1981) also indicated large mass-independent isotope fractionations. As is now known, all oxygen-containing gases in the atmosphere, except water, show deviations from mass-dependent fractionations, all of which are linked to reactions involving stratospheric ozone (Thiemens 2006).

Recent progress in high-precision isotope measurements has led to the determination of fractionation coefficients with significance in the third decimal place. This presents the possibility of distinguishing between equilibrium and kinetic isotope-fractionation mechanisms. For example, isotope equilibrium processes for sulphur are close to a coefficient of 0.515, whereas bacterial sulphate reduction has a lower value, around 0.512.

More recently, significant mass-independent sulphur isotope fractionations have been reported by Farquhar and co-workers (i.e. Farquhar et al. 2000) in sulphides older than 2.4 Ga. These mass-independent fractionations do not occur in measurable amounts in sulphides younger than 2.4 Ga, which is generally interpreted as indicating the transition on Earth from an anoxic to an oxic atmosphere (see further discussion in Holland, this volume).

Multiply substituted isotopologues

In general, bulk isotopic compositions of natural samples are reported in terms of the major isotope (e.g., $\delta^{13}C$, $\delta^{18}O$, etc.). In the measured gases, bulk compositions depend only on abundances of molecules containing one rare isotope (e.g., $^{13}C^{16}O^{16}O$ or $^{12}C^{18}O^{16}O$). However, also present in very low concentration are molecules having more than

Table 7.3 Stochastic abundances of CO_2 isotopologues (Eiler 2007).

Mass	Isotopologue	Relative abundance
44	$^{12}C^{16}O_2$	98.40 %
45	$^{13}C^{16}O_2$	1.11 %
	$^{12}C^{17}O^{16}O$	748 ppm
46	$^{12}C^{18}O^{16}O$	0.40 %
	$^{13}C^{17}O^{16}O$	8.4 ppm
	$^{12}C^{17}O_2$	0.142 ppm
47	$^{13}C^{18}O^{16}O$	44.4 ppm
	$^{12}C^{17}O^{18}O$	1.50 ppm
	$^{13}C^{17}O_2$	1.60 ppb
48	$^{12}C^{18}O_2$	3.96 ppm
	$^{13}C^{17}O^{18}O$	16.8 ppb
49	$^{13}C^{18}O_2$	44.5 ppb

one rare isotope such as $^{13}C^{18}O^{16}O$ or $^{12}C^{18}O^{17}O$. These so-called 'isotopologues' are molecules that differ from one another only in isotopic composition. Table 7.3 gives the stochastic abundances of isotopologues of CO_2.

As early as 1947, it was recognized (Urey 1947; Bigeleisen and Mayer 1947) that multiply-substituted isotoplogues have thermodynamic properties that are different from those for singly-substituted isotopologues of the same molecule. Natural distributions of multiply-substituted isotopologues can thus provide unique constraints on geological, geochemical and cosmochemical processes (Wang et al. 2004).

'Conventional' gas-source mass spectrometers do not allow meaningful abundance measurements of these very rare species. However, their measurement is possible, if some requirements for high abundance sensitivity, precision and mass resolving power are met. J Eiler and co-workers (see e.g. Eiler and Schauble 2004; Affek and Eiler 2006; Eiler 2007) have reported precise (<0.1‰) measurements of CO_2 with mass 47 (Δ_{47}-values) with a specially modified, but normal gas-source, mass spectrometer. In this nomenclature, Δ_{47}-values are defined as the permil (‰) difference between the measured abundance of all molecules with mass 47 relative to the abundance of 47 expected for the stochastic distribution.

This new technique has been termed 'clumped isotope geochemistry' (Eiler 2007) because the respective species are produced by clumping two rare isotopes together. Deviations from stochastic distributions may result from all processes of isotope fractionation observed in nature. Thus, processes that lead to isotope fractionations of bulk compositions also lead to fractionations of multiply-substituted isotopologues, implying that clumped-isotope geochemistry is potentially applicable to many geochemical problems (Eiler 2007). To date, the most frequent application has been a carbonate thermometer based on the formation of the CO_3^{2-} group containing both ^{13}C and ^{18}O. Schauble et al. (2006) calculated an approximate 0.4‰ excess of $^{13}C^{18}O^{16}O$ groups in carbonate groups at room temperature relative to what would be expected in a stochastic mixture of carbonate isotopologues with the same bulk $^{13}C/^{12}C$, $^{18}O/^{16}O$ and $^{17}O/^{16}O$ ratios. The excess amount of $^{13}C^{18}O^{16}O$ decreases with increasing temperature and, therefore, may serve as a thermometer (Ghosh et al. 2006). Potentially the advantage of this thermometer will be that it allows the determination of temperatures of carbonate formation without knowing the isotope composition of the fluid. Came et al. (2007), for example, presented temperature estimates for Early Silurian and Late Carboniferous seawater which are consistent with varying CO_2 concentrations.

FUTURE OUTLOOK

Stable-isotope investigations in geochemistry have grown steadily since their introduction in earth sciences in the late 1940s. This growth has been occurring during the last decade at an ever-increasing pace as a consequence of major advances in mass spectrometry and in microanalytical techniques. Tremendous progress has been achieved during this time in many subfields of stable isotope geochemistry:

(i) The application of multicollector ICP-MS now enables investigations of a wide range of transition and heavy elements that could not be measured before with adequate precision;

(ii) Precise ion-microprobe measurements on the micrometer scale allow the detection of growth and dissolution processes of minerals;

(iii) Analysis of multiple rare isotopes of elements permit the detection and distinction of mass-dependent and mass-independent fractionation processes.

The future of stable-isotope geochemistry is bright indeed. The pattern of technological advance driving new scientific understanding should continue into the near future and yield new insights into both physical and biological processes on Earth, and in so doing, may reveal unexpected phenomena in both domains. Furthermore, the combination of different isotope systems will better constrain the mechanisms of isotope fractionation processes in recent and ancient geologic environments.

ACKNOWLEDGEMENT

Russell Harmon has reviewed an early draft, which is gratefully acknowledged.

FURTHER READING

Bullen, TD, Eisenhauer, A. (2009) Metal Stable Isotopes: Signals in the environment. *Elements* 5: 349–385.

Hoefs, J. (2009) *Stable Isotope Geochemistry*. 6th edn. Springer Berlin, Heidelberg.

Johnson, CM, Beard, BL and Albarede, F. (eds.) (2004) Geochemistry of Non-Traditional Isotopes. *Reviews in Mineralogy & Geochemistry* 55: 1–454.

Sharp, Z. (2007) *Principles of Stable Isotope Geochemistry*. Pearson, Prentice Hall.

Valley, JW, Cole, DR. (eds.) (2001) Stable Isotope Geochemistry. *Reviews in Mineralogy & Geochemistry* 43: 1–662.

REFERENCES

Affek, HP, Eiler, JM. (2006) Abundance of mass 47 CO_2 in urban air, car exhaust and human breath. *Geochim Cosmochim Acta* 70: 1–12.

Balci, N, Bullen, TD, Witte-Lien, K, Shanks, WC, Motelica, M and Mandernack, KW. (2006) Iron isotope fractionation during microbially simulated Fe(II) oxidation and Fe(III) precipitation. *Geochim. Cosmochim. Acta* **70**: 622–39.

Bigeleisen, J, Mayer, MG. (1947) Calculation of equilibrium constants for isotopic exchange reactions. *J. Chem Phys* **15**: 261–267.

Bindeman, IN, Eiler, JN et al. (2005) Oxygen isotope evidence for slab melting in modern and ancient subduction zones. *Earth Planet Sci. Lett.* **235**: 480–96.

Bolhar, R, Kamber, BS, Moorbath, S, Fedo, CM and Whitehouse, MJ. (2004) Characterisation of early Archaean chemical sediments by trace element signatures. *Earth Planet Sci. Lett.* **222**: 43–60.

Came, RE, Eiler, JM, Veizer, J, Azmy, K, Brand, U and Weidman, CR. (2007) Coupling of surface temperatures and atmospheric CO_2 concentrations during the Paleozoic era. *Nature* **449**: 198–201.

Cameron, EM. (1982) Sulphate and sulphate reduction in early Precambrian oceans. *Nature* **296**: 145–8.

Clayton, RN, Grossman, L and Mayeda, TK. (1973) A component of primitive nuclear composition in carbonaceous chondrites. *Science* **182**: 485–8.

Craig, H. (1953) The geochemistry of the stable carbon isotopes. *Geochim. Cosmochim. Acta* **3**: 53–92.

Craig, H. (1954) Geochemical implications of the isotopic composition of carbon in ancient rocks. *Geochim. Cosmochim. Acta* **6**: 186–96.

Craig, H. (1961) Isotopic variations in meteoric waters. *Science* **133**: 1702 3.

Criss, RE, Taylor, HP. (1986) Meteoric hydrothermal systems. In: Stable isotopes in high temperature geological processes. *Rev. Mineral* **16**: 373–424.

Croal, LR, Johnson, CM, Beard, BL and Newman, DK. (2004) Iron isotope fractionation by Fe(II)-oxidizing photoautotrophic bacteria. *Geochim. Cosmochim. Acta* **68**: 1227–42.

Dansgaard, W. (1964) Stable isotopes in precipitation. *Tellus* **16**: 436–68.

Dansgaard, W, Johnsen, S, Clausen, H and Dahl-Jensen, D et al. (1993) Evidence for general instability of past climate from a 250-kyr ice-core record. *Nature* **364**: 218–220.

Dorendorf, F, Wiechert, U and Wörner, G. (2000) Hydrated sub-arc mantle: A source for the Kluchevskoy volcano, Kamchatka/Russia. *Earth Planet Sci. Lett.* **175**: 69–86.

Eiler, JM. (2007) The study of naturally-occurring multiply-substituted isotopologues. *Earth Planet Sci. Lett* **262**: 309–27.

Eiler, JM, Schauble, E. (2004) $^{18}O^{13}C^{16}O$ in earth's atmosphere. *Geochim. Cosmochim. Acta* **68**: 4767–77.

Eiler, JM, Crawford, A, Elliott, T, Farley, KA, Valley, JW and Stolper, EM. (2000) Oxygen isotope geochemistry of oceanic-arc lavas. *J. Petrol.* **41**: 229–56.

Eiler, JM, Farley, KA, Valley, JW, Hauri, E, Craig, H, Hart, S and Stolper, EM. (1997a) Oxygen isotope variations in ocean island basalt phenocrysts. *Geochim. Cosmochim. Acta* **61**: 2281–93.

Eiler, JM, Graham, C and Valley, JW. (1997b) SIMS analysis of oxygen isotopes: matrix effects in complex minerals and glasses. *Chem. Geol.* **138**: 221–34.

Eiler, JM, Farley, KA, Valley, JW, Hofmann, AW and Stolper, EM. (1996) Oxygen isotope constraints on the source of Hawaiian volcanism. *Earth Planet. Sci. Lett.* **144**: 453–68.

Eiler, JM, Valley, JW, Graham, C and Baumgartner, LP. (1995) Ion microprobe evidence for the mechanisms of stable isotope regression in high-grade metamorphic rocks. *Contrib. Mineral. Petrol.* **118**: 365–78.

Eisenhauer, A, Kisakürek, B and Böhm, F. (2009) Marine calcification: An alkali earth metal isotope perspective. *Elements* **5**: 365–8.

Emiliani, C. (1955) Pleistocene temperatures. *J. Geol.* **63**: 538–78.

Emiliani, C. (1978) The cause of the ice ages. *Earth Planet. Sci. Lett.* **37**: 349–54

Fairchild, IJ, Smith, CL, Baker, A, Fuller, L, Spötl et al. (2006). Modification and preservation of environmental signals in speleothems. *Earth Science Reviews* **75**: 105–53.

Farquhar, J, Bao, H and Thiemens, M. (2000) Atmospheric influence of Earth's earliest sulphur cycle. *Science* **289**: 756–9.

Freeman, KH, Hayes, JM, Trendel, JM and Albrecht, P. (1990) Evidence from carbon isotope measurements for diverse origins of sedimentary hydrocarbons. *Nature* **343**: 254–6.

Gat, JR. (1971) Comments on the stable isotope method in regional groundwater investigation. *Water Resources Research* **7**: 980.

Gat, JR, Issar A. (1974) Desert isotope hydrology: water sources of the Sinai desert. *Geochim. Cosmochim. Acta* **38**: 1117–31.

Ghosh, P, Adkins, J, Affek, H, Balta, B et al. (2006) ^{13}C-^{18}O bonds in carbonate minerals: a new kind of paleothermometer. *Geochim. Cosmochim. Acta* **70**: 1439–56.

Giletti, BJ, Shimizu, B. (1989) Use of the ion microprobe to measure natural abundance of isotopes in minerals.

In: *New Frontiers in Stable Isotope Research*, Eds W.C. Shanks III and RE Criss, *US Geol. Soc. Bull.* **1890**: 129–36.

Harmon, RS and Hoefs, J. (1995) Oxygen isotope heterogeneity of the mantle deduced from global 180 systematics of basalts from different geotectonic settings. *Contra. Mineral. Petrol.* **120**: 95–114.

Hattori, K and Muehlenbachs, K. (1982) Oxygen isotope ratios of the Icelandic crust. *Journal of Geophysical Research* **87**: 6559–65.

Hendy, CH, Wilson, AT. (1968) Palaeoclimatic data from speleothems. *Nature* **219**: 48–51.

Hoefs, J. (2009) *Stable Isotope Geochemistry*. 6th edition. Springer Berlin Heidelberg.

Ionov, DA, Harmon, RS, France-Lanord, C, Greenwood, PB and Ashchepkov, IV. (1994) Oxygen isotope composition of garnet and spinel peridotites in the continental mantle: evidence from the Vitim xenolith suite, southern Siberia. *Geochim. Cosmochim. Acta.* **58**: 1463–70.

Johnsen, SJ, Dansgaard, W, Clausen, HB and Langway, CC Jr. (1972) Oxygen isotope profiles through the Antarctic and Greenland ice sheets. *Nature* **235**: 429–34.

Johnson, CM, Beard, BL and Roden, EE. (2008) The iron isotope fingerprints of redox and biogeochemical cycling in modern and ancient Earth. *Annual Review Earth Planet. Sci.* **36**: 457–93.

Jouzel, J. (2003) Water Stable Isotopes: Atmospheric Composition and Applications. In: *Polar Ice Core Studies*. In: Treatise on Geochemistry, RF Keeling (ed.) Executive Editors: HD Holland and KK Turekian, **4**: 213–43.

Kelley, SP, Fallick, AE. (1990) High precision spatially resolved analysis of $\delta^{34}S$ in sulfides using a laser extraction technique. *Geochim. Cosmochim. Acta* **54**: 883–8.

Kolodny, Y, Bar-Matthews, M, Ayalon, A and McKeegan, KD. (2003) A high spatial resolution $\delta^{18}O$ profile of a speleothem using an ionmicroprobe. *Chem. Geol.* **197**: 21–28.

Markl, G, von Blanckenburg, F and Wagner, T. (2006a) Iron isotope fractionation during hydrothermal ore deposition and alteration. *Geochim. Cosmochim. Acta* **70**: 3011–3030.

Markl, G, Lahaye, Y and Schwinn, G. (2006b) Copper isotopes as monitors of redoc processes in hydrothermal mineralization. *Geochim. Cosmochim. Acta* **70**: 4215–28.

Masson, V, Vimeux, F, Jouzel, J, Morgan, V, Delmotte, M, Ciais, P, Hammer, C, Johnsen, S, Lipenkov, V,

Mosley-Thompson, E, Petit, J-R, Steig, EJ, Stievenard, M and Vaikmae, R. (2000) Holocene Climate Variability in Antarctica Based on 11 Ice-Core Isotopic Records. *Quat. Res.* **54**: 348–58.

Mattey, D, Lowry, D and Macpherson, C. (1994) Oxygen isotope composition of mantle peridotite. *Earth Planetary Science Lett.* **128**: 231–41.

Mauersberger, K. (1981) Measurements of heavy ozone in the stratosphere. *Geophysical Research Letters* **8**: 935–7.

McKay, DS, Gibson, EK, Thomas-Kprta, KL, Vali, H et al. (1996) Search for past life on Mars: possible relic biogenic activity in martian meteorite ALH 84001. *Science* **273**: 924–30.

Mojzsis, SJ, Arrhenius, G, McKeegan, KD, Harrison, TM, Nutman, AP and Friend, CRL. (1996) Evidence for life on Earth before 3,800 million years ago. *Nature* **384**: 55–9.

Monster, J, Appel, PW, Thode, HG, Schidlowski, M et al. (1979) Sulphur isotope studies in early Archean sediments from Isua, West Greenland: implications for the antiquity of bacterial sulphate reduction. *Geochim. Cosmochim. Acta.* **43**: 405–13.

Nier, AO, Ney, EP and Inghram, MG. (1947) A null method for the comparison of two ion currents in a mass spectrometer. *Review Science Instruments* **18**: 294.

Ohmoto, H. (1972) Systematics of sulphur and carbon isotopes in hydrothermal ore deposits. *Economic Geology* **67**: 551–78.

Ohmoto, H. (1986) Stable isotope geochemistry of ore deposits. *Reviews in Mineralogy* **16**: 491–559.

Ohmoto, H, Rye, RO. (1979) Isotopes of sulphur and carbon. In: *HL Barnes Geochemistry of Hydrothermal Ore Deposits*, 2nd edn. Holt Rinehart and Winston, New York.

Ono, S, Wing, BA, Johnston, D, Farquhar, J and Rumble, D. (2006) Mass-dependent fractionation of quadruple sulphur isotope systems as a new tracer of sulphur biogeochemical cycles. *Geochim. Cosmochim. Acta* **70**; 2238–52.

Peckmann, J, Thiel, V. (2005) Carbon cycling of ancient methane-seeps. *Chem. Geol.* **205**: 115–33.

Rankama, K. (1954) Early pre-Cambrian carbon of biogenic origin from the Canadian Shield. *Science* **119**: 506.

Sakai, H. (1968) Isotopic properties of sulfur compounds in hydrothermal processes. *Geochemical Journal* **2**: 29–49.

Schauble, EA, Ghosh, P and Eiler, JM. (2006) Preferential formation of $^{13}C-^{18}O$ bonds in carbonate minerals, estimated using first-principles lattice dynamics. *Geochim. Cosmochim. Acta* **70**: 2510–519.

Schrag, DP. (1999) Effects of diagenesis on the isotope record of late Paleogene tropical sea surface temperature. *Chemical Geology* **161**: 2265–78.

Shackleton, NJ. (1986) Paleogene stable isotope events. *Palaeogeography, Palaeoclimatology, Palaeoecology* **57**: 91–102.

Shackleton, NJ. (1997) The deep-sea sediment record and the Pliocene-Pleistocene boundary. *Quaternary International* **40**: 33–5.

1, NJ, Opdyke, ND. (1973) Oxygen isotope and palaeo-magnetic stratigraphy of equatorial Pacific core V28-238: oxygen isotope temperatures and ice volumes on a 10^5 and 10^6 year scale. *Quaternary Research* **3**: 39–55.

Sharp, ZD. (1990) A laser-based microanalytical method for the in situ determination of oxygen isotope ratios of silicates and oxides. *Geochim. Cosmochim. Acta* **54**: 135–7.

Sharp, ZD. (2007) *Principles of Stable Isotope Geochemistry*. Pearson Prentice Hall.

Shen, Y, Buick, R. (2004) The antiquity of microbial sulfate reduction. *Earth Science Reviews* **64**: 243–72.

Skulan, JL, Beard, BL and Johnson, CM. (2002) Kinetic and equilibrium Fe isotope fractionation between aqueous Fe(III) and hematite. *Geochim. Cosmochim. Acta* **66**: 2995–3015.

Taran, YA, Kliger, GA and Sevastianov, VS. (2007) Carbon isotope effect in the open system Fischer Tropsch synthesis. *Geochim. Cosmochim. Acta* **71**: 4474–87.

Taylor, HP. (1987) Comparison of hydrothermal systems in layered gabbros and granites and the origin of low ^{18}O magmas. In: Magmatic processes: Physicochemical Principles. *Geochemical Soc. Special Publication* **1**: 337–57.

Taylor, HP. (1997) Geochemistry of hydrothermal ore deposits. In: HL. Barnes (ed.) *Geochemistry of Hydrothermal Ore Deposits*, J Wiley & Sons: 229–88.

Thiemens, MH. (2006) History and applications of mass-independent isotope effects. *Annual Review Earth Planet. Sci.* **34**: 217–62.

Thiemens, MH, Heidenreich, JE. (1983) The mass independent fractionation of oxygen- A novel isotope effect and its cosmochemical implications. *Science* **219**: 1073–5.

Thode, HG, Macnamara, J and Collins, CB. (1949) Natural variations in the isotopic content of sulphur

and their significance. *Canadian Journal of Research* **27**: 361.

Thompson, LG, Hastenrath, S and Morales Arnao, B. (1979) Climatic ice core records from the tropical Quelccaya Ice Cap. *Science* **203**, 1240–1243.

Thompson, LG, Mosley-Thompson, E, Davis, ME, Henderson, KA et al. (2002) Kilimanjaro Ice Core Records: Evidence of Holocene climate change in tropical Africa. *Science* **298**: 589–93.

Treble, PC, Chappell, J, Gagan, MK, McKeegan, KD and Harrison, TM. (2005) In situ measurement of seasonal $\delta^{18}O$ variations and analysis of isotopic trends in a modern speleothem from southwest Australia. *Earth Planet. Sci. Lett.* **233**: 17–32.

Treble, PC, Schmitt, AK, Edwards, RL, McKeegan, KD et al. (2007) High resolution Secondary Ionisation Mass Spectrometry (SIMS) $\delta^{18}O$ analyses of Hulu Cave speleothem at the time of Heinrich Event 1: *Chem. Geol.* **238**: 197–212.

Ueno, Y, Ono, S, Rumble, D and Maruyama, S. (2008) Quadruple sulfur isotope analysis of ca 3.5 Ga Dresser Formation: New evidence for microbial sulfate reduction in the early Archean. *Geochim. Cosmochim. Acta* **72**: 5675–91.

Urey, HC. (1947) The thermodynamic properties of isotopic substances. *Journal Chemical Society* **1947**: 562–81.

Valley, JW. (2001) Stable isotope thermometry at high temperatures. *Reviews in Mineralogy & Geochemistry* **43**: 365–413.

Valley, JW, Graham, CM. (1993) Cryptic grain-scale heterogeneity of oxygen isotope ratios in metamorphic magnetite. *Science* **259**: 1729–33.

Valley, JW, Graham, CM, Harte, B, Eiler, JM and Kinney, PD. (1998) Ion microprobe analysis of oxygen, carbon and hydrogen isotope ratios. In: MA McKibben, WC Shanks III and WI Ridley (eds.) Applications of microanalytical techniques to understanding mineralization processes. *Review Economic Geology* **7**: 73–98.

Wang ZR, Eiler JM (2008) Insights into the origin of low-delta O-18 basaltic magmas in Hawaii revealed from in situ measurements of oxygen isotope compositions of olivines. *Earth Planet. Sci. Lett.* **269**: 376–86.

Wang, Z, Schauble, EA and Eiler, JM. (2004) Equilibrium thermodynamics of multiply substituted isotopologues of molecular gas. *Geochim. Cosmochim. Acta* **68**: 4779–97.

Wiechert, U, Hoefs, J. (1995) An excimer laser-based microanalytical preparation technique for *in situ* oxygen isotope analysis of silicate and oxide minerals. *Geochim. Cosmochim. Acta* **59**: 4093–101.

Xiao, Y, Zhang, Z, Hoefs, J and van den Kerkhof, AM. (2006) Ultrahigh pressure rocks from the Chinese continental Scientific Drilling Project: II. Oxygen isotope and fluid inclusion distribution through vertical sections. *Contributions Mineralogy Petrology* **152**: 443–58.

Zheng, Y, Fu, B, Li, Y, Xiao, Y and Li, S. (1998) Oxygen and hydrogen isotope geochemistry of ultra-high pressure eclogites from the Dabie mountains and the Sulu terrane. *Earth Planet. Sci. Lett.* **155**: 113–29.

Part 2
Frontiers in Geochemistry

8 Geochemistry of Geologic Sequestration of Carbon Dioxide

YOUSIF K. KHARAKA[1] AND DAVID R. COLE[2]

[1]U.S. Geological Survey, Menlo Park, CA, USA
[2]School of Earth Sciences, The Ohio State University, Columbus, OH, USA

ABSTRACT

Geologic sequestration of carbon dioxide (CO_2) is considered one plausible option to reduce greenhouse gases (GHGs) emissions to mitigate global warming and related climate changes. Geochemical studies are essential for providing improved understanding of CO_2-brine-mineral interactions, which strongly affect storage injectivity and security. Reservoir capacity and integrity are strongly dependent on the four main CO_2-trapping mechanisms: 'structural', 'residual', 'solution' and 'mineral'. The bulk of CO_2 will be stored initially as a supercritical fluid, with some rapidly dissolving in formation water, but mineral and additional solution-trapping will be slower, yet more permanent. Results from natural analogues indicate that dissolution in brine is the major sink for CO_2. Geochemical results from Frio tests (Texas) proved powerful in: (i) tracking the successful injection of CO_2; (ii) detecting CO_2 in the overlying 'B' sandstone; (iii) showing mobilization of metals and organics, and major changes in chemical and isotopic compositions of brine. Modelling and Fe isotopes indicate dissolution of calcite and Fe-oxyhydroxides, and well-pipe corrosion.

Geochemical changes, including pH-lowering, alkalinity increases and metals mobilization, were also observed in shallow groundwater following CO_2 injection at the zero-emission research and technology (ZERT) site, Montana. Results show highly sensitive chemical and isotopic tracers that can be used to monitor injection performance, and provide early detection of CO_2 and brine leakages into groundwater.

INTRODUCTION

Fossil fuels – coal, oil and natural gas – are essential sources of primary energy, currently supplying about 85% of the global energy (EIA 2009). Combustion of these fuels, however, releases large amounts of greenhouse gases (GHGs), primarily CO_2, to the atmosphere. Currently, close to 30 Gt/yr of CO_2 are being added to the atmosphere from these sources, and the amount is projected to increase to 43 Gt/yr by 2030 (Broecker and Kunzig 2008; EIA 2009; Haszeldine 2009). Increased anthropogenic emissions of CO_2 have raised its atmospheric concentrations from about 280 ppmv during pre-industrial times to approximately 390 ppmv today, and based on several defined scenarios CO_2 concentrations are projected to increase up to 1100 ppmv by 2100 (White et al. 2003; IPCC 2005, 2007). There is now a broad scientific consensus that current global warming (average temperature is

Frontiers in Geochemistry: Contribution of Geochemistry to the Study of the Earth, First edition. Edited by Russell S. Harmon and Andrew Parker.
© 2011 Blackwell Publishing Ltd. Published 2011 by Blackwell Publishing Ltd.

approximately 0.8°C higher than during pre-industrial times) and related climate changes are caused mainly by these atmospheric CO_2 increases; average global temperatures are projected to increase by 2–6°C by 2100 (White et al. 2003; IPCC 2007; Oelkers and Cole 2008). Related climate changes with potential adverse impacts may include sea-level rise from the melting of alpine glaciers and ice sheets, and from the ocean warming; increased frequency and intensity of wildfires, floods, droughts and tropical storms; and changes in the amount, timing, and distribution of rain, snow and runoff. Rising atmospheric CO_2 is also increasing the amount of CO_2 dissolved in ocean water, increasing its acidity (lowering its pH from 8.1 to 8.0), with potentially disruptive effects on coral reefs, marine plankton and ecosystems (Adams and Caldeira 2008; Sundquist et al. 2009).

Carbon-dioxide sequestration, in addition to energy conservation and increased use of renewable and lower-carbon, intensity fuels, is now considered an important component of the portfolio of options for reducing greenhouse gas emissions to stabilize atmospheric levels of these gases and global temperatures at acceptable values that would not severely impact global economic growth (Benson and Cook 2005; Holloway et al. 2007; IPCC 2007; Lackner 2010; Oelkers and Cole 2008; Schrag 2009; Sundquist et al. 2009). Sedimentary basins in general, and deep saline aquifers in particular, are being investigated as possible repositories for large volumes of anthropogenic CO_2 that must be sequestered to mitigate global warming and related climate changes (Hitchon 1996; Benson and Cole 2008; Kharaka et al. 2006, 2009).

Currently there are a total of five commercial projects operating world-wide that capture and inject about seven million tons of CO_2 annually and that provide valuable experience for assessing the efficacy of carbon capture and sequestration (CCS). There are also more than 100 commercial 'enhanced oil recovery' (EOR) projects world-wide with approximately 80% in the USA, primarily in the prolific oil reservoirs of the Permian Basin of Texas and New Mexico. In the USA, enhanced oil-recovery started in 1973 and 5% of oil production is currently from EOR operations, and US CO_2 sales for EOR reached approximately 3 billion ft^3/d in 2008 (DOE/NETL 2008; Moritis 2009). In addition, there are approximately 25 geologic sequestration field-demonstration projects in the USA at various stages of planning and deployment, and an equal number of tests are being implemented in other countries to investigate the storage of CO_2 in various rock formations (clastic, carbonate or basalt) using different injection schemes, monitoring methods, hazards assessment protocols and mitigation strategies (Litynski et al. 2008; Cook 2009; Haszeldine 2009; Matter et al. 2009; Michael et al. 2009).

Considerable uncertainties and scientific gaps, however, exist in understanding CO_2-brine-mineral interactions at reservoir conditions, because supercritical CO_2 is buoyant, displaces huge volumes of formation water, and becomes reactive when dissolved in the formation water (Kharaka et al. 2009; Haszeldine 2009). Dissolved CO_2 is likely to react with the reservoir and cap rocks, causing dissolution, precipitation and transformation of minerals, and changing the porosity, permeability and injectivity of the reservoir, as well as impacting the extent of CO_2 and brine leakage that, as noted by Kharaka et al. (2006, 2009) and Benson and Cole (2008) could contaminate underground sources of drinking water (USDW). Reservoir capacity, performance and integrity are strongly affected by the four possible CO_2 trapping (Fig. 8.1) mechanisms (Benson and Cook 2005; Friedmann 2007; Benson and Cole 2008): (1) 'structural and stratigraphic trapping', where the injected CO_2 is stored as a supercritical and buoyant fluid below a cap rock or adjacent to an impermeable barrier; (2) 'residual trapping' of CO_2 by capillary forces in the pores of reservoir rocks away from the supercritical plume; (3) 'solution trapping', where the CO_2 is dissolved in formation water forming aqueous species such as $H_2CO_3^0$, HCO_3^-, and CO_3^{2-}; and (4) 'mineral trapping', with the CO_2 precipitated as calcite, magnesite, siderite and dawsonite (Gunter et al. 1993; Palandri et al. 2005; Bénézeth et al. 2007, 2009; Oelkers et al. 2008).

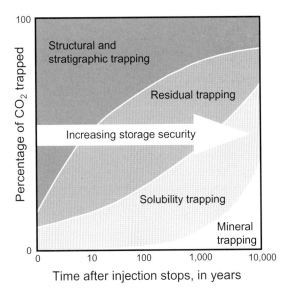

Fig. 8.1 Schematic diagram showing the proportions of CO_2 trapped over extended periods of time by various processes following injection. Note the general increases with time for solubility and mineral trapping (from Benson and Cook 2005).

This review discusses the geochemistry of CO_2 sequestration in subsurface reservoirs, emphasizing CO_2-brine-mineral interactions in sedimentary basins at reservoir conditions. We discuss CO_2 phase relations and geochemical results obtained from natural CO_2 analogues (e.g. Gilfillan et al. 2009), and from CO_2-injection field sites, especially our reported results from the Frio Brine Pilot tests, an American DOE-funded multilaboratory field experiment located near Dayton, Texas (Hovorka et al. 2006; Kharaka et al. 2006). The Frio tests were carried out in 2004–2008 to investigate the potential for geologic storage of CO_2 in deep (approximately 1500 m) saline aquifers, and to develop geochemical, geophysical, simulation and other tools to map the flow of CO_2 in the injection zones (Daley et al. 2007; Kharaka et al. 2009). We emphasize temporal changes in the chemical and isotopic composition of formation brine and gases that were used for 'deep monitoring' of fluid leakage from the injection sandstone 'C' into the overlying 'B' sandstone.

Significant isotopic and chemical changes, including the lowering of pH, increases in alkalinity and mobilization of metals, were also observed in samples obtained from shallow groundwater following CO_2 injection at the ZERT site, located in Bozeman, Montana (Spangler et al. 2009; Kharaka et al. 2010b). Results from both the deep and shallow field-tests show that geochemical methods provide highly sensitive chemical and isotopic tracers for tracking changes resulting from water-CO_2-sediment interactions; these methods are recommended for CO_2 injection sites to monitor injection performance, and for early detection of any CO_2 and brine leakages.

CO_2 PHASE RELATIONS

In the subsurface, CO_2 can exist as a dissolved component in groundwater, basinal brines and geothermal fluids as well as in gaseous and liquid phases. The physical state of CO_2 varies with temperature and pressure as shown in Fig. 8.2 (Hu

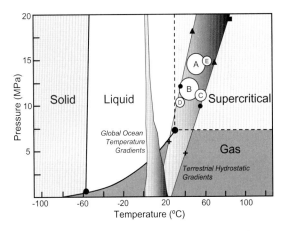

Fig. 8.2 Phase behaviour of CO_2 as a function of pressure and temperature, with stability fields representing ocean and hydrostatic pressure and temperature gradients (modified after Sundquist et al. 2009). Legend: (A) Alberta, Canada; (B) Gippsland Basin, Australia; (C) Nagaoka, Japan; (D) Sleipner, North Sea; (E) Frio Pilot, Texas; (+) 500 m depth; (•) 1000 m depth; (▲) 1500 m depth; (■) 2000 m depth.

et al. 2007; Sundquest et al. 2009). At ambient T and P conditions, CO_2 is a gas. At T of $-55°$ to $31°C$ and P greater than $7.4\,MPa$, CO_2 is a liquid, but at T greater than $31°C$ and P greater than $7.4\,MPa$ (the T and P at the critical point) CO_2 is classified as a supercritical fluid. In the supercritical state, CO_2 behaves like a gas, but under high P the density of the CO_2 can be very high, approaching or even exceeding the density of liquid water. In most sequestration scenarios, CO_2 is injected in liquid form, but transforms to a supercritical fluid as it enters the formation and is subjected to the formation T and P.

The calculated density-volume relationship of CO_2 in the subsurface is shown in Fig. 8.3, based on the National Institute of Standards and Technology 14 database (NIST 1992) for a typical geothermal gradient of $30°C/km$. The density decreases markedly from supercritical conditions at about $700\,kg/m^3$ down to gaseous conditions below $250\,kg/m^3$. This smooth transition occurs at a depth of between 1000 and $800\,m$. At depths below about $800\,m$, CO_2 density is high enough to allow efficient pore filling and to decrease the buoyancy difference compared with *in situ* fluids (Benson and Cole 2008). Below this depth interval the volume of the CO_2 relative to gas at 1 bar pressure is roughly 0.002. If CO_2 migrates vertically through a system, the phase behaviour is dependent on the pressure-temperature path (Fig. 8.4a) – i.e. the geothermal gradient (Oldenburg 2007). CO_2 rising along a normal geothermal gradient of, say, $15°C/km$ would more than likely undergo two phase transitions – from supercritical to liquid, and liquid to gas, as it ascends to the surface. Alternatively, CO_2 rising along a steep geothermal gradient such as $30°C/km$ could experience a phase transition directly from a supercritical fluid to a gas. For sequestration in saline aquifers and oil reservoirs, CO_2 is less dense than the original *in situ* fluids, causing it to rise towards the base of the cap rock. The viscosity is from 10 to 20 times less than for water. Conversely, CO_2 tends to sink in gas reservoirs and displaces native CH_4 because it is denser and more viscous (Oldenburg 2007).

SOLUBILITY BEHAVIOUR

Injection of CO_2 of any form into the pore water of a reservoir leads to 'solubility trapping'. The solubility depends on several factors, most notably pressure, temperature and brine salinity (Spycher et al. 2003; Lagneau et al. 2005; Rosenbauer et al. 2005; Koschel et al. 2006; Raistruck et al. 2006). The chemical equilibria associated with the dissolution of CO_2 in formation brine are expressed in equations (1)–(3):

$$CO_{2\,(g)} + H_2O \leftrightarrow H_2CO_3 \tag{8.1}$$

$$H_2CO_3 \leftrightarrow HCO_3^- + H^+ \tag{8.2}$$

$$HCO_3^- \leftrightarrow CO_3^{2-} + 2H^+ \tag{8.3}$$

At pH values lower than approximately 6, carbonic acid (H_2CO_3) is the dominant carbonate species, with bicarbonate (HCO_3^-) and carbonate (CO_3^{2-}) becoming dominant in brine at intermediate and high pH values respectively. At condi-

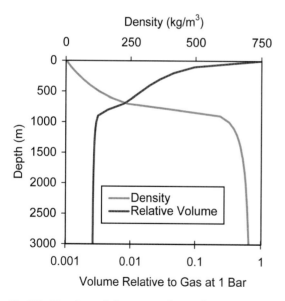

Fig. 8.3 Density and change in volume of CO_2 as a function of depth below ground surface for a typical gradient of approximately $30°C/km$ (from Benson and Cole 2008). See Plate 8.3 for a colour version of this image.

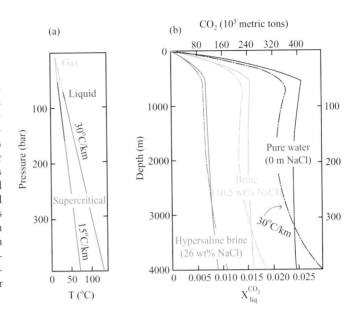

Fig. 8.4 (a) Phase behaviour of CO_2 as a function of temperature and pressure for two different geothermal gradients. (b) Solubility, in mole fraction, of CO_2 in NaCl solution as a function of depth and salinity at two different geothermal gradients. Both figures modified after Oldenburg (2007) based on results presented in Spycher et al. (2003) and Spycher and Pruess (2005). An example of the mass of CO_2 (in metric tons) trapped is illustrated using a simple scenario where CO_2 is injected into a 20 m-thick formation with 10% of its void space available for a CO_2 dissolution process extending 1 km out from the well in all directions. A pure water system can dissolve five times the amount of CO_2 compared to a hypersaline brine. See Plate 8.4 for a colour version of these images.

tions expected for most geological sequestration sites (50° to 150°C, and up to a few hundred bar total pressure), the solubility of CO_2 increases with increasing pressure, but decreases with increasing temperature and salinity (Fig. 8.4b). The salinity effect, known as the salting-out effect, is large, lowering the solubility of CO_2 by a factor of five in hypersaline brine (26% salinity) compared with pure water. As observed in Fig. 8.4(b), the effects of geothermal gradient are complicated by fluid pressure related to depth. For example, at depths greater than approximately 2 km, CO_2 solubility increases slowly for the 15°C/km geothermal gradient, but the increase is much faster for the 30°C/km geothermal gradient (Oldenburg 2007).

A number of experimental and modelling studies have focused on the mutual solubility of H_2O and CO_2 at conditions relevant to geologic storage (e.g. King et al. 1992; Spycher et al. 2003; Qin et al. 2008; Pappa et al. 2009). Mole fraction solubilities of H_2O in CO_2 tend to increase with increasing temperature and pressure above the critical pressure of CO_2 and are quite low at the conditions expected in geologic storage with

values ranging from slightly over 0.004 at 40°C to nearly 0.01 at 80°C, both at 100 bars pressure. For temperatures 80°C and below, the H_2O solubility exhibits a general pattern of decreasing magnitude below about 80–90 bars, then an abrupt increase above this pressure range which tends to start gradually flattening out above approximately 200–250 bars.

At the onset of injection, the fluid in the vicinity of the well is quite acidic, leading to increased reactivity with the host minerals, and perhaps the well casing. These reactions generally lead to a pH increase (rock buffering) that promotes a shift from $CO_{2(aq)}$-rich fluids to fluids dominated by either HCO_3^- or CO_3^{2-}. This feedback process leads to a total dissolved CO_2 concentration much larger than the initial concentration observed at lower pH. Bench-scale experiments (Czernichowski-Lauriol et al. 1996) have demonstrated that dissolution is rapid when the formation water and CO_2 share the same pore space, but once the formation fluid is saturated with CO_2, the uptake rate slows and is controlled by diffusion and convection rates. The higher density of CO_2-saturated formation water, however,

induces instability and convective mixing, helping to accelerate CO_2 dissolution (Lindeberg and
Wessel-Berg 1997; Xu et al. 2005). However, in a
real injection system, CO_2 dissolution may be
rate-limited owing to the variability of the contact
area between the CO_2 and the fluid phase. The
principal benefit of 'solubility trapping' is that
once CO_2 is dissolved, it no longer exists as a
separate phase, thereby eliminating the buoyant
forces that drive it upwards. Conversely, the
implication of decreasing solubility as CO_2 rises
to the near-surface environment is that 'CO_2-
boiling' is likely to occur. Reservoir engineering
entailing injection into a horizontal well at the
bottom of the reservoir and the extraction of formation fluid could accelerate the rate of dissolution (Leonenko and Keith 2008).

CO₂ TRAPPING MECHANISMS

Results from general (Fig. 8.1 Benson and Cook
2005) and site-specific (Fig. 8.5, Han 2008; Han
and McPherson 2008; Xu et al. 2010) simulations
indicate that initially the bulk of CO_2 will be
stored as supercritical fluid, because the target
reservoirs are likely to have temperature and pres-

sure values higher than 31°C and 74 bar, the critical values for CO_2 (Fig. 8.2). The amount of CO_2
residually trapped is likely to be small until the
injection is terminated and the CO_2 plume
migrates away from the injection wells (Han and
McPherson 2008); it will also be smaller in structural traps, relative to saline formations (Burruss
et al. 2009). As the injected CO_2 contacts the
formation water, it rapidly dissolves reaching
saturation, with the dissolved CO_2 comprising
1–5% of brine weight (Fig. 8.4) at the likely injection reservoir conditions at 1–4 km below ground
level (Spycher and Pruess 2005; Benson and Cole
2008; Burruss et al. 2009). Equations of state for
CO_2 indicate that its solubility increases with
decreasing salinity of formation water (Fig. 8.4),
increasing the proportion of dissolved Ca to Na
and increasing pressure of CO_2; solubility
decreases with increasing temperature to about
150°C, with solubility increasing at higher temperatures (Duan and Sun 2003; Li and Duan 2007;
Rosenbauer et al. 2005). Results from natural gas
fields dominated by a CO_2 phase that provide a
natural analogue for assessing the geological
storage of anthropogenic CO_2 over millennial
timescales (Ballentine et al. 2001) indicated that
dissolution in formation water was the sole or the
major sink for CO_2 (Gilfillan et al. 2009).

Computer simulations, laboratory experiments,
and field tests indicate that mineral trapping
would be slower (Figs. 8.1 and 8.5) but more permanent, and the amount sequestered dependent
on the chemical composition of the formation
water and reservoir pressure and temperature, but
primarily on the reactivity of the reservoir minerals (Hitchon 1996; Knauss et al. 2005; Marini
2007; Parry et al. 2007; Han and McPherson 2008).
Injection of CO_2 into limestone reservoirs will
result in relatively rapid dissolution of carbonate
minerals to saturation with calcite and an approximate 2% increase in porosity (Rosenbauer et al.
2005; Emberley et al. 2005). Field results and geochemical modelling indicated that injection of
CO_2 into the Frio sandstone beds resulted in initially lowered pH values (as low as around 3.0)
and dissolution of calcite and iron oxyhydroxides;
brine pH then would increase as a result of the

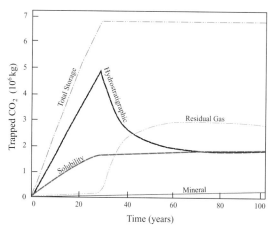

Fig. 8.5 Trapping of CO_2 injected for EOR at the
SACROCK oil field western Texas calculated by Han
(2008). See Plate 8.5 for a colour version of this image.

slower dissolution of feldspar minerals (oligoclase), and calcite, dawsonite and other minerals would ultimately precipitate (Bénézeth et al. 2007, 2009; Hellevang et al. 2005; Xu et al. 2010). More carbonate minerals would ultimately precipitate from sandstones with increased feldspar content, particularly where the feldspars are more calcic (e.g. anorthite).

Injections of flue gas with SO_2 and H_2S into ferric-iron-bearing sediments such as redbeds have been proposed as a mineral trap for CO_2 (Palandri and Kharaka 2004, 2005; Xu et al. 2007). The sulphur gases react to reduce the ferric to ferrous iron, which precipitates as siderite (eqn. 4).

$$Fe_2O_3 + 2CO_{2(g)} + SO_{2(g)} + H_2O \leftrightarrow \\ 2FeCO_3 + H_2SO_4 \quad (8.4)$$

Laboratory experiments show this reaction proceeding at 150°C, 300 bar (Palandri and Kharaka 2004; Palandri et al. 2005). However, other laboratory experiments (Rosenbauer et al. 2005) and reactive transport modeling (Knauss et al. 2005) point out that the presence of SO_2 in Ca-rich brines and oxidizing environments could significantly alter the quantity of CO_2 trapped in carbonate minerals owing to the precipitation of anhydrite and formation of very low pH pore fluids.

General and site-specific simulations indicate that the proportions of CO_2 stored as supercritical fluid in structural and stratigraphic traps will decrease with time relative to the other three trapping mechanisms (Figs. 8.1 and 8.5) during the period required (up to 10,000 years) for sequestration (Benson and Cook 2005; Han 2008; Han and McPherson 2008). This decrease would likely lead to increased storage security, especially with regard to the upward leakage of the buoyant (density 0.5–0.8 g/cm³) supercritical CO_2; the slightly increased density (up to 1%) of brine saturated with CO_2, and the more stable CO_2 residually trapped and stored as carbonate minerals add to the increased overall stability of the system (Benson and Cook 2005). Because the density of CO_2 stored as carbonate or in solution is higher than that of the supercritical CO_2, the storage

capacity should increase with time and will be higher in reservoirs with high proportions of carbonate precipitating minerals (e.g. arkosic sandstones high in Ca-plagioclase feldspar).

CO_2 INJECTION INTO BASALTS AND ULTRAMAFIC ROCKS

Geological sequestration of CO_2 as stable carbonate minerals is favoured if the CO_2 is injected into basalts and peridotites, which are high in reactive Mg, Fe and Ca silicates (Goldberg et al. 2008; Gysi and Stefansson 2008; Lackner 2002; Flaathen et al. 2009). Carbonation reactions are exothermic, and reaction rates are favourable over experimental time-scales. In addition, basalts and ultramafic rocks are abundant and widely distributed in accessible and suitable continental and marine systems (Broecker 2008; Matter and Kelemen 2009). Natural carbonation reactions are common in many basalts and most ultramafic rocks, which have sufficient permeability and pore space to react with CO_2-rich fluids (Oelkers et al. 2008; Matter et al. 2009), forming calcite, dolomite, magnesite, siderite and Mg-Fe carbonate solid solution (Brown et al. 2009; Flaathen et al. 2009). During *in-situ* mineral sequestration, CO_2 is injected into an aquifer where it dissolves in water and produces carbonic acid that can react with fosterite (eqn. 5). The dissociation into bicarbonate (eqn. 6) releases a proton (H^+), which is free to react with the host rock, here shown with basaltic glass from Stapafell, Iceland (Oelkers et al. 2008):

$$2H_2CO_3 + Mg_2SiO_4 \leftrightarrow 2MgCO_3 + \\ SiO_{2(am)} + 2H_2O \quad (8.5)$$

$$SiAl_{0.36}Ti_{0.02}Fe^{III}_{0.02}Ca_{0.26}Mg_{0.28}Fe^{II}_{0.17}Na_{0.08}K_{0.008}O_{3.45} \\ + 2.82H^+ \leftrightarrow SiO_{2(aq)} + 0.36Al^{3+} + 0.02TiO_{2(aq)} \\ + 0.02Fe^{3+} + 0.26Ca^{2+} + 0.28Mg^{2+} + 0.17Fe^{3+} \\ + 0.08Na^+ + 0.008K^+ + 1.41H_2O \\ (8.6)$$

This reaction releases carbonate-forming cations, Ca, Fe and Mg, into the solution at the same time as it consumes protons (H[+]), which leads to a pH increase and promotes the precipitation of secondary minerals including carbonates.

The carbonation reaction of CO_2 with basalt and ultramafic rocks, however, results in substantial volume increases that occur during the transformation from Mg- and Ca-silicate minerals to carbonates and the associated by-products of the reactions, including secondary silicate minerals. The large volume change and precipitation of SiO_2 minerals will result in sealing of rock porosity and reduction in the efficiency of carbonation reactions. A pilot study investigating the possibility of sequestering CO_2 captured from a geothermal power plant as carbonate minerals in Icelandic basalt is being performed by the CarbFix research group (Gislason et al. 2009; Matter et al. 2009).

GEOCHEMISTRY OF NATURAL CO$_2$ ANALOGUES

The use of natural CO_2-bearing geologic systems as proxies for understanding the behaviour of CO_2 in the subsurface has gained considerable momentum in recent years as various developing countries wrestle with the issue of storage of anthropogenic CO_2 in geological formations (e.g. Pearce et al. 2003; Lewicki et al. 2007; Fessenden et al. 2009; Wilkinson et al. 2009). The study of these natural analogue systems can provide valuable insights into: (i) potential leakage or migration pathways and CO_2 fluxes to the shallow subsurface: (ii) long-term consequences of CO_2-brine-rock interactions on reservoir rocks and cap rock seals: (iii) relative contribution by various storage mechanisms to the security of CO_2 storage for time frames in excess of those currently available to us via existing CO_2 injection demonstration tests, and (iv) the robustness of computer modelling of water-rock interactions to predict behaviour well beyond the limits of either bench-scale or field tests. One distinct advantage natural analogues provide is a snap-shot of long-lived systems that contained large amounts of CO_2. A major difference, however, with proposed injection scenarios associated with stationary point sources of CO_2 (i.e. power plants) is that the injection rates in the industrial systems will be much greater than what we typically predict for natural settings.

The sources of CO_2 can be categorized into four main groups, based largely on their $^{13}C/^{12}C$ isotope compositions: (i) atmospherically-derived CO_2 in open groundwater systems, $\delta^{13}C$ = approximately −7‰; (ii) organic sources including fermentation, microbial oxidation, kerogen and coal decarboxylation and microbial or thermochemical sulphate reduction of hydrocarbons, $\delta^{13}C$ approximately −40 to −20‰; (iii) burial diagenesis of limestones or carbonate-cemented clastics, $\delta^{13}C$ approximately −20 to +3‰; and (iv) deep sources such as crustal, thermal metamorphism or juvenile volcanic typically associated with active tectonic regions; $\delta^{13}C$ approximately −6 to +3‰. Interestingly, it has been shown that indigenous reservoir CO_2 can be distinguished from the injectate CO_2 depending on differences among the carbon and oxygen isotope compositions of each end member (Kharaka et al. 2006; Cole et al. 2008). It is important to note that there can be considerable overlap in the $\delta^{13}C$ values of CO_2 such that differentiating different sources can be difficult (Fig. 8.6). This is particularly true for natural gas fields containing high concentrations of CO_2 (greater than 70%) where $\delta^{13}C$ compositions generally fall within the overlapping range between magmatic degassing and carbonate breakdown (Gilfillan et al. 2008). Compounding this issue is the fact that isotopic compositions of CO_2 can be altered by a number of different reaction pathways such as dissolution into brine or oil, and reaction with host rock particularly carbonates. Furthermore, it is unlikely that just one source of CO_2 will dominate any given geologic system, so the challenge is to deploy a set of multiple geochemical tools to sort out the various potential sources.

Recently, it has been demonstrated that noble gas isotopic and abundance measurements (He, Ne, Ar, Kr and Xe) can be used to constrain origins and subsurface interaction of CO_2 with the reservoir fluid (e.g. Gilfillan et al. 2008; 2009). Noble

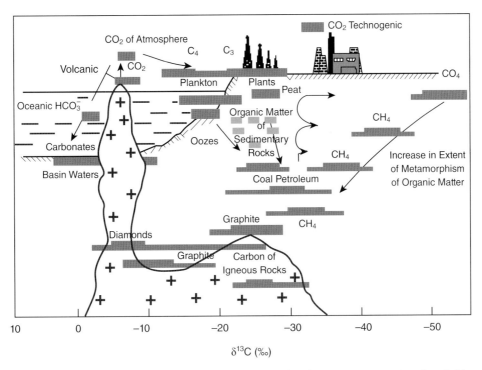

Fig. 8.6 Ranges in carbon isotope compositions for most major carbon-bearing reservoirs. See Plate 8.6 for a colour version of this image.

gas isotopes and chemical signatures from the terrestrial atmosphere, dissolved in groundwater can be distinguished from both the primordial signature of the mantle and signals produced by the radioactive decay of U, Th and K within the crust (e.g. Kennedy et al. 1997). According to Gilfillan et al. (2008), it is possible to determine the relative noble gas input from each source to any crustal fluid by coupling the isotopic compositions with the distinct elemental abundance patterns of the different sources. This method allows the extent of crustal, mantle and atmospheric contributions to the fluid to be quantified (Kennedy et al. 1997; Ballentine et al. 2002; Uysal et al. 2009). Additionally, noble gas chemistry and isotopes can be used to quantify the residence times of gas species in fluids as well as the degree of interaction of a crustal fluid with a groundwater system (e.g. Ballentine and Sherwood Lollar 2002; Zhou et al. 2005).

Although CO_2 accumulations occur globally, those found in the western US, particularly in the Colorado Plateau and adjoining Rocky Mountain regions (Fig. 8.7), have received a great deal of interest in terms of application of various geochemical tools to assess the origin of CO_2 and its interaction with surrounding host rocks (e.g. Moore et al. 2005; Shipton et al. 2004, 2005; White et al. 2005; Gilfillan et al. 2008). There are at least nine producing or abandoned gas fields within or near the plateau region containing up to $2800 \times 10^9 \, m^3$ of natural gas; a number of these are used for EOR (Allis et al. 2005). The dominant producing lithologies of the region are either limestone or sandstone. Mineralogically, it is noteworthy that Moore et al. (2005) documented the formation of late-stage dawsonite [$NaAlCO_3(OH)_2$] and kaolinite from dissolution of carbonate cements and detrital feldspars in the Springerville-St.Johns CO_2 field along the

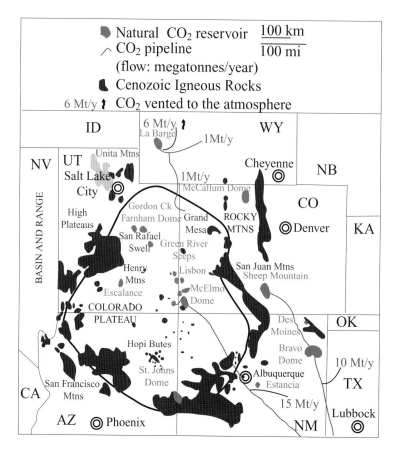

Fig. 8.7 Map of the Colorado Plateau illustrating the sites of major Cenozoic igneous provinces, location of the natural CO_2 reservoirs sampled and other CO_2 reservoirs within the region (from Gilfillan et al. 2008) (reproduced with permission from Elsevier: Gilfillan et al. 2008, Geochim. Cosmochim. Act, 72; 1174–1198). See Plate 8.7 for a colour version of this figure.

Arizona-New Mexico border. Dawsonite is a phase predicted by various thermodynamic water-rock interaction codes (e.g. Johnson et al. 2004) and is thought to be one of the key carbonate phases to form during CO_2 injection into feldspar-bearing sandstones (Palandri et al. 2005; Benezeth et al. 2007).

Detailed studies of mineral paragenesis during diagenesis in select sandstone and limestone-hosted CO_2 systems such as the Bravo Dome of New Mexico, the Miller field in the North Sea, and the Da Nang basin of Vietnam provide a number of important insights into the link between the generation of CO_2 and carbonate mineral formation during burial (Baines and Worden 2004). These include: (i) carbonate minerals, either in limestone or as secondary cements

in otherwise clean quartz sandstone are not capable of sequestering secondary CO_2-, they will dissolve in CO_2-charged low pH fluids; (ii) CO_2 added to sandstones will induce combined aluminosilicate dissolution and carbonate precipitation if the silicates contain Ca, Mg and/or Fe; and (iii) minimal evidence of carbonate sequestration in long-lived high CO_2 palaeo-systems (approximately 10^4 to 10^5 yr), indicates that either dissolution of reactive aluminosilicates and/or formation of carbonate minerals is relatively slow. Therefore, the impact of carbonate sequestration on the large volumes of CO_2 injected in any given site may not be meaningful even for cases where there is a source of Ca and/or Mg. This means that other mechanisms such solubilization of CO_2 into the reservoir brine or residual-gas trapping will tend

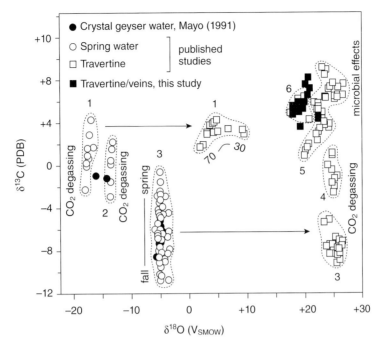

Fig. 8.8 Stable carbon (δ^{13}C PDB) and oxygen (δ^{18}O SMOW) isotope data from travertines, calcite/aragonite veins and springwaters (from Shipton et al. 2004 and references therein). Samples of veins and travertines (solid squares) are from Little Grand Wash and Salt Wash faults (east central Utah). Published data are from six active travertine deposits: (1) Mammoth Hot Springs, Yellowstone, spring temperature in °C; (2) Durango, Colorado; (3) Oklahoma; (4) Coast Range, California; (5) near Florence Italy; (6) central Italy. Differences between veins and travertine at individual sites could be due to either: (i) a decrease in temperature when fluids exit onto at the surface, or (ii) different sources. Vertical data patterns of travertine and travertine-forming spring waters are common and result from downstream degassing of spring water, seasonal variations in isotopic values of source waters, and variable microbial influence in facilitating precipitation of travertine carbonates (reproduced with permission from the Geological Society and the authors: Shipton et al. 2004, Geol. Soc. London Spec. Pub. 233. 43–58).

to dominate storage sites for the decade to century time frame (Benson and Cole 2008).

As noted above, noble gas and CO_2 carbon isotopes (Fig. 8.8) are powerful tracers of crustal processes that influence the behavior of subsurface CO_2 (e.g. Gilfillan et al. 2009). In particular, they are especially useful in helping delineate the phase behaviour of CO_2 and the relative importance of different trapping mechanisms other than carbonate mineralization highlighted above. Based on previously reported He-isotope data from east central Utah (Heath et al. 2009), which suggested only a minor mantle-derived source and the preponderance of heavier C-isotope values

(greater than 0‰), it was concluded that CO_2 was derived from either microbially mediated hydrocarbon reactions or thermally-induced decarbonation of carbonates. Conversely, Gilfillan et al. (2008) reported CO_2/^{3}He ratios from Bravo Dome, Sheep Mountain, McCallum Dome, St. John's Dome and McElmo Dome consistent with a dominantly magmatic source (Fig. 8.9). Age of nearest magmatic activity varies depending on the area: e.g., 8–10 Ka for the Bravo Dome versus 70–42 Ma for the McElmo Dome, which implies that CO_2 can be stored within the subsurface on a millennia timescale. More importantly the results described by Gilfillan et al. (2008; 2009) provide

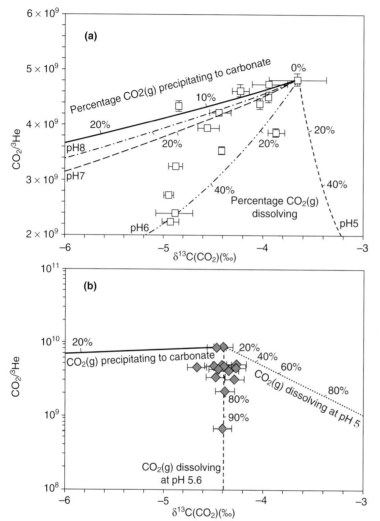

Fig. 8.9 Plot of $\delta^{13}C(CO_2)$ against $CO_2/^3He$ for Bravo dome and McElmo Dome (from Gilfillan et al. 2009). (a) Bravo Dome. The solid line shows the predicted trend for carbonate-mineral precipitation and the broken lines show $CO_2(g)$ dissolution trends for the indicated formation-water pH. These data limit the maximum effect of CO_2 precipitation in samples to approximately 18%. (b) McElmo Dome. Invariant $\delta^{13}C(CO_2)$ with a change in $CO_2/^3He$ of over an order of magnitude in McElmo dome gases cannot be accounted for by precipitation (solid line). Dissolution of reservoir CO_2 into formation water at pH 5.6 is consistent with observed results. Error bars are 1σ (reprinted by permission from Macmillan Publishers, Ltd: Gilfillan et al. 2009, Nature, 458, 614-618).

detailed documentation for the complex two-stage re-dissolution of CO_2 involving coherent fractionation of groundwater $^{20}Ne/^{36}Ar$ with crustal radiogenic noble gases (^4He, ^{21}Ne, ^{40}Ar). Stage one involved the injection of magmatic CO_2 into the groundwater system that strips dissolved air-derived noble gases and accumulated crustal/radiogenic noble gas via CO_2/water phase partitioning. In stage two, noble gases re-dissolve into any available gas-stripped groundwater, which can be modelled using a Rayleigh distillation

process; the model provides the volume of both the originally stripped groundwater and the groundwater involved in the subsequent re-dissolution interaction. These new studies reinforce the observation that CO_2 storage will be dominated by trapping mechanisms other than carbonate formation for geologically significant durations.

Present-day CO_2 leakage from these systems is considered small, but there are significant outflows of high bicarbonate water in many areas, suggesting continuous migration of CO_2 in the aqueous phase from depth. There are also localized CO_2-flux anomalies, and the outflow of ground water containing dissolved CO_2 present challenges for effective, long-term monitoring of CO_2 leakage (Allis et al. 2005; Fessenden et al. 2009; Heath et al. 2009). Measurement of CO_2 and CH_4 fluxes over the Rangely oil field, Colorado, where there has been CO_2 injection for EOR since 1986, is also considered small (Klusman 2003a,b).

SEQUESTRATION OF CO_2 IN SEDIMENTARY BASINS

Sedimentary basins in general, and deep saline aquifers in particular, are being investigated as possible repositories for large volumes of anthropogenic CO_2 that must be sequestered to mitigate global warming and related climate changes (White et al. 2005; Rubin 2008; Benson and Cole 2008; Sundquist et al. 2009). Currently there are a total of three commercial projects operating world-wide that capture and sequester close to three million tons of CO_2 annually, and that provide valuable experience for assessing the efficacy of CCS. There are also two commercial projects (Weyburn-Midale, Canada and Cranfield, MS) that capture and inject close to three million tons of CO_2 annually for EOR, but these differ from the more than 100 EOR projects world-wide, because they aim to integrate EOR with CO_2 sequestration (Cantucci et al. 2009; Moritis 2009; Hovorka et al. 2010). In addition there are approximately 25 geologic-sequestration field demonstration projects in the US at various stages of planning and deployment, and an equal number of projects in other countries (including the Minami-Nagaoka in Japan, the Otway and Gorgon in Australia, and the Ketzin in Germany) that are investigating the storage of CO_2 in various clastic and carbonate rock formations using different injection schemes, monitoring methods, hazards assessment protocols, and mitigation strategies (Torp and Gale 2003; Litynski et al. 2008; Cook 2009; Haszeldine 2009; Hitchon 2009; Matter

et al. 2009). Sedimentary basins are attractive for CO_2 storage because: (i) they generally have large estimated local and global (greater than 10,000 Gt CO_2) storage capacities (Holloway 1997; Bradshaw et al. 2007; US Department of Energy 2008); (ii) they often have advantageous locations close to power plants and other major CO_2 sources (Hitchon 1996; Friedmann 2007; Burruss et al. 2009); and (iii) there is a great deal of geologic and other relevant information gained from petroleum exploration and production, and disposal of produced water and other wastes (Benson and Cook 2005; Kharaka et al. 2006; Bachu et al. 2007).

The five commercial projects that currently capture and inject close to seven million tons of CO_2 annually are the Sleipner project, offshore Norway, the Weyburn-Midale EOR project in the Williston Basin of Saskatchewan, Canada, and the In Salah gas field project in Algeria; two projects started in 2008: the Snøhvit field in the Barents Sea offshore Norway and the Cranfield Oil field sequestration-EOR project in Mississippi. The Sleipner Project, operated by the StatoilHydro since 1996, is the world's first industrial-scale operation. Approximately one million tons of CO_2 are being extracted annually from the produced gas that has approximately 10% CO_2 to meet quality specifications, and stored approximately 1000 m below sea level in the Utsira Formation (Chadwick et al. 2004, 2009; Hermanrud et al. 2009). The Utsira Formation consists of thick and poorly consolidated sandstones of high porosity (35–40%) and permeability (1–8.3 darcies), and is overlain by the thick Nordland shale caprock (Fig. 8.10, Bickle 2009). The sand comprises 75% quartz , 13% K-feldspar, and 3% calcite (Chadwick et al. 2004). The thickness of the formation varies between 150 and 250 m and the top is located approximately 800 m beneath the bottom of the North Sea (Bickle 2009).

The gas, though injected into the bottom of the Utsira sandstone, has migrated through intervening shale beds into nine different sandstone layers (Fig. 8.10). However, results from the seismic investigations show that no CO_2 is leaking out of the formation (Chadwick et al. 2004; Bickle 2009).

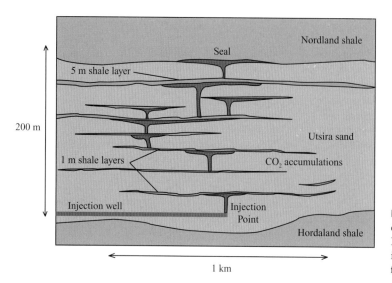

Fig. 8.10 Schematic diagram of carbon dioxide storage at the Sleipner Field, Norway based on seismic images (Bickle 2009). See Plate 8.10 for a colour version of this image.

Two-dimensional reactive-transport modelling of CO_2 injected at the Sleipner site indicates that after 10,000 years, only 5% of the injected gas will be trapped into minerals, while 95% will be dissolved in the brine (Audigane et al. 2007). The relatively small percentage of mineral trapping is due mainly to the low concentrations of divalent cations in the reservoir sandstones.

The Weyburn-Midale project (Williston Basin, Canada) started injecting CO_2 into the Weyburn field in 2000, combining CO_2 sequestration with EOR by injecting the gas and brine into an almost-depleted oil field to increase fluid pressure and oil production (Emberley et al. 2005). The CO_2 used is obtained from the Dakota gasification plant near Beulah, North Dakota, where the gas is captured after coal gasification, liquefied by compression and piped 320 km north to the oil fields. It is the first man-made source of CO_2 being used for EOR. The oil field consists of shallow marine carbonate (depth of approximately 1500 m), and the injection rate is now greater than three million tons of CO_2/yr. The chemical and C-isotope composition of both brine and H_2O exhibit significant changes resulting from CO_2 injection; geochemical modelling and monitoring of the aquifer indicate that the injected CO_2 reacts with the brine and the host rock, dissolving car-

bonate minerals, increasing alkalinity, and lowering the pH of produced water; slow dawsonite precipitation is also predicted (Emberley et al. 2005; Cantucci 2009).

The In Salah Gas Joint Venture CO_2 storage project, located in the Algerian Sahara, is currently the largest onshore project in the world (Iding and Ringrose 2009). Approximately one million tons/yr of CO_2 are injected into the water leg of the Carboniferous Krechba sandstone gas-reservoir (20 m thick) via three horizontal wells at a depth of approximately 1900 m. The CO_2 is obtained from several local natural-gas fields that contain up to 10% CO_2, which has to be reduced to 0.3% before the gas is sold (Rutqvist et al. 2009). CO_2 injection started in August 2004 into sandstones with relatively low porosity (11–20%) and permeability (approximately 10–md) with some fractures and small faults in both the reservoir unit and in the Carboniferous mudstone and siltstone beds of the caprock. Despite the evidence of fractures, the thick mudstone caprock sequence provides an effective mechanical seal for containing the CO_2 (Iding and Ringrose 2009). A number of key technologies to monitor the injection, and the subsurface movement and storage of CO_2, have been, and will continue to be, deployed to provide long-term assurance of

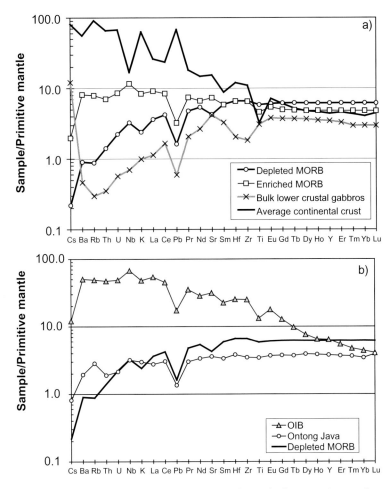

Plate 4.5 (a) Variation of incompatible elements in depleted and enriched MORB (Sun and McDonough 1989) as well as a bulk lower oceanic crust estimate (Godard et al. 2009) compared to average continental crust (Rudnick and Fountain 1995). (b) Average ocean island basalt, depleted MORB (both from Sun and McDonough 1989) and a primitive Ontong Java Plateau basalt glass (Tejada et al. 2004). The data are normalized to the estimated concentrations of primitive mantle (Sun and McDonough 1989).

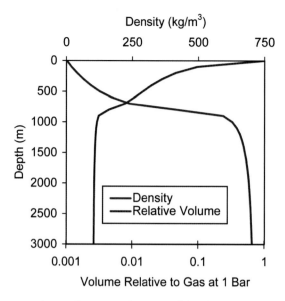

Plate 8.3 Density and change in volume of CO_2 as a function of depth below ground surface for a typical gradient of approximately 30°C/km (from Benson and Cole 2008).

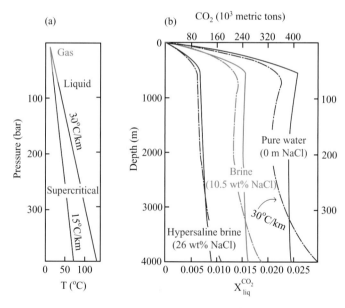

Plate 8.4 (a) Phase behaviour of CO_2 as a function of temperature and pressure for two different geothermal gradients. (b) Solubility, in mole fraction, of CO_2 in NaCl solution as a function of depth and salinity at two different geothermal gradients. Both figures modified after Oldenburg (2005) based on results presented in Spycher et al. (2003) and Spycher and Pruess (2004). An example of the mass of CO_2 (in metric tons) trapped is illustrated using a simple scenario where CO_2 is injected into a 20 m thick formation with 10% of its void space available for a CO_2 dissolution process extending 1 km out from the well in all directions. A pure water system can dissolve five times the amount of CO_2 compared to a hypersaline brine.

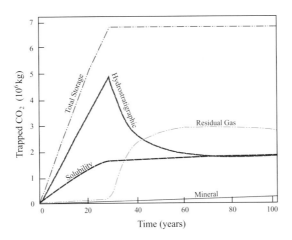

Plate 8.5 Trapping of CO_2 injected for EOR at the SACROCK oil field, western Texas, calculated by Han (2008).

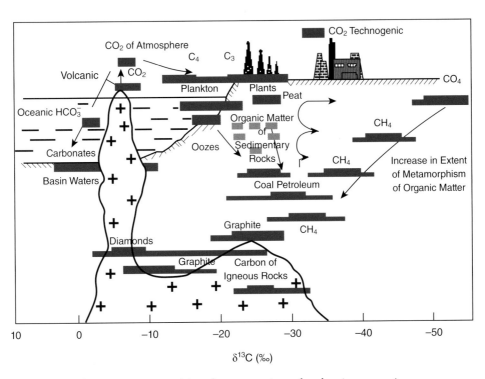

Plate 8.6 Ranges in carbon-isotope compositions for most major carbon-bearing reservoirs.

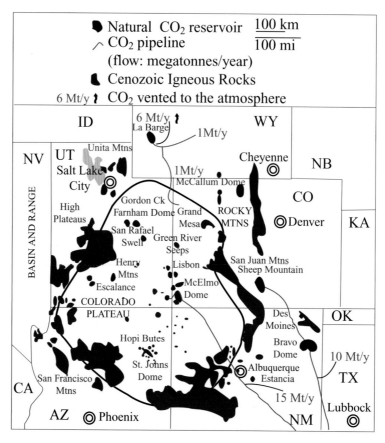

Plate 8.7 Map of the Colorado Plateau illustrating the sites of major Cenozoic igneous provinces, location of the natural CO_2 reservoirs sampled and other CO_2 reservoirs within the region (from Gilfillan et al. 2008).

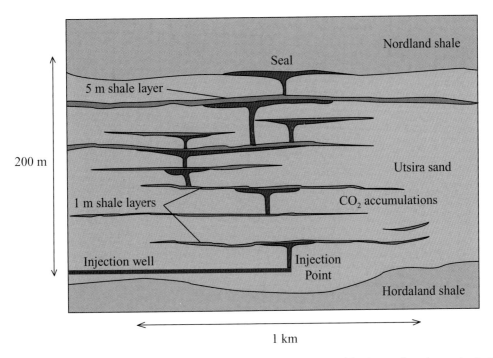

Plate 8.10 Schematic diagram of carbon dioxide storage at the Sleipner Field, Norway based on seismic images (Bickle 2009).

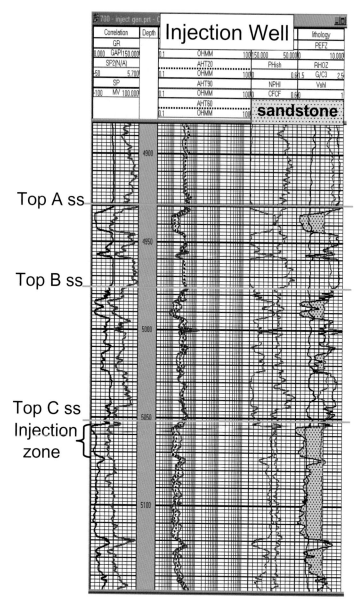

Plate 8.11 Open-hole logs of the injection well. Note the relatively thick beds of shale and siltstone between the injection zone, Frio 'C', and the overlying monitoring sandstone, Frio 'B' (modified from Kharaka et al. 2009).

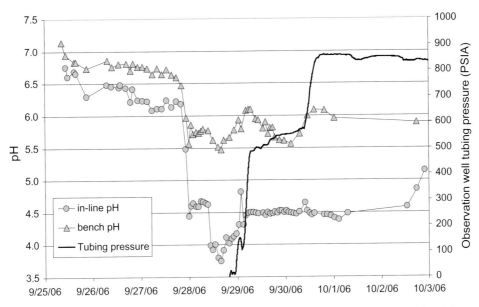

Plate 8.14 Bench and in-line pH values obtained from Frio II brines before and following CO_2 breakthrough at the observation well. Note the sharp drops of pH, especially values from in-line probe following the breakthrough of CO_2.

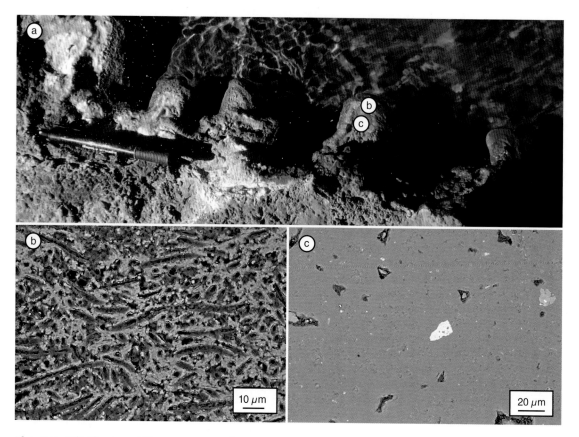

Plate 9.7 Silicification of filamentous cyanobacteria at El Tatio Geyser Field, Chile. (a) Nodules forming at edge of main geyser pool, showing (b) moderate silicification of bacterial sheaths, and (c) complete mineralization of the microbial community and subsequent biosignature preservation.

sequestration. Time-lapse satellite images (using PSInSAR Technology), which measure ground deformation, show a surface uplift of the order of 5 mm/yr above active CO_2 injection wells, and the uplift pattern extends several kilometres from the injection wells. The observed surface uplift is used to constrain the coupled reservoir-geomechanical model, and to show that surface deformations from InSAR can be useful for tracking the fluid pressure and for detection of a leakage path (e.g. in a permeable fault) through the overlying caprock (Rutqvist et al. 2009).

CO₂-BRINE-MINERAL INTERACTIONS IN THE FRIO FORMATION, TEXAS

The Frio Brine pilot was the first multi-laboratory field study of its kind, funded by US DOE, to investigate the potential for geologic storage of CO_2 in saline aquifers and to develop geological, geochemical and geophysical tools, and multi-phase simulation programs to track the injected CO_2 and predict its interactions with reservoir brine and minerals (Hovorka et al. 2006; Doughty et al. 2008; Kharaka et al. 2009). For Frio-I, approximately 1600 tons of CO_2, were injected during October 2004 into a 24-m thick 'C' sandstone of the Oligocene Frio Formation, a regional brine reservoir in the Gulf Coast region of the southern US (Hovorka et al. 2006). Using a variety of tools, fluid samples were obtained before CO_2 injection for baseline geochemical characterization, during the CO_2 injection to track its breakthrough into the observation well, and after injection to investigate changes in fluid composition and leakage into the overlying 'B' sandstone (Fig. 8.11). New geophysical (Daley et al. 2007, 2008) and geochemical tools were deployed and additional detailed tests were carried out during the Frio II Brine test (September 2006 to October 2008), where approximately 300 tons of CO_2 were injected into a 17-m thick Frio 'Blue' sandstone, located approximately 120 m below the Frio 'C'. Geochemical tools and methods deployed for Frio II included online pH, EC and temperature probes, field determinations

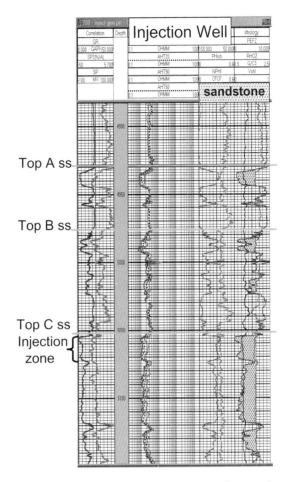

Fig. 8.11 Open-hole logs of the injection well. Note the relatively thick beds of shale and siltstone between the injection zone, Frio 'C', and the overlying monitoring sandstone, Frio 'B' (modified from Kharaka et al. 2009). See Plate 8.11 for a colour version of this figure.

of Fe^{2+} and Fe^{3+}, and collection and analyses for a large metals using ICP-MS and organic compounds (Kharaka et al. 2010a).

Regional setting and experimental methods

The Frio Brine site is located within the South Liberty oil field, near Dayton, Texas, a region of the Gulf Coast where large industrial sources of CO_2 are present. Wells in this field were drilled in the 1950s, with production from the Eocene Yegua Formation at depths of approximately 2900-m. The Frio Formation has a dip of 16° to the south, and comprises several reworked fluvial sandstone and siltstone beds that are separated by transgressive marine shale. The sandstones are generally subarkosic fine-grained, moderately sorted quartz and feldspar sandstone, with minor amounts of illite/smectite, calcite and iron oxyhydroxides. The sandstones have a high mean porosity greater than 30%, and very high permeability of 2–3 Darcies. Approximately 1600 and 300 tons of CO_2 were respectively injected into the Frio 'C' and Frio Blue sandstones at a depth of around 1500 m. Situated above the 'C', the 'B' sandstone has an approximately 4 m thick reworked fluvial sandstone bed at the top, but is separated from 'C' by approximately 15 m of shale, muddy sandstone and siltstone beds (Fig. 8.11). However, the regional cap – the main barrier to CO_2 leakage to shallow groundwater – is expected to be the overlying thick marine shale of the Miocene-Oligocene Anahuac Formation (Hovorka et al. 2006).

During 2004–8, more than 200 brine and gas samples were collected from the Frio I and II tests, using a variety of tools and methodologies, including the Kuster sampler that can be evacuated and lowered to the desired sampling depth (Kharaka et al. 2006), and the novel downhole U-tube system developed for this field experiment (Freifeld et al. 2005). The bulk of samples were obtained from an observation well located 30 m updip of the injection well. The samples were subjected to detailed organic and inorganic chemical and isotope analyses of brine, associated gases and added tracers (Freifield et al. 2005; Kharaka et al. 2009).

Results and discussion

Results of chemical analyses for Frio-I samples prior to CO_2 injection show that the brine is a Na-Ca-Cl type water, with a salinity of approximately 93,000 mg/L TDS (Fig. 8.12), with relatively high concentrations of Mg and Ba, but low values for SO_4, HCO_3, DOC and organic-acid

anions. The high salinity and the low Br/Cl ratio (0.0013) relative to seawater, indicate dissolution of halite from the nearby salt dome (e.g. Kharaka and Hanor 2007). The brine has 40–45 mM dissolved CH_4, which is close to saturation at reservoir conditions (60°C and 150 bar), and CH_4 comprises approximately 95% of total gas, but the CO_2 content is low at approximately 0.3% (Table 8.1).

During CO_2 injection, 4–14 October, 2004, on-site measurements of electrical conductance (EC) exhibited only a small increase from a pre-injection value of approximately 120 mS/cm at around 22°C. However, there were major changes in some chemical parameters (Fig. 8.13) when the CO_2 reached the observation well, including a sharp drop in pH from 6.5 to 5.7 (measured in a 150 mL bottle at surface conditions), and large increases in alkalinity, from 100 to 3000 mg/L, as HCO_3^- (Kharaka et al. 2006). Additionally, laboratory determinations showed major increases in dissolved Fe (from 30 to 1100 mg/L) and Mn, and significant increases in the concentration of Ca. The most dramatic changes in chemistry occurred at CO_2-breakthrough 51 hours after injection (Fig. 8.14), as evidenced also by PFT tracer analysis (Phelps et al. 2006) and on-site analysis of gas samples from the U-tube system (Freifeld et al. 2005) that showed CO_2 concentrations increasing from 0.3% to 3.6% and then quickly to 97% of total gas (Table 8.1).

The variations in the field-measured pH proved the most sensitive early warning parameter for tracking the arrival of CO_2 at the observation well, especially when the online pH probe was successfully deployed during Frio II (Fig. 8.14). The pH values obtained from the online probe in particular, but also from bench measurement at ambient conditions, could be used also to indicate the arrival of the dissolved CO_2, which is expected to reach the observation well before the supercritical phase. The sharp drop in pH and increases in alkalinity, dissolved iron, measured CO_2 (Freifeld et al. 2005) and other constituents in the sample collected at 15:45 on 6 October, marked the time of breakthrough of the supercritical CO_2 plume (Kharaka et al. 2009).

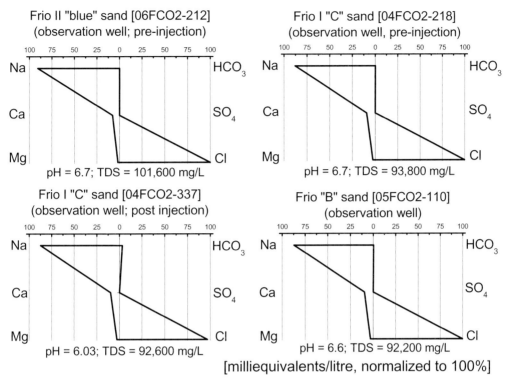

Fig. 8.12 Modified Stiff diagrams showing concentrations (equivalent units normalized to 100%) of major cations and anions, together with salinity and pH of Frio I brine from 'C' and 'B' sandstones before and after CO₂ injection (Kharaka et al. 2009).

Table 8.1 Composition (volume %) of gas obtained from Frio 'C' and 'B' sandstones, during Frio I field test near Houston, Texas, USA (from Kharaka et al., 2009).

Gas	¹"C"	²"C"	³"B"
He	0.0077	0	0.01
H₂	0.040	0.19	0.92
Ar	0.041	0	0.13
CO₂	0.31	96.8	2.86
N₂	3.87	0.037	1.51
CH₄	93.7	2.94	94.3
C₂H₆⁺	1.95	0.0052	0.12

(¹) From the injection well before CO₂ injection; (²) from the observation well after CO₂ breakthrough; (³) from the observation well 6 months after CO₂ injection.

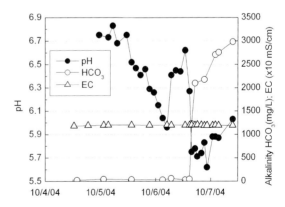

Fig. 8.13 Electrical conductance (EC), pH and alkalinity of Frio I brine from monitoring well, determined on site during CO₂ injection. Note the sharp drop of pH and alkalinity increase with the breakthrough of CO₂ (from Kharaka et al. 2006).

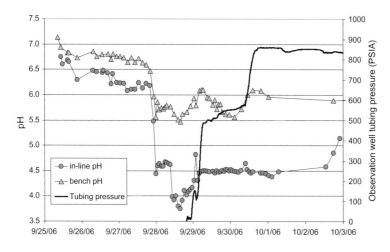

Fig. 8.14 Bench and in-line pH values obtained from Frio II brines before and following CO_2 break-through at the observation well. Note the sharp drops of pH, especially values from in-line probe following the breakthrough of CO_2. See Plate 8.14 for a colour version of this image.

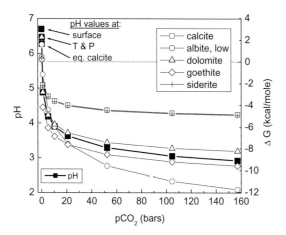

Fig. 8.15 Computed pH values and saturation states of selected minerals in Frio I brine as a function of CO_2 partial pressure at subsurface conditions. Note the sharp initial drop of pH from around 6.4, the average value computed at T and P, and calcite saturation before CO_2 injection (modified from Kharaka et al. 2006).

Results of geochemical modelling, using updated SOLMINEQ (Kharaka et al. 1988) indicate that the Frio brine in contact with the supercritical CO_2 would have a pH of about 3 at subsurface conditions, and this low pH causes the brine to become highly undersaturated with respect to carbonate, aluminosilicate and most other minerals present in the Frio Formation (Fig.

8.15). Because mineral dissolution rates are generally higher by orders of magnitude at such low pH values, the observed increases in concentrations of HCO_3^- and Ca^{2+} probably result from the rapid dissolution of calcite via the reaction:

$$H_2CO_3^\circ + CaCO_{3(s)} \leftrightarrow Ca^{2+} + 2HCO_3^-. \quad (8.7)$$

The large increases observed in concentrations of dissolved Fe and equivalent bicarbonate alkalinity could result from dissolution of siderite, but no siderite was observed in the retrieved core. Alternatively, these increases could be caused by dissolution of the observed iron oxyhydroxides, depicted in this redox-sensitive reaction:

$$2Fe(OH)_{3(s)} + 4H_2CO_3^\circ + H_2^\circ = 2Fe^{2+} + 4HCO_3^- + 6H_2O. \quad (8.8)$$

However, some of the increase in dissolved Fe and equivalent bicarbonate could also result from corrosion of pipe and well casing that contact low pH brine (Kharaka et al. 2009; Hitchon 2000), as indicated by the redox-sensitive reaction:

$$Fe_{(s)} + 2H_2CO_3^\circ = Fe^{2+} + 2HCO_3^- + H_{2(g)} \quad (8.9)$$

Similarly, a redox-sensitive reaction may be written for Mn, which increased from 3 to 18 mg/L. There were also increases in the concen-

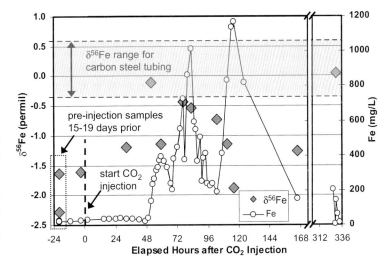

Fig. 8.16 The concentrations of Fe and δ⁵⁶Fe values for selected brine samples from Frio II. Samples with high Fe values following CO₂ injection have δ⁵⁶Fe values that fall between those of the carbon steel tubing and preinjection brine, indicating a mixed origin for dissolved Fe (Kharaka et al. 2010a).

tration of other metals, including Zn, Pb and Mo, which are generally associated (sorbed and coprecipitated) with iron oxyhydroxides, but could also be present in the low-carbon steel pipe used in petroleum wells (Celia et al. 2006).

Even higher concentrations of Fe (up to 1200 mg/L as Fe⁺⁺) and other metals were obtained in brines collected from Frio II, the Blue sandstone (Fig. 8.16). For selected samples from Frio II, the δ⁵⁶Fe was determined and compared with three values obtained from carbon steel tubing used in the wells. The steel pipe, as expected, has δ⁵⁶Fe values of 0±0.5‰, whereas brine samples before CO₂ injection had δ⁵⁶Fe values of –2.0±0.5‰. Brine samples following CO₂ injection have δ⁵⁶Fe values that fall between those of the steel pipe and brine before CO₂ injection (Fig. 8.16). Geochemical modelling, and other chemical data, support results of Fe isotopes that dissolution of Fe-oxyhydroxides and corrosion of well pipe contribute to the very high dissolved-Fe concentrations.

Geochemical modelling indicates that the brine pH increases from dissolution of carbonate and iron oxyhydroxide minerals discussed above, as well as from dissolution of oligoclase and other aluminosilicate minerals present in the Frio. Aluminosilicate-mineral dissolution generally is not congruent, but probably follows the incongru-

ent reaction described by Eqn. 10 below, in which dawsonite, gibbsite and amorphous silica are precipitated, at the expense of Ca-plagioclase dissolution (White et al. 2003).

$$0.4H^+ + Ca_{.2}Na_{.8}Al_{1.2}Si_{2.8}O_{8(s)} + 0.8CO_{2(g)} +$$
$$1.2H_2O \leftrightarrow 0.2Ca^{2+} + 0.8NaAlCO_3(OH)_{2(s)} +$$
$$0.4Al(OH)_{3(s)} + 2.8SiO_{2(s)}$$
(8.10)

As the pH increases from mineral interactions and the mixing of CO₂-saturated and pristine brines, modelling indicates that mineral saturations reverse the trend shown in Fig. 8.15, resulting in precipitation of carbonate and other minerals. The overall result is that the brine gradually evolves toward its pre-injection composition, but additional fluid sampling is planned to further investigate gas-water-rock interactions in this system.

Dissolved organics

Dissolved organic carbon (DOC) values obtained in Frio-I samples before CO₂ injection are expectedly low (1–5 mg/L) and the values obtained from the 'C' sandstone during the CO₂ injection

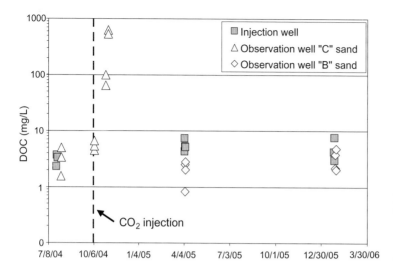

Fig. 8.17 Concentration of dissolved organic carbon (DOC) in Frio I brine. Note the extremely high values obtained on November, 2004, approximately 20 days after the end of CO_2 injection (Kharaka et al. 2009).

increased moderately to 5–6 mg/L. However, DOC values increased unexpectedly by a factor of 100 on samples collected approximately 20 days after injection stopped (Fig. 8.17). The concentrations of organic acid anions and BTEX (benzene, toluene, ethylbenzene and xylene) in these samples were low (less than 1 mg/L), but values of formate, acetate and toluene were generally higher in the enriched DOC samples. Results from a more detailed sampling protocol for samples collected six and fifteen months after injection show oil and grease values below detection limit, low levels (up to 30 ppb phenol) of volatile organic compounds (VOCs) and semi-VOCs (30 ppb naphthalene), and only slightly elevated DOC values (5–8 mg/L) values (Fig. 8.17).

It is difficult to rule out contamination from well operations for the very high DOC values, but they probably represent a 'slug' of organic matter mobilized by the injected CO_2, as reported in laboratory experiments simulating carbon dioxide storage in deep coal beds (Kolak and Burruss 2006) and as happens in EOR operations (Shiraki and Dunn 2000). Additional investigations are required in this area, but these results suggest that mobilization of organics from depleted oil reservoirs and non oil-bearing aquifers could have major implications for the environmental aspects

of CO_2 storage and containment. The concern here is warranted, as high concentrations of toxic organic compounds, including benzene, toluene (up to 60 mg/L for BTEX), phenols (less than 20 mg/L), and polycyclic aromatic hydrocarbons (up to 10 mg/L for PAHs), have been reported in oil-field waters (Kharaka and Hanor 2007).

Isotopic composition of water and gases

Significant shifts were observed in the isotopic compositions of H_2O (Fig. 8.18) and DIC (Fig. 8.19) following CO_2 injection, but only subtle changes in the δD and $\delta^{13}C$ values of CH_4. The $\delta^{13}C$ values of DIC became profoundly lighter, shifting from –3 to –33‰, reflecting the fact that the injected CO_2 is the dominant C source and is depleted, with $\delta^{13}C$ = –44 to –34‰, depending on the mixing proportions of the two gas sources. The $\delta^{18}O$ values of brine became isotopically lighter with time, shifting from 0.80 to –11.1‰, and there was a corresponding increase in the $\delta^{18}O$ values of CO_2, from 9 to 43‰. Because water and CO_2 rapidly exchange oxygen isotopes even at low temperature, it is possible to use their $\delta^{18}O$ values in mass-balance equations to estimate the brine-to-CO_2 mass and volume ratios in the reservoir. The equation for a closed system and no

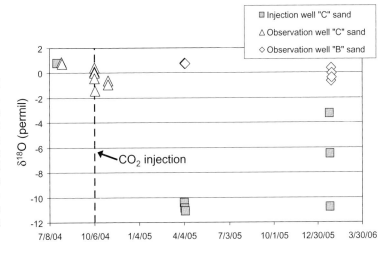

Fig. 8.18 The $\delta^{18}O$ values of Frio I brine samples collected before, during and after CO_2 injection. Note the depleted values obtained from the 'C' sandstone following CO_2 injection relative to those before injection . The $\delta^{18}O$ values obtained from the 'B' sandstone show some depletion, especially in samples collected approximately 15 months after completion of injection (Kharaka et al. 2009).

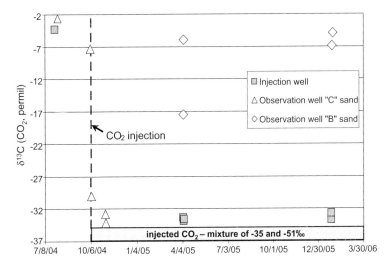

Fig. 8.19 The $\delta^{13}C$ values of DIC in Frio I samples obtained before, during and after CO_2 injection. Note the highly depleted values obtained from the 'C' sandstone following CO_2 injection relative to those before injection. The $\delta^{13}C$ values obtained from the 'B' sandstone show some depletion, especially in samples collected approximately 6 months after completion of injection (Kharaka et al. 2009).

isotopic exchange with minerals is given (Clark and Fritz 1997) by:

$$X_{brine}/X_{CO2} = \frac{\delta^{18}O^f_{CO2} - \delta^{18}O^i_{CO2}}{\delta^{18}O^i_{H2O} - \delta^{18}O^f_{H2O}}. \quad (8.11)$$

where the superscripts 'i' and 'f' are the initial and final δ values for brine and CO_2, respectively, and X is the atomic oxygen in the subscripted component.

Results from the observation well (Table 8.2) show that initially the system is brine-dominated, with CO_2 comprising approximately 10% of the fluid at reservoir conditions from one day after the CO_2 breakthrough on October 7, 2004 through November 3, 2004. However, samples collected from the injection well on 4–6 April, 2005 yield a value of approximately 50% for the volume of CO_2 at reservoir conditions. The initial brine-dominated system could indicate that the injected

Table 8.2 Calculated brine/CO_2 volume ratios in the Frio 'C' sandstone following CO_2 injection, based on the $\delta^{18}O$ values for brine and CO_2 (from Kharaka et al. 2006).

Date	Brine	CO_2	vol. ratio*
	$\delta^{18}O$ shift	$\delta^{18}O$ shift	Brine/CO_2
10-5-04	0	0	→∞
10-6-04	0.37	32	43
10-6-04	0.69	32	23
10-6-04	0.77	32	21
10-6-04	1.22	32	13
10-7-04	2.24	32	7.1
11-3-04	1.43	32	11
11-3-04	1.74	32	9.1
4-4-05	11.2	22	0.97
5-4-05	11.7	22	0.93
6-4-05	11.9	22	0.92

*To convert from mole oxygen basis (eq. 4 in text) to brine/CO_2 vol. ratio at reservoir conditions, we multiply by 0.495, using a density (gm/cc) of CO_2 = 0.60 and brine = 1.0 (Kharaka et al., 2006).

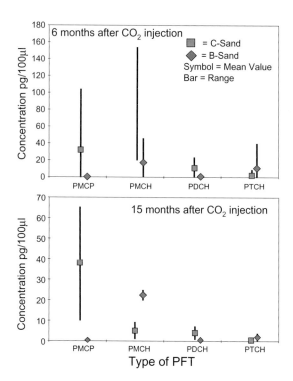

Fig. 8.20 Concentration of perfluorocarbon tracers (PFT) in Frio I brine samples Mean (dot) and range (bar) for each perfluorocarbon tracers in the Frio 'C' sandstone (squares) and the 'B' sandstone (diamonds). Data from samples collected 6 (top) and 15 months (bottom) after CO_2 injection (modified from Phelps et al. 2006).

CO_2 acts like a piston, pushing the pore water out with minimal mixing and isotopic exchange. Contact and isotopic equilibration with a larger volume of injected CO_2 is indicated from data of April 2005. These results are comparable to 'residual' CO_2-saturation values obtained with the reservoir saturation (RST) and other geophysical tools (Hovorka et al. 2006), indicating the usefulness of this isotopic approach.

Subsurface monitoring

Monitoring at and close to the ground level for CO_2 and brine leakage into soil gas and shallow (approximately 30 m depth) groundwater was not effective at the Frio site, primarily because of perturbation caused by injection operations (Nance et al. 2005). Because of such difficulties, and the long time that could be required for a potential CO_2 and/or brine leakage to reach the surface, a rigorous program for deep subsurface monitoring was carried out. Results of brine and gas analyses of fluid samples obtained from the

'B' sandstone, first perforated and sampled six months after CO_2 injection, showed: (i) Slightly elevated concentrations of bicarbonate, Fe and Mn; (ii) significantly depleted $\delta^{13}C$ values (−17.5 to −5.9‰ versus approximately −4‰) of DIC (Fig. 8.19); and (iii) somewhat depleted $\delta^{18}O$ values of brine (Fig. 8.18), relative to pre-injection values obtained for the 'C' samples. A more definitive proof of the migration of injected CO_2 into the 'B' sandstone was obtained from the presence (Fig. 8.20) of the two PFT tracers (PMCH and PTCH) Perfluoromethyl cyclohexane and perfluorotrimethyl cyclohexane that were added to the injected CO_2, and migrated with the initial CO_2 breakthrough (Phelps et al. 2006). Additional proof of the migration of injected CO_2 into the 'B' sandstone is obtained from the high concentration (2.9 versus approximately 0.3%) of CO_2 in dissolved gas obtained from one of the two downhole Kuster samples (Table 8.1).

Results of samples collected from Frio 'B' in 23–27 January, 2006 (approximately 15 months after CO_2 injection) gave brine and gas compositions that are approximately similar to those obtained from the 'C' sandstone before CO_2 injection (Fig. 8.12). The $\delta^{13}C$ values of DIC (Fig. 8.19) and the $\delta^{18}O$ values of brine (Fig. 8.18) are slightly depleted in the heavy isotopes relative to preinjection values obtained for the 'C' samples. These overall results indicate the absence of significant amounts of injected CO_2 in the 'B' fluids sampled. However, a contrary conclusion is indicated from the fact that PMCH and PTCH were measured (Fig. 8.20) in the six samples also analysed for PFT tracers (Phelps et al. 2006). It is possible that the PMCH and PTCH measured in January 2006 represent desorbed PFT tracers that were introduced into 'B' earlier and do not require migration of additional injected CO_2 into the 'B' sandstone.

Results from the 'B' sandstone clearly show some CO_2 migration from the 'C' sandstone after about six months following injection. However, these results cannot be used to estimate the volume of CO_2 that migrated to 'B', or the path of migration, but they highlight the importance of subsurface monitoring for detecting early leaks (Kharaka et al. 2009). These results are comparable to those observed at Sleipner, where CO_2 is injected into the bottom of the Utsira sandstone, but it has migrated through intervening shale beds (one extensive and approximately 5 m thick) into nine different sandstone layers (Fig. 8.10). Results from the seismic investigations show that no CO_2 is leaking out of the cap rocks at Sleipner (Chadwick et al. 2004, Gaus et al. 2005; Bickle 2009). Leakage of CO_2 injected into the Shatuck sandstone (Permian) in a pilot study similar to Frio was measured at the surface at the West Pearl Queen field, NM, where 2090 tons of CO_2, tagged with PFT tracers, were injected at a depth of 1400 m. Using an array of PFTs capillary absorption tubes placed 2 m below ground, a leakage rate

of 0.009% of injected CO_2 per year was measured by Wells et al. (2007).

NEAR-SURFACE MONITORING AT THE ZERT SITE, MONTANA

The Zero Emission Research and Technology (ZERT) facility has been developed at a field site in Bozeman, Montana, USA, to allow controlled studies of near-surface CO_2 transport and detection technologies. A slotted well divided into six zones was installed horizontally 2–2.3 m deep. Controlled releases of CO_2 tagged with PFTs and other tracers were performed in the summers of 2007, 2008 and 2009. A wide variety of detection techniques, including soil-gas flux, composition and isotopes, eddy-covariance measurements, hyperspectral and multispectral imaging of plants, and differential absorption measurements using laser-based instruments were deployed by collaborators from many institutions. Even at relatively low CO_2 fluxes, most techniques were able to detect elevated levels of CO_2 in the soil or atmosphere. Additionally, modelling of CO_2 transport and concentrations in the saturated soil and in the vadose zone was successfully conducted. (See Spangler et al. 2009, for program details and list of references).

As part of this multidisciplinary ongoing research project, 80 samples of water were collected during the 2008 season from 10 shallow monitoring wells (1.5 or 3.0 m deep) installed 1–6 m from the injection pipe, and from two distant monitoring wells (Kharaka et al. 2010b). Approximately 300 kg/day of food-grade CO_2 was injected through the perforated pipe during 9 July to 7 August, 2008 at the field test. Samples were collected before, during and following CO_2 injection. The main objective of the study was to investigate changes in the concentrations of major, minor and trace inorganic and organic compounds during and following CO_2 injection.

The ZERT site is located on a relatively flat 12-hectare agricultural plot close to the Montana State University campus in Bozeman. At this site, the topsoil of organic-rich silt and clay, with some sand, ranges in thickness from 0.2–1.2 m, and a caliche layer, high in calcite (approximately 15%), is observed at depths of approximately 50–80 cm. Beneath the topsoil layer is a cohesionless deposit of coarse sandy gravel extending to 5 m, the maximum depth investigated (Spangler et al. 2009). Gravels comprise approximately 70% of rock volume, and andesite is the chief rock fragment among the gravels and coarse sands, but minor amounts of detrital limestone and dolostone are also observed. The sand- and silt-sized fraction of this sediment consists of approximately 40% quartz, 40% magnetite and magnetic rock fragments, and 20% grains of amphibole, biotite/chlorite and feldspar.

Dissolved inorganic chemicals

The chemical data for samples collected prior to CO_2 injection show that the groundwater is a Ca-Mg-Na-HCO_3 type water, with a fresh-water salinity of about 600 mg/L TDS (Fig. 8.21, sample Z-109, Table 8.3). The groundwater has a pH of approximately 7.0, and HCO_3^- is the dominant anion, but the concentrations of Cl^- and SO_4^{2-} are relatively low. The concentrations of Fe, Mn, Zn, Pb and other trace metals are expectedly low, at ppb levels.

Following CO_2 injection, field-measured pH values decreased systematically to values around 6.0; early response (within one day) occurred in well 2B, only 1 m from the injection pipe and in the direction of groundwater flow, with pH decreasing to 5.7 (Fig. 8.22a). The pH of water for samples from well 4B (6 m from pipe) started decreasing only after the three days following CO_2 injection, but pH values remained above pH 6.0 (Fig. 8.22a). The measured pH values of groundwater were controlled primarily by pCO_2, which was measured by Strazisar et al. (2008) in capped wells, and was computed with SOLMINEQ (Kharaka et al. 1988) using the measured temperature, pH, alkalinity and other chemical parameters. The pCO_2 values measured and computed for the ZERT samples were 0.035 bar before CO_2 injection; they increased to values close to 1.0 bar following CO_2 injection. It is evident from these

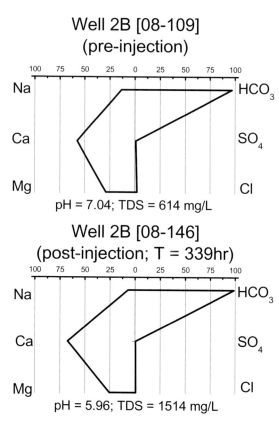

Fig. 8.21 Modified Stiff diagrams showing concentrations (equivalent units normalized to 100%) of major cations and anions, together with salinity and pH of groundwater from well 2B at the ZERT site, Montana, before and during CO_2 injection (From Kharaka et al. 2010b).

results that pH is an excellent early indicator for detection of CO_2 intrusion into groundwater at this and similar sites.

The alkalinity of groundwater increased from about 400 mg/L as HCO_3 to values of up to 1330 mg/L, following CO_2 injection (Fig. 8.22b). Alkalinity values for different wells, as with the pH values, show variable trends reflecting distance from the CO_2 injection pipe, impacts from precipitation events, and possibly local variations in the mineral composition of soils and sediments. Values for the electrical conductance, also

ble 8.3   Chemical composition of water samples from the ZERT monitoring well 2B, Bozeman, Montana (Kharaka et al. 2010b).

mple	Z-109	Z-118	Z-132	Z-136	Z-146	Z-150	Z-154	Z-161	Z-165	Z-169	Z-172	Z-177
ate	7/9/08	7/12/08	7/17/08	7/19/08	7/23/08	7/26/08	7/29/08	8/4/08	8/7/08	8/8/08	8/11/08	8/13/08
me	9:00	10:00	11:15	11:15	13:15	11:35	13:45	12:15	10:45	11:25	10:20	12:30
μS/cm)	651	952	119	1342	1424	1339	1235	1195	1201	732	615	606
H	7.04	6.4	5.91	5.97	5.96	5.87	5.82	5.78	5.74	5.95	5.76	6.42
(°C)	12.2	9.1	9.4	9.6	10.1	10.4	10.1	10.9	11.2	10.8	11.3	11.2
ajor solutes (mg/L)												
CO$_3$	434	664	924	1120	1150	1050	967	916	884	511	451	389
a	9.1	9.7	9.5	9.9	10.2	9.6	8.8	8.5	8.8	7.1	7.2	7.8
	5.4	71	8.0	7.4	7.4	7.1	7.6	7.7	7.7	5.9	5.4	5.2
g	28.0	40.8	48.8	54.6	54.9	47.0	40.0	35.9	34.9	20.3	17.4	16.4
a	91.9	142	191	223	239	241	216	219	212	125	106	94.1
	0.30	0.45	0.57	0.69	0.73	0.68	0.58	0.52	0.50	0.29	0.25	0.23
	0.10	0.19	0.26	0.23	0.24	0.22	0.25	0.27	0.26	0.15	0.12	0.10
n	0.028	0.19	0.14	0.011	0.014	0.015	0.13	0.19	0.21	0.13	0.090	0.052
	<0.01	0.075	0.53	<0.025	<0.025	<0.025	0.78	1.1	1.2	0.87	0.35	0.15
	0.17	0.14	0.055	<0.05	0.13	0.050	0.064	0.10	0.074	0.18	0.24	0.27
	5.35	5.31	5.55	5.54	5.59	5.63	5.66	5.80	6.12	6.31	6.53	6.88
	0.041	0.048	0.049	0.056	0.051	0.047	0.055	0.052	0.062	0.065	0.070	0.073
O$_3$	0.26	0.12	0.20	0.25	0.35	0.41	0.46	0.64	0.77	1.0	1.1	1.4
O$_4$	0.10	0.046	<0.015	0.24	0.26	<0.015	<0.015	<0.015	<0.015	<0.015	0.023	0.061
$_4$	7.17	7.39	7.77	8.35	8.60	8.49	8.00	7.81	8.02	7.98	7.89	7.84
O$_2$	32	40	37	38	39	30	29	38	38	31	29	29
OS	173	246	302	340	358	342	310	318	310	197	174	162
ace solutes (μg/L)												
	3.3	5.2	5.8	6.0	8.2	7.0	5.1	10	8.4	3.5	3.0	2.3
	1.3	1.0	0.42	0.88	1.6	1.5	1.2	0.42	055	0.44	0.49	0.65
	19	27	22	22	21	26	20	20	18	21	18	20
	0.29	0.45	0.43	0.22	0.22	0.19	0.18	0.32	0.23	0.19	0.14	0.15
	0.4	1.2	1.2	0.5	0.6	<0.5	1.2	1.3	1.5	0.8	0.6	0.5
	12	54	21	7.2	1.2	<1.0	<1.0	41	6.6	12	13	7.4
	2.5	2.3	2.2	2.4	2.4	2.0	1.5	1.9	2.0	1.6	1.7	1.6
	7.0	9.1	8.2	7.7	7.5	7.9	6.0	6.7	5.3	4.8	4.2	4.4
o	0.66	0.51	0.51	<0.5	0.68	<0.75	0.21	0.40	0.31	0.58	0.52	0.67
	0.06	0.08	0.06	0.05	0.06	<0.05	<0.05	<0.05	0.03	0.03	0.02	0.04
	5.0	5.9	<3.0	3.0	4.3	<5.0	<5.0	<3.0	2.8	<2.0	<2.0	2.5
	4.3	3.8	4.3	4.1	4.4	4.0	4.0	4.0	4.1	4.1	4.4	4.3
	3.8	9.0	2.3	2.8	3.5	4.0	4.4	3.5	4.2	2.8	2.5	5.9

measured at the site, show similar trends to the alkalinity, increasing from approximately 600 μS/cm before CO$_2$ injection to approximately 1800 μS/cm following CO$_2$ injection (Fig. 8.22c). Electrical conductance and alkalinity, as with pH, are easily measured field parameters that could provide early detection of CO$_2$ leakage into shallow groundwater from deep storage operations.

The alkalinity increases following CO$_2$ injection are balanced primarily by increases in the

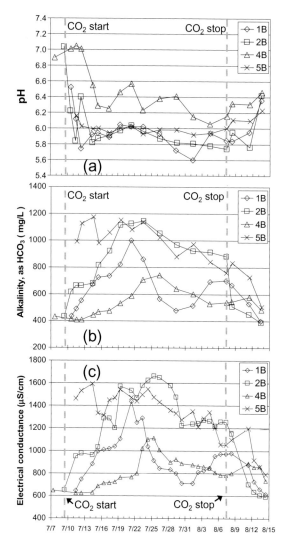

concentrations of Ca and Mg, whereas the concentrations of Na (10 ±2 mg/L) are relatively constant (Fig. 8.23). The concentrations of Ca increase from approximately 80 to 240 mg/L and those for Mg from approximately 25 to 70 mg/L (Fig. 8.23). Dissolutions of both calcite (eqn. 12) and dolomite (eqn. 13) are required to explain the changes in alkalinity and concentrations of Ca and Mg (Table 8.3).

$$CO_{2(g)} + H_2O + CaCO_{3(s)} \leftrightarrow Ca^{2+} + 2HCO_3^- \quad (8.12)$$

$$2H^+ + CaMg(CO_3)_{2(s)} \leftrightarrow Ca^{2+} + Mg^{2+} + 2HCO_3^- \quad (8.13)$$

These conclusions are supported by the initial characterization of minerals in core samples that show that calcite is abundant in a caliche layer observed at depths of approximately 50–80 cm. Small traces of carbonates were also observed in fines above 2.5 m, and minor amounts of detrital limestone and dolostone were observed in the gravel section. Results of geochemical modelling with SOLMINEQ (Kharaka et al. 1988) also support dissolution of calcite and disordered dolomite as possible reactions at all pH values. Desorption-ion exchange reactions on clay minerals with H⁺ have been suggested as alternative explanation (Zheng et al. 2009) for the increases in the concentrations of Ca and Mg.

The concentrations of Fe and Mn, the two most abundant trace metals in groundwater, also increase following CO_2 injection (Fig. 8.24). Increased concentrations of Fe could reflect dissolution of several Fe(II) and Fe(III) minerals, including siderite and ferrihydrite, depicted in Eqns. (14) and (15), respectively. The concentration of Fe in groundwater is a strong function of Eh, which was measured in water from the deeper wells that was not impacted by CO_2 injection (150 to 200 mV) indicating oxidizing conditions that account for the low concentration of dissolved Fe; much higher Fe values are possible under reducing conditions because of the higher solubility of Fe(II) minerals (Hem 1985). The Fe concentrations (Fig. 8.24a) increase from around 5 to 1200 ppb, but show very low values during

Fig. 8.22 Field-measured groundwater pH values (4a), alkalinities (4b), and electrical conductance (4c), obtained from selected ZERT wells as a function of time of sampling. Note the systematic decrease in pH values from approximately 7.0 before CO_2 injection to values as low as 5.6 during injection, and subsequent pH increases after CO_2 injection was terminated. Alkalinities increased from about 400 mg/L before CO_2 injection to values close to 1200 mg/L as HCO_3, and electrical conductance also increased from about 600 μS/cm before CO_2 injection to values higher than 1600 μS/cm during injection (From Kharaka et al. 2010b).

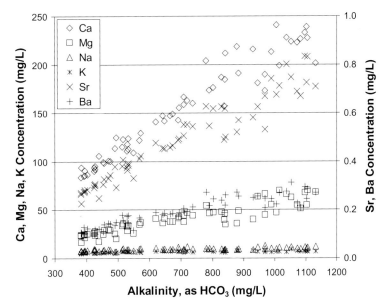

Fig. 8.23 Concentrations of major cations in groundwater from the ZERT wells plotted as a function of water alkalinities. Note the relatively constant concentrations of Na and K, but the general increases in the concentrations of divalent cations with water alkalinities, possibly indicating dissolution of carbonate minerals (From Kharaka et al. 2010b).

20–26 July following significant precipitation events, even when pH values were low. Dilution alone cannot explain the low Fe concentrations during 20–26 July, but the low values can be attributed to the oxidizing conditions possibly caused by increased dissolved O_2 content in groundwater transported with percolating water from precipitation events. Ion-exchange reactions on clays with H^+ and major dissolved cations, such as Ca, Mg and Na, are other possible controls on Fe concentrations (Birkholzer et al. 2008; Zheng et al. 2009), and these will be investigated after the completion of leaching experiments.

Geochemical modelling indicates that the large increases observed in concentrations of Fe (from around 5 to 1200 ppb) could result from dissolution of siderite (Eqn. 14), but are most likely caused by dissolution of iron oxyhydroxides, depicted in redox-sensitive reactions (Eqns. 15–17).

$$Fe(CO_3)_{(s)} + H^+ \leftrightarrow Fe^{2+} + HCO_3^- \quad (8.14)$$

$$Fe(OH)_{3(s)} + 3H^+ \leftrightarrow Fe^{3+} + 3H_2O \quad (8.15)$$

$$4FeOOH_{(s)} + 8H^+ \leftrightarrow 4Fe^{2+} + 6H_2O + O_2 \quad (8.16)$$

$$2Fe(OH)_{3(s)} + 4H_2CO_3^\circ + H_{2(g)} = 2Fe^{2+} +$$
$$4HCO_3^- + 6H_2O. \quad (8.17)$$

The concentrations of Mn show similar trends to those of Fe, increasing from approximately 5 to 1400 ppb following CO_2 injection, but also show low values during 20–26 July (Fig. 8.24b).

The concentrations of Pb, As, Zn and other trace metals (Table 8.3) generally show an increase with increasing alkalinity (lower pH value) following CO_2 injection. The values reported, however, carry high uncertainties as they are, in some cases, close to the analytical detection limits. The concentration increases are probably caused by desorption-ion exchange reactions with H^+, Ca and Mg resulting from lowered pH values. The concentrations, it should be noted, are all significantly below the maximum contaminant levels (MCLs) for the respective trace metals (e.g. 15 ppb for Pb, 6 ppb for As) (US Environmental Protection Agency 2009). The initial values and the increases in concentrations of these trace metals, although small, are readily measured by the sampling and analytical methods used in this study.

The chemical changes observed in the ZERT groundwater are similar in trends, though much lower in actual concentrations, to the changes observed in the Frio Brine Pilot tests discussed earlier in this chapter. The differences between pH results from Frio (pH as low as 3) and ZERT

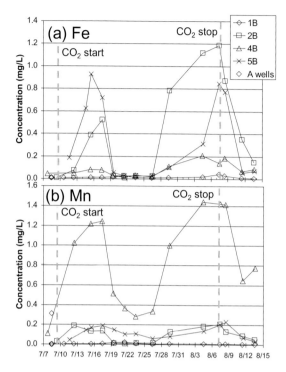

Fig. 8.24 Concentrations of Fe (a) and Mn (b) in ground-water from selected ZERT wells plotted as a function of time of sampling. Note the low Fe and Mn concentrations during July 20 to July 26, which we are attributing to the oxidizing conditions possibly caused by increased dissolved O_2 content in groundwater transported with percolating water from precipitation events (From Kharaka et al. 2010b).

tests are related to several geochemical parameters, but an important reason relates to the subsurface pCO_2 value for Frio, which was approximately 150 bar (Kharaka et al. 2009). The pCO_2 value measured (Stratizar et al. 2008) and computed with SOLMINEQ (Kharaka et al. 1988) for the ZERT samples ranged from 0.035 to approximately 1.0 bar. The maximum amount of CO_2 dissolved in water is a strong function of fluid pressure. Field determinations showed rapid and systematic changes in pH (7.0 to 5.6), alkalinity (400 to 1300 mg/L as HCO_3^-) and electrical conductance (600 to 1800 μS/cm). These easily monitored and sensitive chemical changes could

provide early detection of CO_2 leakage into shallow groundwater from deep storage operations; laboratory results could be used to confirm any such leaks.

Carbon isotopes at ZERT

The isotopic composition of carbon was measured by Fessenden et al. (2010) on dominant plants, soil cores (from 0 cm to 70 cm depth), DIC in groundwater, and in the local and regional CO_2. The $\delta^{13}C$ values were measured in these reservoirs during several seasons, before and during CO_2 releases. Results show that the dominant vegetation has a C_3 C-isotope signature, with values for grasses and alfalfa as low as −25‰ and −29‰, respectively. Once the CO_2 was released in the 2007 experiment, vegetation growing over the pipe and as far as 7 m to the north of the pipe experienced a change in their $\delta^{13}C$ signatures. The $\delta^{13}C$ of the CO_2 released from the tank was −52‰ and the $\delta^{13}C$ of the leaf tissues became increasingly depleted over the course of the experiment. The soils remained constant through each of the experiments, but the $\delta^{13}C$ signature of the soil column did become enriched with depth. Groundwater showed a CO_2 signature from the tank with a depletion of ^{13}C in the DIC that was observed within a day of the release and as far as 6 m north and down gradient from the pipe. The chamber CO_2 showed a CO_2 tank signature within 1 day of the release. This experiment showed that C-isotope analysis in the carbon reservoirs was a useful tool to detect CO_2 seepage either within days or weeks of release (Spangler et al. 2009; Fessenden et al. 2010).

POTENTIAL IMPACTS AND RISKS OF GEOLOGIC STORAGE OF CO_2

The major environmental risks associated with a CO_2 storage site may potentially include induced seismicity in vulnerable locations, and leakage of CO_2 and brine that may damage USDW and other natural resources, or cause harm to humans, animals and ecosystems (US EPA 2008). Fluid leakage could occur along unmapped fracture systems and faults, improperly sealed abandoned and orphaned wells (Nordbotten et al. 2009; Celia et al. 2006), corroded well casings and cements (Carey et al. 2007) or even via pathways created in the rock seals as a result of CO_2-brine-rock interactions (Kharaka et al. 2006; 2009). Maintaining reservoir integrity that limits CO_2 leakage to very low levels (less than 0.01%) is essential to the long-term success of injection operations (Hepple and Benson 2005). Preventing brine migration into overlying drinking water supplies is equally important, because dissolution of minerals following CO_2 injection would mobilize Fe, Mn and other metals, as well as toxic organics, including BTEX, phenols and other toxic compounds observed in the Frio Brine tests (Kharaka et al. 2009).

To evaluate these risks requires a detailed geologic site characterization, and an improved understanding of formation properties and how the injected CO_2 spreads and interacts with the rock matrix and reservoir fluids (Bachu 2003; Friedmann 2007). Geologic formations typically consist of layers of rock with different porosities, thicknesses, and brine and mineral compositions. All of these factors, together with the presence of faults and fracture systems, affect the suitability of the formation as a site for CO_2 sequestration. Porosity and thickness determine the storage capacity of the formation, and chemical composition determines the interaction of CO_2 with the minerals in place. Also, an impervious cap rock and absence of high-permeability faults are necessary to prevent the sequestered buoyant CO_2 from migrating to the surface. Finally, if the geological section consists of a series of aquifers, it is necessary to ensure that CO_2 stored in a saline formation does not migrate to a potable aquifer.

ENVIRONMENTAL IMPACTS

From an environmental point of view, leakage is the most serious potential problem. Leakage to the atmosphere negates the original environmental benefit and economic effort expended in sequestering the CO_2 (Friedman 2007; Stenhouse et al. 2009). Another potential problem is accumulation of CO_2 in pockets in soil and on the surface of the Earth, where it could present a health hazard to humans, animals and ecosystems. Furthermore, CO_2 could migrate into other strata, with the potential for contaminating USDW or causing other problems. If the formation into which CO_2 is being injected is below the ocean, leakage of CO_2 into the marine environment could affect ocean pH and chemistry, and have potentially serious consequences for marine life.

Results from the Frio Brine tests indicate that the injected supercritical CO_2, which is a very effective solvent for organic compounds (Kolak and Burruss 2006), could mobilize and transport organics, including BTEX, phenols, PAHs and other toxic organic compounds that have been reported in relatively high concentrations (10–60 mg/L) in oil-field brines (Kharaka and Hanor 2007; Kharaka et al. 2009). Preventing brine migration into overlying USDW is equally important, because dissolution of minerals following CO_2 injection would mobilize Fe, Mn and other metals, in addition to the high concentration of metals and other chemicals (salinity 5000 to more than 200,000 mg/L TDS) present in the original brine (Kharaka and Hanor 2007). Additional risks could arise from potential damage to nearby hydrocarbon resources caused by the displacement of fluids by the injected CO_2, such as saline water production at wells that had been producing oil or gas. Finally, if a project does not operate within prescribed injection rates and pressures, there is some potential for initiating seismic activity (Friedmann 2007). On the other hand, injection sites should not be located near major active faults, such as the San Andreas fault in California, because results from recent simulations indicate that large earthquakes along such faults would cause intense ground motions leading to rock fracturing and possible CO_2 leakage over large areas (up to 300 km by 50 km) (US Geological Survey, 2008, http://www.shakeout.org/).

Health and safety impacts

Information on the responses of humans, animals and plants to elevated and potentially hazardous levels of CO_2 and low levels of oxygen can be found in the literature (Deel, et al. 2006; Bergfeld et al. 2006; NIOSH/OSHA 1981; US EPA 2008). The National Institute of Occupational Safety and Health (NIOSH/OSHA) has determined that human time-weighted average (8 hours TWA) exposure limits to CO_2 of 1% (10,000 ppm), and short-term exposure limit (15 minute STEL) of 3% are appropriate (NIOSH/OSHA 1981, 1989). Human exposure to elevated levels of CO_2 can be hazardous in two ways – by a reduction in the oxygen content of the ambient air, causing hypoxia, or through direct CO_2 toxicity. In most cases of hazardous CO_2 exposure, the gas is presumed to act as a simple asphyxiant. However, extensive research indicates that exposure to elevated CO_2 concentrations higher than 3% has significant effects, including kidney damage, before oxygen dilution becomes physiologically significant. Concentrations higher than 10% have caused difficulty in breathing, impaired hearing, nausea, vomiting, a strangling sensation, sweating and loss of consciousness within 15 minutes. As O_2 concentration drops below 17%, increasingly severe physiological effects occur until below 6% O_2, loss of consciousness is rapid, and death takes place within minutes.

Another potential safety problem could result from accidents while working around the facilities required to capture, condense, transport and inject the CO_2. If H_2S is sequestered along with CO_2, health risks are significantly increased, as H_2S is highly toxic. Also, the health impacts would increase substantially if significant amounts of BTEX, PAHs and other mobilized organics are carried with the leaking CO_2 (Kharaka et al. 2009).

In order to minimize the environmental hazards and risks associated with CO_2 storage, a rigorous program should be implemented for the measurement, monitoring and verification (MMV) of the injected CO_2 and associated brine (Benson and Cook 2005; Friedmann 2007;

Doughty et al. 2008). Measurement, monitoring and verification is concerned with the capability to measure the amount of CO_2 stored at a specific sequestration site, map its spatial disposition through time, develop techniques for surface and subsurface monitoring for the early detection of leakage, and verify that the CO_2 is stored or isolated as intended and will not adversely impact the host ecosystem, including USDW. Measurement, monitoring and verification for geologic sequestration consists of three areas: (i) modelling and analysis of the geology and hydrology of the total injection system before injection occurs; (ii) tracking and monitoring the movement of the CO_2 plume; and (iii) above-surface measurements that verify that the CO_2 remains sequestered (US EPA 2008). Optimization of the MMV technologies is needed to ensure that the full range of spatial and temporal scales are covered. Results from pilot studies highlight the importance of using the more sensitive geochemical markers (Kharaka et al. 2006, 2009; Wells et al. 2007) and the importance of subsurface monitoring for detecting any early leakage of injected CO_2 and brine to minimize damage to groundwater and the local environment.

GEOCHEMICAL AND MULTI-PHASE REACTIVE TRANSPORT MODELLING

Mathematical models and numerical simulation tools of various types and complexity play an important role in evaluating the feasibility of geologic storage of CO_2. Geochemical codes such as an updated version of SOLMINEQ (Kharaka et al. 1988) or PHREEQC (Parkhurst and Appelo 1999) use formation-water analyses, together with formation mineralogy, to assess their equilibrium status. Once these equilibrium criteria are established, then kinetic codes such as PATH (Perkins and Gunter 1995) or EQ3/6 (Wolery and Daveler 1992) can be used to assess the reactivity and geochemical trapping potential over time when CO_2 is injected into a depleted oil reservoir or deep saline aquifer. Transport codes, such as GEOCHEMIST WORKBENCH (Bethke 2007) and

TOUGHREACT (Xu 2008; Xu et al. 2010) can be used to assess the distribution of CO_2 in time and space over different phases. Finally, hydrogeological codes such as MODFLOW (Harbaugh 2005), or STOMP (White and Oostrom 2006) can model the regional implications due to displacement of large volumes of saline water in the aquifers. These codes can be initially used sequentially to assess implications for dynamic CO_2 storage, followed by code improvement and possibly code integration.

Numerous studies of water-rock interaction have shown that without the close connection between experimental results and well-constrained natural systems, geochemical models can be highly inaccurate because of the complex interactions between mineral dissolution, mineral precipitation, aqueous chemistry (e.g. organics) and transport processes (Maher et al. 2006, 2009; Navarre-Sitchler et al. 2009). There are several challenges in modelling systems with high ionic strength (Zhang et al. 2009) and mixtures of CO_2 and brine (Lu et al. 2009). The credibility of theoretical models for injecting CO_2-rich solutions into the subsurface with the intent of successful hydrodynamic, solubility or mineralogical sequestration (Benson and Cole 2008; Oelkers et al. 2008; Bickle 2009) is dependent on the thermodynamic and transport parameters adopted, as well as the theoretical algorithms, initial and boundary conditions, and related assumptions and approximations (Birkholzer and Zhou 2009). Furthermore, any credible prediction for the reactivity of CO_2 in the subsurface requires the temperature and pressure dependence of: (i) the standard-state thermodynamic properties for minerals and solvent components, as well as all organic and inorganic solute species; (ii) non-standard composition-dependent mixing approximations, presented in terms of fugacity and activity coefficients; and (iii) kinetic rate laws describing reaction rates as a function of reactive surface area, reaction affinity, pH, and any additional catalytic or inhibitory effects (e.g. Al^{3+} or organic acids). Modelling and simulations should be carried out at all stages of a storage project, and the models used should be calibrated continuously as new data become available.

CONCLUDING REMARKS

Since supercritical CO_2 is buoyant and reactive in water, it will have a tendency to flow upward in the geologic section towards the Earth's surface. Therefore, despite the fact that many deep geological formations may be suitable for geological carbon sequestration, supercritical and gaseous CO_2 carries the possibility of leakage, which would negate the benefits of sequestration, and introduce elements of risk to groundwater and biota. Contamination of groundwater that provides approximately 21% of all water and 50% of US drinking water by the leakage of injected CO_2, and displaced brine should be prevented, because once contaminated, groundwater remediation is very difficult and expensive (Reilly et al. 2008). Importantly, CO_2 leakage risk will not be uniform across all sites, thus CO_2 storage sites will have to demonstrate minimal risk potential in their site characterization plans (Bradshaw et al. 2007; Oldenburg et al. 2008). A small percentage of sites might end up having significant leakage rates during injection, which will require subsurface monitoring (Kharaka et al. 2009) and/or near-surface monitoring using the variety of geophysical and geochemical tools such as those deployed at the ZERT site (Spangler et al. 2009; Kharaka et al. 2010b); storage sites would require validation as well as mitigation plans. Based on analogous experience in CO_2 injection such as acid-gas disposal and enhanced oil recovery, these risks are relatively minor, possibly comparable to those of current oil and gas operations (Benson and Cook 2005). If storage projects are not sited near seismically active faults, and operate within prescribed injection rates and pressures, then the potential for initiating significant seismic activity would probably be small. Finally, the likelihood and extent of any potential CO_2 leakage should slowly decrease with time after injection stops, because the formation pressure will begin to drop to pre-injection levels, as more of the injected CO_2 dissolves into the pore brine and begins the long-term process of forming chemically stable carbonate minerals.

There is a strong need for development of an integrated multiphase transport code capable of predicting fluid flow and mineral-brine-petroleum-gas interactions in the subsurface. To accomplish this difficult task, geochemists need to: (i) develop kinetic rate laws for mineral-dissolution/precipitation reactions under the full range of CO_2 saturation conditions; (ii) quantify rates and identify mechanisms that control transient reaction kinetics; (iii) determine the thermodynamic properties of complex solutions with dissolved organics, and solids, including the effects of nanoscale confinement and trace components; (iv) clarify the role of bacteria, other micro-organisms and catalysis on reaction kinetics; and (v) quantify the reactive surface area for minerals and organic matter in the subsurface.

ACKNOWLEDGEMENTS

This review is based on several of our referenced publications that have many of our colleagues as co-authors. We thank these as well as our co-workers, who contributed greatly to field sampling and chemical and isotope analyses. We thank Robert Rosenbauer, Robert Michel and William Herkelrath for reviewing an early version of this manuscript and suggesting important changes and modifications. We are grateful to James Thordsen and Dina Drennan for logistical support on this chapter. The Frio project was managed by Sue Hovorka (Bureau of Economic Geology, University of Texas, Austin), with financial support from DOE-NETL. The ZERT research was conducted within the ZERT project directed by Lee Spangler and managed by Laura Dobeck, MSU, Bozeman, MT. The ZERT research was funded primarily by the Electric Power Research Institute, EPRI, but funds were also obtained from EPA, DOE, LBNL and USGS.

REFERENCES

Adams, EE and Caldeira, K. (2008) Ocean storage of CO_2. *Elements* **4**; 319–24.

Allis, R, Bergfeld, D, Moore, J, McClure, K, Morgan, C, Chidsey, T Heath, J and Macpherson, B. (2005) Implications of results from CO_2 flux surveys over known CO_2 systems for long-term monitoring. Proc. Fourth Annual Conference on Carbon Capture and Sequestration DOE/NETL, May 2–5, 2005, CD-ROM, pp. 1367–88.

Audigane, P, Gaus, I, Czernichowski-Lauriol, I, Pruess, K and Xu, TF. (2007) Two-dimensional reactive transport modeling of CO_2 injection in a saline Aquifer at the Sleipner site, North Sea. *Am. J. Sci.* **307**: 974–1008.

Bachu, S. (2003) Screening and ranking of sedimentary basins for sequestration of CO_2 in geological media in response to climate change. *Environmental Geology* **44**: 277–89.

Bachu, S, Bonijoly, D, Bradshaw, J, Burruss, R, Holloway, S, Christensen, NP and Mathiassen, OM. (2007) CO_2 storage capacity estimation: Methodology and gaps. *International Journal of Greenhouse Gas Control* **1**: 430–43.

Baines, SJ and Worden, RH. (2004) The long-term fate of CO_2 in the subsurface: natural analogues for CO_2 storage. In: SJ Baines and WH Worden (eds.) Geological Storage of Carbon Dioxide. *Geol. Soc. London Spec. Pub.* **233**: 59–85.

Ballentine, CJ and Sherwood Lollar, B. (2002) Regional groundwater focusing of nitrogen and noble gases into the Hugoton-Panhandle giant gas field, USA. *Geochim. Cosmochim. Acta* **66**: 2483–97.

Ballentine, CJ, Burgess, R and Marty, B. (2002) Tracing fluid origin: transport and interaction in the crust. In: DR Porcelli, CJ Ballentine and R Weiler (eds.) Noble Gases *Geochim. Cosmochim. Acta* **47**: 539–614.

Ballentine, CJ, Schoell, M, Coleman, D and Cain, BA. (2001) 300-Myr-old magmatic CO_2 in natural gas reservoirs of the west Texas Permian basin. *Nature* **409**: 327–31.

Bénézeth, P, Ménez, B and Noiriel, C. (2009) CO_2 geological storage: Integrating geochemical, hydrodynamical, mechanical and biological processes from the pore to the reservoir scale. *Chem. Geol.* **265**: 1–2.

Bénézeth, P, Palmer, AD, Anovitz, LM and Horita, J. (2007) Dawsonite synthesis and re-evaluation of its thermodynamic properties from solubility measurements: Implications for mineral trapping of CO_2. *Geochim. Cosmochim. Acta* **71**: 4438–55.

Benson, SM and Cole, DR. (2008) CO_2 sequestration in deep sedimentary formations. *Elements* 4 (5), In: DR Cole and EH Oelkers (eds.) Carbon Dioxide Sequestration. pp. 305–10.

Benson, SM and Cook, P. (2005) Underground Geological Storage. In: *Carbon Dioxide Capture and Storage: Special Report of the Intergovernmental Panel on Climate Change (IPCC)*. Cambridge University Press, Interlachen, Switzerland, pp. 5–1 to 5–134.

Bergfeld, D, Evans, WC, Howle, JF and Farrar, CD. (2006) Carbon dioxide emissions from vegetation-kill zones around the resurgent dome of Long Valley caldera, Eastern California, USA. *Journal of Volcanology and Geothermal Research* 152: 140–56.

Bethke, CM. (2007) The Geochemist's Workbench ® (Version 7.0). Hydrogeology Program, University of Illinois.

Bickle, MJ. (2009) Geological carbon storage. *Nature Geoscience* 2(12): 815–818.

Birkholzer, JT and Zhou, Q. (2009) Basin–scale hydrogeologic impacts of CO2 storage: Capacity and regulatory implications. *International Journal of Greenhouse Gas Control* 3: 745–56.

Birkholzer, J, Apps, JA., Zheng, L, Zhang, Y, Xu, T and Tsang, C-F. (2008) Research project on CO_2 geological storage and groundwater resources: water quality effects caused by CO_2 intrusion into shallow groundwater. Lawrence Berkeley National Laboratory Technical Report, LBNL-1251E, pp. 450.

Bradshaw, J, Bachu, S, Bonijoly, D, Burruss, R, Holloway, S, Christensen, NP and Mathiassen, OM. (2007) CO_2 storage capacity estimation: Issues and development of standards. *International Journal of Greenhouse Gas Control* 1: 62–8.

Broecker, WS. (2008) CO_2 capture and storage: Possibilities and perspectives. *Elements* 4: 295–7.

Broecker, WS and Kunzig, R. (2008) CO_2, Fixing Climate: What Past Climate Changes Reveal about the Current Threat – and How to Counter it. Hill and Wang, *New York*, pp. 253.

Brown Jr, GE, Bird, DK, Kendelewicz, T, Maher, K, Mao, W, Johnson, N, Rosenbauer, RJ and García Del Real, P. (2009) Geological Sequestration of CO_2: Mechanisms and Kinetics of CO_2 Reactions with Mafic and Ultramafic Rock Formations. 2009 Annual Report to the Global Climate and Energy Project, Stanford University, Stanford, CA. (http:// www.stanford.edu/~gebjr/).

Burruss, R, Brennan, ST, Freeman, PA, Merrill, MD, Ruppert, LF, Becker, MF, Herkelrath, WN, Kharaka, YK, Neuzil, CE, Swanson, SM, Cook, TA, Klett, TR, Nelson, PH and Schenk, CJ. (2009) Development of a Probabilistic Assessment Methodology for Evaluation of Carbon Dioxide Storage. USGS Open-File Report No 09–1035.

Cantucci, B, Montegrossi, G, Vaselli, O, Tassi, F, Quattrocchi, F and Perkins, EH (2009) Geochemical modeling of CO_2 storage in deep reservoirs: The Weyburn Project (Canada) case study. *Chemical Geology v.* 265: pp. 181–97.

Carey, JW Wigand, M, Chipera, SJ, Wolde, G, Pawar, R, Lichtner, PC, Wehner, SC, Raines, MA and Guthrie, GD. (2007) Analysis and performance of oil well cement with 30 years of CO_2 exposure from the SACROC Unit, West Texas, USA. *International Journal of Greenhouse Gas Control* 1: 75–85.

Celia, MA, Kavetski, D, Nordbotten, JM, Bachu, S and Gasda, SE. (2006) Implications of abandoned wells for site selection, In: CO2SC 2006 International Symposium on Site Characterization for CO_2 Geological Storage, March 20–22, 2006. Proceedings: Berkeley, CA, Lawrence Berkeley National Laboratory, 157–9.

Chadwick, RA, Noy, D, Arts, R and Eiken, O. (2009) Latest time-lapse seismic data from Sleipner yield new insights into CO_2 plume development. *Energy Procedia* 1: 2103–10.

Chadwick, RA, Zweigel, P, Gregersen, U, Kirby, GA, Holloway, S and Johannessen, PN. (2004) Geological reservoir characterization of a CO_2 storage site: The Utsira Sand, Sleipner, northern North Sea. *Energy* 29: 1371–81.

Clark, I.D and Fritz, P. (1997) *Environmental Isotopes in Hydrogeology*. CRC Press, New York.

Cole, DR, Phelps, TJ, Kharaka, YK, Horita, J, Hovorka, SD, Knauss, KG, Thordsen, JJ and Nance, HS. (2008) Application of gas and fluid chemistry, stable isotopes and perfluorocarbon tracers as MMV tools in assessing water-rock interaction during the Frio CO_2 injection test. Proceedings of the 7th Annual Conf. Carbon Capture and Sequestration, Pittsburgh, PA May 5–8, 2008.

Cook, P.J (2009) Demonstration and Deployment of Carbon Dioxide Capture and Storage in Australia. *Energy Procedia* 1: 3859–66.

Czernichowski-Lauriol, I, Sanjuan, B, Rochelle C, Bateman K, Pearce J and Blackwell P. (1996) The underground disposal of carbon dioxide. *Inorganic Geochemistry J* 27: 183–276.

Daley, TM, Myer, LR, Peterson, JE, Majer, EL and Hoversten, GM. (2008) Time-lapse crosswell seismic and VSP monitoring of injected CO2 in a brine aquifer. *Environmental Geology* 54: 1657–65.

Daley, TM, Solbau, RD, Ajo-Franklin, JB and Benson, SM. (2007) Continuous active-source monitoring of CO2 injection in a brine aquifer. *Geophysics* 72 (5): A57–A61.

Deel, D, Mahajan, K, Mahoney, CR, McIlvried, HG and Srivastava, RD. (2006) Risk assessment and management for long-term storage of CO_2 in geologic formations. United States Department of Energy R&D. *Journal of Systemics, Cybernetics and Informatics* **5**: 79–85.

DOE/NETL (2008) Storing CO_2 with Enhanced Oil Recovery. DOE/NETL-402/1312/01-070-08, Advanced Resources International Inc., Feb. 7, 2008.

Doughty, C, Freifeld, BM and Trautz, RC. (2008) Site characterization for CO_2 geologic storage and vice versa: the Frio brine pilot, Texas, USA as a case study. *Environmental Geology* **54**(8): 1635–56.

Duan, A and Sun, R. (2003) An improved model calculating CO_2 solubility in pure water and aqueous NaCl solutions from 273 to 533 K and from 0 to 2000 bar. *Chemical Geology* **193**: 257–71.

Energy Information Administration (EIA) (2009) Annual Energy Outlook 2010 Early Release Overview. Report #:DOE/EIA-0383. Washington DC. http://www.eia.doe.gov/oiaf/ieo/pdf/emissions_tables.pdf

Emberley, S, Hutcheon, I., Shevalier, M, Durocher, K, Mayer, B, Gunter, WD and Perkins, EH. (2005) Monitoring of fluid-rock interaction and CO_2 storage through produced fluid sampling at the Weyburn CO_2-injection enhanced oil recovery site, Saskatchewan, Canada. *Applied Geochemistry* **20**: 1131–57.

Fessenden, JE, Clegg, SM, Rahn, TA, Humphries, SD and WS. Baldridge W.S. (2010) Novel MVA tools to track CO_2 seepage, tested at the ZERT controlled release site in Bozeman, MT. *Environmental Earth Sciences* DOI: 10.1007/s12665-010-0489-3.

Fessenden, JE, Stauffer, PH and Viswanathan, HS (2009) Natural Analogs of Geologic CO_2 Sequestration: Some General Implications for Engineered Sequestration. In: BJ McPherson and ET Sundquist (eds.). *Carbon Sequestration and Its Role in the Global Carbon Cycle*. American Geophysical Union, Geophysical Monograph 183, Washington, D.C., pp. 135–46.

Flaathen, TK, Gislason, SR, Oelkers, EH and Sveinbjörnsdóttir, ÁE. (2009) Chemical evolution of the Mt. Hekla, Iceland, groundwater: A natural analogue for CO_2 sequestration in basaltic rocks. *Applied Geochemistry* **24**: 463–74.

Freifeld, BM, Trautz, RC, Kharaka, YK, Phelps, TJ, Myer, LR, Hovorka, SD and Collins DJ (2005) The U-tube: A novel system for acquiring borehole fluid samples from a deep geologic CO_2 sequestration experiment. *Journal of Geophysical Research* **110**: B10203.

Friedmann, SJ (2007) Geological carbon dioxide sequestration. *Elements* **3**: 179–84.

Gaus, I, Azaroual, M and Czernichowski-Lauriol, I. (2005) Reactive transport modeling of the impact of CO_2 injection on the clayey cap rock at Sleipner (North Sea). *Chem. Geol.* **217**: 319–37.

Gilfillan, SMV, Sherwood-Lollar, B, Holland, G, Blagburn, D, Stevens, S, Schoell, M, Cassidy, M, Ding, Z, Zhou, Z, Lacrampe-Couloume, G and Ballentine, CJ. (2009) Solubility trapping in formation water as dominant CO_2 sink in natural gas fields. *Nature* **458**: 614–618.

Gilfillan, SMV, Ballentine, CJ, Holland, G, Blagburn, D, Sherwood-Lollar, B, Stevens, S, Schoell, M and Cassidy, M. (2008) The noble gas geochemistry of natural CO_2 gas reservoirs from the Colorado Plateau and Rocky Mountain provinces, USA. *Geochim. Cosmochim. Acta* **72**: 1174–98.

Gislason, SR, Wolff-Boenisch, D, Stefansson, A, Oelkers, E, Gunnlaugsson, E, Sigurdardóttir, H, Sigfússon, Broecker, W, Matter, J, Stute, M, Axelsson, G and Fridriksson, T. (2009) Mineral sequestration of carbon dioxide in basalt: The CarbFix project. *International Journal of Greenhouse Gas* (submitted).

Goldberg, DS, Takahashi, T and Slagle, AL. (2008) Carbon dioxide sequestration in deep-sea basalt. *Proc. Nat. Acad. Sci. U.S.A.* **105**(29): 9920–5.

Gunter, WD, Perkins, EH and McCann, TJ (1993) Aquifer disposal of CO_2-rich gases: reaction design for added capacity. *Energy Conversion and Management* **34**: 941–8.

Gysi, AP and Stefánsson, A. (2008) Numerical modeling of CO_2-water-basalt interaction. *Mineralogical Magazine* **72**: 55–9.

Han, WS. (2008) Evaluation of CO_2 trapping mechanisms at the SACROC northern platform: Site of 35 years of CO_2 injection: Socorro. The New Mexico Institute of Mining and Technology. Ph.D. dissertation, 426.

Han, WS and McPherson, B. (2008) Comparison of two different equations of state for application of carbon dioxide sequestration. *Advances in Water Resources* **31**: 877–90.

Harbaugh, AW. (2005) MODFLOW-2005, the U.S. Geological Survey modular ground-water model: the Ground-Water Flow Process. U.S. Geological Survey Techniques and Methods 6-A16, various pages.

Haszeldine, SR. (2009) Carbon Capture and Storage: How Green Can Black Be? *Science* **325**: 1647–52.

Heath, JE, Lachmar, TE, Evans, JP, Kolesar, PT and Williams, AP. (2009) Hydrogeochemical

Characterization of Leaking Carbon Dioxide: Charged Fault Zones in East–Central Utah, With Implications for Geological Carbon Storage. In: BJ McPherson and ET Sundquist (eds.) *Carbon Sequestration and Its Role in the Global Carbon Cycle.* American Geophysical Union, Geophysical Monograph 183, Washington, D.C., pp. 147–58.

Hellevang, HP, Aagaard, P, Oelkers, EH and Kvamme, B. (2005) Can dawsonite permanently trap CO_2? *Environmental Science and Technology* **39**: 8281–7.

Hem, JD. (1985) Study and Interpretation of the Chemical Characteristics of Natural Water. U.S. Geological Survey Water-Supply Paper 2254, 264 pp.

Hepple, RP and Benson, SM. (2005) Geologic storage of carbon dioxide as a climate change mitigation strategy: performance requirements and the implications of surface seepage. *Environ Geol* **47**: 576–85.

Hermanrud, C, Andresen, T, Eiken, O, Hansen, H, Janbu, A, Lippard, J, Bolas, HN, Simmenes, TH, Teige, GMG and Ostmo, S. (2009) Storage of CO_2 in saline aquifers: lessons learned from 10 years of injection into the Utsira Formation in the Sleipner area. *Energy Procedia* **1**: 1997–2004.

Hitchon, B (ed.) (2009) Pembina Cardium CO_2 Monitoring Pilot: A CO2-EOR Project, Alberta, Canada. Geoscience Publishing, Box 79088, Sherwood Park, Alberta, 360, pp.

Hitchon, B. (2000) 'Rust' contamination of formation waters from producing wells. *Appl. Geochem* **15**: 1527–33.

Hitchon, B. (ed.) (1996) *Aquifer Disposal of Carbon Dioxide.* Geoscience Publishing Ltd., Sherwood Park, Alberta, Canada, pp. 165.

Hitchon, B (ed.), (1996) *Aquifer Disposal of Carbon Dioxide.* Geoscience Publishing Ltd., Sherwood Park, Alberta, Canada, 165 pp.

Holloway, S, Pearce, JM, Hards, VL, Ohsumi, T and Gale, J. (2007) Natural emissions of CO_2 from the geosphere and their bearing on the geological storage of carbon dioxide. *Energy* **32**: 1194–201.

Holloway, S. (1997) An overview of the underground disposal of carbon dioxide. *Energy Conversion and Management* **38**: 193–8.

Hovorka, SD, Benson, SM, Doughty, CK, Freifeld, BM, Sakurai, S Daley, TM, Kharaka, YK, Holtz, MH, Trautz, RC, Nance, HS, Myer, LR and Knauss, KG. (2006) Measuring permanence of CO_2 storage in saline formations: the Frio experiment. *Environmental Geosciences* **13** 105–121.

Hovorka, SD, Choi, J-W, Menckel, TA, Trevino, RH, Zeng, H, Kordi, M, Wang, FP and Nicot, J-P. (2010)

Measured and modeled CO2 flow through heterogeneous reservoir – early results of SECARB test at Cranfield Mississippi. Abstract, Ninth Annual Conference on Carbon Capture and Sequestration DOE/NETL, May, 2010 (in press).

Hu, J, Duan, Z, Zhu, C and Chou, I-M. (2007) PVTx properties of the CO_2–H_2O and CO_2–H_2O–NaCl systems below 647 K: Assessment of experimental data and thermodynamic models. *Chem. Geol.* **238**: 249–67.

Iding, M and Ringrose, P. (2009) Evaluating the impact of fractures on the long-term performance of the In-Salah CO2 storage site. *Energy Procedia* **1**: 2021–28.

Intergovernmental panel on Climate Change (IPCC). (2005) *Special Report on Carbon Dioxide Capture and Storage. Prepared by the Working Group* III *of the Intergovernmental Panel on Climate Change* [B Metz, O Davidson, HC de Coninck, M Loos and LA Meyer (eds.)], Cambridge University Press, Cambridge, United Kingdom and New York, NY, USA, 442 pp.

Intergovernmental panel on Climate Change (IPCC). 2007 *Climate Change 2007: The Physical Science Basis. Contribution of Working Group I to the Fourth Assessment Report of the Intergovernmental Panel on Climate Change* [S Solomon, D Qin, M Manning, Z Chen, M Marquis, KB Averyt, M Tignor and HL Miller (eds.)]. Cambridge University Press, Cambridge, United Kingdom and New York, NY, USA, 996 pp.

Johnson, JW, Nitao, JJ and Knauss, K. G. (2004) Reactive transport modeling of CO_2 storage in saline aquifers to elucidate fundamental processes, trapping mechanisms and sequestration partitioning. In: SJ Baines and RH Worden (eds.). *Geological Storage of Carbon Dioxide. Geol, Soc. London Spec*. Pub. **233**, 107–28.

Kennedy, BM, Kharaka, YK, Evans, WC, Ellwood, A, DePaolo, DJ, Thordsen, JJ, Ambats, G and Mariner, RH. (1997) Mantle fluids in the San Andreas fault system, California. *Science* **278**: 1278–81.

Kharaka, YK and Hanor, JS. (2007), Deep fluids in the continents I: sedimentary basins. In: JI Drever (ed.) *Surface and Ground Water, Weathering and Soils: Treatise on Geochemistry*, v. **5**, pp. 1–48.

Kharaka, YK, Thordsen, JJ, Bullen, TD, Cole, DR, Phelps, TJ, Birkholzer, JT and Hovorka, SD. (2010a) Near surface and deep subsurface monitoring for successful geologic sequestration of CO_2. Proc. Of WRI-13 , Birkle, P & Torres-Alvarado, IS (eds.) A Balkema, p. 867–70.

Kharaka, YK, Thordsen, JJ, Kakouros, E, Ambats, G, Herkelrath, WN, Birkolzer, JT, Apps, JA, Spycher, NF, Zheng, L, Trautz, RC, Rauch, HW and Gullickson,

KS. (2010b) Changes in the chemistry of shallow groundwater related to the 2008 injection of CO_2 at the ZERT field site, Bozeman, Montana. *Environmental Earth Sciences*. **60**: 273–84.

Kharaka, YK, Thordsen, JJ, Hovorka, SD, Nance, HS, Cole, DR, Phelps, TJ and Knauss, KG. (2009) Potential environmental issues of CO_2 storage in deep saline aquifers: Geochemical results from the Frio-I brine pilot test, Texas, USA. *Applied Geochem.* **24**: 1106–12.

Kharaka, YK, Cole, DR, Hovorka, SD, Gunter, WD, Knauss, KG and Freifeld, BM. (2006) Gas-water-rock interactions in Frio Formation following CO_2 injection: implications to the storage of greenhouse gases in sedimentary basins. *Geology* **34**: 577–80.

Kharaka, YK, Gunter, WD, Garwal, PK, Perkins, EH and DeBrall, JD. (1988) SOLMINEQ.88: A computer program for geochemical modeling of water-rock interactions. U.S. Geological Survey Water Resources Invest. Rep. 88–4227.

King, MB, Murbarak, A, Kim, JD and Bott, TR. (1992) The mutual solubilities of water with supercritical and liquid carbon dioxide. *J. Supercrit. Fluids* **5**: 296–302.

Klusman, RW. (2003a) A geochemical perspective and assessment of leakage potential for a mature carbon dioxide-enhanced oil recovery project and as a prototype for carbon dioxide sequestration; Rangely field, Colorado. *Amer. Assoc. Pet. Geol.* **87**: 1485–508.

Klusman, RW. (2003b) Rate measurements and detection of gas microseepage to the atmosphere from an enhanced oil recovery/sequestration project, Rangely, Colorado, USA. *App. Geochem.* **18**: 1825–38.

Knauss, KG, Johnson, JW and Steefel, CI. (2005) Evaluation of the impact of CO_2, co-contaminant gas, aqueous fluid and reservoir-rock interactions on the geologic sequestration of CO_2. *Chemical Geology* **217**: 339–50.

Kolak, JJ and Burruss, RC. (2006) Geochemical investigation of the potential for mobilizing non-methane hydrocarbons during carbon dioxide storage in deep coal beds. *Energy Fuels* **20**(2): 566–74.

Koschel, D, Coxam, J-Y, Rodier, L, Majer, V. (2006) Enthalpy and solubility data of CO_2 in water and NaCl (aq) at conditions of interest for geological sequestration. *Fluid Phase Equil.* **247**: 107–20.

Lackner, KS. (2010) Carbon Dioxide Capture from Ambient Air. In: W Blum, M Keilhacker, U Platt, W Roether (eds.) *The Physics Perspective on Energy Supply and Climate Change – A Critical Assessment, v. 2009*. Springer Verlag: Bad Honnef.

Lackner, KS. (2002) Carbonate chemistry for sequestering fossil carbon. *Annu. Rev. Energ. Env.* **27**: 193–232.

Lagneau, V, Pipart, A and Catalette, H. (2005) Reactive transport modeling of CO_2 sequestration in deep saline aquifers. *Oil & Gas Science and Technology* **60**: 231–47.

Leonenko, Y and Keith, DW. (2008) Reservoir Engineering to Accelerate the Dissolution of CO_2 Stored in Aquifers. *Environ. Sci. Technol.* **42**: 2742–7.

Lewicki, J, Oldenburg, C, Dobeck, L and Spangler, L. (2007) Surface CO_2 leakage during the first shallow subsurface CO_2 release experiment. *Geophys. Res. Lett.* **34**: L24402.

Li, D and Duan, Z. (2007) The speciation equilibrium coupling with phase equilibrium in the H_2O–CO_2–NaCl system from 0 to 250°C, from 0 to 1000 bar, and from 0 to 5 molality of NaCl. *Chem. Geol.* **244**: 730–51.

Lindeberg, E and Wessel-Berg, D. (1997) Vertical convection in an aquifer column under a gas cap of CO_2, *Energy Conversion and Management* **38**: S229–S234.

Litynski, JT, Plasynski, S, McIlvried, HG, Mahoney, C and Srivastava, RD. (2008) The United States Department of Energy's Regional Carbon Sequestration Partnerships Program Validation Phase. *Environment International* **34**: 127–38.

Lu, C, Han, WS, Lee, S-Y, McPherson, BJ, Lichtner, PC. (2009), Effects of density and mutual solubility of a CO_2-brine system on CO_2 storage in geological formations: 'Warm' vs. 'cold' formations. *Advances in Water Resources* **32**(12): 1685–702.

Maher, K, Steefel, CI, White, AF and Stonestrom, DA. (2009) The Role of Secondary Minerals and Reaction Affinity in Regulating Weathering Rates at the Santa Cruz Marine Terrace Chronosequence. *Geochim. Cosmochim. Acta* **73**: 2804–31.

Maher, K, Steefel, CI, Depaolo, DJ and Viani, BE. (2006), The mineral dissolution rate conundrum: Insights from reactive transport modeling of U isotopes and pore fluid chemistry in marine sediments. *Geochim. Cosmochim. Acta* **70**(2): 337–63.

Marini, L. (2007) *Geological Sequestration of Carbon Dioxide: Thermodynamics, Kinetics and Reaction Path modeling*. Elsevier, Amsterdam, pp.470.

Matter, JM and Kelemen, PB. (2009) Permanent storage of carbon dioxide in geological reservoirs by mineral carbonation. *Nature Geoscience* **2**(12): 837–41.

Matter, JM, Broecker, WS, Stute, S, Gislason, SR, Oelkers, EH, Stefánsson, A, Wolff-Boenisch, D,

Gunnlaugsson, E, Axelsson, G and Björnsson, G. (2009) Permanent Carbon Dioxide Storage into Basalt: The CarbFix Pilot Project, Iceland. *Energy Procedia* **1**: 3641–6.

Michael, K, Arnot, M, Cook, P, Ennis-King, J, Funnell, R, Kaldi, J, Kirste, D and Paterson, L. (2009) CO_2 storage in saline aquifers I: current state of scientific knowledge. *Energy Procedia* **1**: 3197–204.

Moore, J, Adams, M, Allis, R, Lutz, S and Rauzi, S. (2005) Mineralogical and geochemical consequences of the long-term presence of CO_2 in natural reservoirs: An example from the Springerville–St. Johns Field, Arizona, and New Mexico, *U.S.A. Chem. Geol.* **217**: 365–85.

Moritis, G. (2009) Special Report: More CO_2-EOR projects likely as new CO_2 supply sources become available. Dec. 2009, *Oil & Gas J.*

Nance, HS, Rauch, H, Strazisar, B, Bromhal, G, Wells, A, Diehl, R, Klusman, R, Lewicki, JL, Oldenberg, CM, Kharaka, YK and Kakouros, E. (2005), Surface environmental monitoring at the Frio Test site. Proceedings of the Fourth Annual Conference on Carbon Capture and Sequestration. Alexandria, VA., May 2–5, 16 p, CD-ROM.

Navarre-Sitchler, A, Steefel, CI, Yang, L, Tomutsa, L and Brantly, S. (2009) Evolution of porosity and diffusivity associated with chemical weathering of a basalt clast. *J. of Geophys. Res.* **114**: 1–14.

NIOSH/OSHA (1989) Toxicologic review of selected chemicals: Carbon dioxide. http://www.cdc.gov/niosh/pel88/124-38.html

NIOSH/OSHA (1981) Occupational Health Guidelines for Chemical Hazards: Department of Health and Human Services (National Institute for Occupational Safety and Health) Publication No. 81–123, United States Government Printing Office, Washington, DC.

NIST (1992) National Institute of Standards and Technology Database 14, Mixture Property Database, version 9.08. US Dept. of Commerce, Washington, DC.

Nordbotten, JM, Kavetski, D, Celia, MA and Bachu, S. (2009), Model for CO_2 Leakage including Multiple Geological Layers and Multiple Leaky Wells. *Environmental Science and Technology* **43**: 743–9.

Oelkers, EH and Cole, DR. (2008) Carbon dioxide sequestration: A solution to a global problem. *Elements* **4**: 305–10.

Oelkers, EH, Gislason, SR and Matter, J. (2008) Mineral Carbonation of CO_2. *Elements* **4**(5): 333–7.

Oldenburg, CM (2007) Migration mechanisms and potential impacts of CO_2 leakage and seepage. In: EJ Wilson and D Gerard (eds.) *Carbon Capture and Sequestration: Integrating Technology, Monitoring and Regulation.* Blackwell Publishing pp. 127–46.

Oldenburg, CM., Nicot, J and Bryant, SL. (2008) *The certification framework: Risk assessment for safety and effectiveness of geologic carbon sequestration [abs].* American Geophysical Union, San Francisco, CA, Fall Meeting 2008.

Palandri, J and Kharaka, YK. (2004) A compilation of rate parameters of water-mineral interaction kinetics for application to geochemical modeling. US Geol. Surv. Open File Report 2004-1068, pp. 64.

Palandri, JL and Kharaka, YK. (2005) Ferric iron-bearing sediments as a mineral trap for CO_2 sequestration: iron reduction using sulfur-bearing waste gas. *Chem. Geol.* **217**: 351–64.

Palandri, JL, Rosenbauer, RJ and Kharaka, YK. (2005) Ferric iron in sediments as a novel CO_2 mineral trap: CO_2-SO_2 reaction with hematite. *Applied Geochemistry* **20**: 2038–48.

Pappa, GD, Perakis, C, Tsimpanogiannis, IN and Voutsas, EC. (2009) Thermodynamic modeling of the vapor-liquid equilibrium of the CO_2/H_2O mixture. *Fluid Phase Equil.* **284**: 56–63.

Parkhurst, DL and Appelo, CAJ. (1999) User's Guide to PHREEQC (Version 2): A computer program for speciation, batch-reaction, one-dimensional transport, and inverse geochemical calculations. USGS Water-Resources Investigations Report 99-4259, Denver, Colorado.

Parry, WT, Forester, CB, Evans, JP, Bowen, BB and Chan, MA. (2007) Geochemistry of CO_2 sequestration in the Jurassic Navajo Sandstone: Colorado Plateau, Utah. *Environmental Geosciences* **14**: 91–109.

Pearce, JM, Baker, J, Beaubien, S, Brune, S, Czernichowski-Lauriol, I, Faber, E, Hatziyannis, G, Hildebrand, A, Krooss, BM, Lombardi, S, Nador, A, Pauwels, H and Schroot, BM. (2003) Natural CO_2 accumulations in Europe: understanding long-term geological processes in CO_2 sequestration. In: J Gale and Y Kaya (eds.) *Greenhouse Gas Control Tech.* Vol. **1**, pp. 417–22.

Perkins, EH and Gunter, WD (1995) A user's manual for β PATHARCH.94: A reaction path-mass transfer program. Alberta Research Council Report, ENVTR pp. 95–11, 179.

Phelps, TJ, McCallum, SD, Cole, DR, Kharaka, YK and Hovorka, SD. (2006) Monitoring geological CO_2 sequestration using perfluorocarbon gas tracers and isotopes. Proc. Fifth Annual Conference on Carbon Capture and Sequestration, Alexandria, VA., May 8–11, p. 8 CD-ROM.

Qin, J, Rosenbauer, RJ and Duan, Z. (2008) Experimental Measurements of Vapor-Liquid Equilibria of the $H_2O + CO_2 + CH_4$ Ternary System. *J. Chem. Eng. Data* 53: 1246–9.

Raistruck, M, Mayer, B, Shevalier, M, Perez, RJ, Hutcheon, I, Perkins, E and Gunter, B. (2006) Using Chemical and Isotopic Data to Quantify Ionic Trapping of Injected Carbon Dioxide in Oil Field Brines. *Environ. Sci. Technol.* 40: 6744–9.

Reilly, TE, Dennehy, KF, Alley, WM and Cunningham, WL. (2008) Ground-Water Availability in the United States. U.S. Geological Survey Circular 1323, 70 available at http://pubs.usgs.gov/circ/1323/

Rosenbauer, RJ, Koksalan, T and Palandri, JL. (2005) Experimental investigation of CO_2-brine-rock interactions at elevated temperature and pressure: Implications for CO_2 sequestration in deep-saline aquifers. *Fuel Processing Technology* 86: 1581–97.

Rubin, ES. (2008) CO_2 Capture and Transport. *Elements* 4: 311–317.

Rutqvist, J, Vasco, DW and Myer, L. (2009) Coupled reservoir-geomechanical analysis of CO_2 injection at In Salah, Algeria. *Energy Procedia* 1: 1847–54.

Schrag, DP. (2009) Storage of carbon dioxide in offshore sediments. *Science* 325: 1658–9.

Shipton, ZK, Evans, JP, Heath, J, Williams, A, Kirchner, D and Kolesar, PT. (2005) Chapter 4. Natural leaking CO2-charged systems as analogs for failed geologic storage reservoirs. In: DC Thomas and SM Benson (eds.) *Carbon Dioxide Capture for Storage in Deep Geologic Formations.* Vol. 2, pp. 699–712.

Shipton, ZK, Evans, JP, Kirshner, D, Kolesar, PT, Williams, AP and Heath, J. (2004) Analysis of CO_2 leakage through 'low permeability' faults from natural reservoirs in the Colorado Plateau, east-central Utah. In: SJ Baines and RH Worden (eds.) Geological Storage of Carbon Dioxide. *Geol. Soc. London Spec. Pub.* 233: pp. 43–58.

Shiraki, R. and Dunn, TL. (2000) Experimental study on water-rock interactions during CO2 flooding in the Tensleep Formation, Wyoming, USA. *Applied Geochemistry* 15: 265–79.

Spangler, LH, Dobeck, LM, Dobeck, LM, Repasky, KS, Nehrir, A, Humphries, S, Barr, J, Keith, C, Shaw, J, Rouse, J, Cunningham, A, Benson, S, Oldenburg, CM, Lewicki, JL, Wells, A, Diehl, R, Strazisar, B, Fessenden, J, Rahn, T, Amonette, J, Barr, J, Pickles, W, Jacobson, Silver, E, Male, E, Rauch, H, Gullickson, K, Trautz, R, Kharaka, YK, Birkholzer, J and Wielopolski, L. (2009) A controlled field pilot for testing near surface

CO_2 detection techniques and transport models. *Energy Procedia* 1: 2143–50.

Spycher, N and Pruess, K (2005) CO_2-H_2O mixtures in the geological sequestration of CO_2. II. Partitioning in chloride brines at 12–100°C and up to 600 bar: *Geochimica et Cosmochimica Acta* 69: 3309–20.

Spycher, N, Pruess, K and Ennis-King, J. (2003) CO_2-H_2O mixtures in geological sequestration of CO_2, I: Assessment and calculation of mutual solubilities from 12 to 100°C and up to 600 bar. *Geochim. Cosmochim. Acta* 67: 3015–31.

Stenhouse, M, Arthur, R and Zhou, W. (2009) Assessing environmental impacts from geological CO_2 storage. *Energy Procedia* 1: 1895–902.

Strazisar, BR, Wells AW and Diehl, JR. (2008) Soil gas monitoring for the ZERT shallow CO_2 injection project. *Prepr. Pap.-Am. Chem. Soc., Div. Fuel Chem.* 52(2).

Sundquist, ET, Ackerman, KV, Parker, L and Huntzinger, DN. (2009) An Introduction to Global carbon Cycle. In: BJ McPherson and ET Sundquist (eds.). *Carbon Sequestration and Its Role in the Global Carbon Cycle.* American Geophysical Union, Geophysical Monograph 183, Washington, D.C., pp. 1–23.

Torp, TA and Gale, J. (2003) Demonstrating storage of CO_2 in geological reservoirs: the Sleipner and SACS projects. In: J Gale and Kaya, J. (eds.) Proceedings of the 6th International Conference on Greenhouse Gas Control Technologies (GHGT-6), 1–4 October 2002, Kyoto, Japan. Pergamon, Amsterdam, v. I, pp. 311–316.

US Department of Energy, Office of Fossil Energy, National Energy Technology Laboratory (2008) Carbon Sequestration Atlas II of the United States and Canada, p. 142.

US Environmental Protection Agency (2009) National Primary Drinking Water Regulations: <http://www.epa.gov/safewater/consumer/pdf/mcl.pdf> (accessed September 18, 2009).

US Environmental Protection Agency (2008) Vulnerability evaluation framework for geologic sequestration of carbon dioxide. EPA430-R-08-009, p. 85.

US Geological Survey (2008) The Great Southern California ShakeOut. At: http://www.shakeout.org/

Uysal, I, Feng, Y-X, Zhao, J-X, Isik, V, Nuriel, P and Golding, SD. (2009) Hydrothermal CO_2 degassing in seismically active zones during the late Quaternary. *Chem. Geol.* 265: 442–54.

Wells, AW, Diehl, JR, Bromhal, G, Strazisar, BR, Wilson, TH and White, CM. (2007) The use of tracers to assess leakage from the sequestration of CO_2 in a depleted oil reservoir, New Mexico, USA. *Applied Geochemistry* **22**: 996–1016.

White, MD and Oostrom, M. (2006) STOMP Subsurface Transport Over Multiple Phases, Version 4.0: User's Guide, PNNL-15782. Pacific Northwest National Laboratory, Richland, Washington.

White, SP, Allis, RG and Moore, J. (2005) Simulation of reactive transport of injected CO_2 on the Colorado Plateau, Utah, USA. *Chemical Geology* **217**: 387–405.

White, CM, Strazisar, BR and Granite, EJ. (2003) Separation and capture of CO_2 from large stationary sources and sequestration in geological formations: coalbeds and deep saline aquifers. *Journal of Air & Waste Management Association* **53**: 645–715.

Wilkinson, M, Hazeldine, RS, Fralick, AE, Odling, N, Stoker, SJ and Gatliff, RW. (2009) CO_2-mineral reaction in a natural analogue for CO_2 storage: implications for modeling. *J. Sed. Res.* **79**: 486–94.

Wolery, TJ and Daveler, SA. (1992) A computer program for reaction path modeling of aqueous geochemical solutions: Theoretical manual, user's guide, and related documentation (version 7.0). Report UCRL-MA-110662 PT IV Lawrence Livermore National Laboratory, Livermore, CA.

Xu, T. (2008) TOUGHREACT User's Guide: A Simulation Program for Non-isothermal Multiphase Reactive Geochemical Transport in Variably Saturated Geologic Media: V1.2.1. Lawrence Berkeley National Laboratory Paper LBNL-55460-2008, accessed 12/31/08 at http://repositories.cdlib.org/lbnl/LBNL-55460-2008.

Xu, T, Apps, JA and Pruess, K. (2005) Mineral sequestration of carbon dioxide in a sandstone-shale system. *Chemical Geology* **217**: 295–318.

Xu, T, Apps, JA, Pruess, K and Yamamoto, H. (2007) Numerical modeling of injection and mineral trapping of CO_2 with H_2S and SO_2 in a sandstone formation. *Chem. Geol.* **242**: 319–46.

Xu, T, Kharaka, YK, Doughty, CBM and Daley, TM. (2010) Reactive Transport Modeling to Study Changes in Water Chemistry Induced by CO_2 Injection at the Frio-I Brine Pilot. *Chemical Geology* **222**: 153–64.

Zhang, W, Li, Y, Xu, T, Cheng, H, Zheng, Y and Xiong, P. (2009) Long-term variations of CO_2 trapped in different mechanisms in deep saline formations: A case study of the Songliao Basin, China. *Greenhouse Gas Control Technologies* **3**(2): 161–80.

Zheng L, Apps J, Spycher N, Birkholzer J, Kharaka YK, Thordsen J, Kakouros E, Trautz R, Rauch H,

Gullickson K (2009) Changes in shallow groundwater chemistry at the 2008 ZERT CO_2 injection experiment: II- Modeling analysis. Abstract, Eight Carbon Capture and Sequestration Conference. Pittsburgh, PA, May 4–7, 2009.

Zhou, Z, Ballentine, CJ, Kipfer, R, Schoell, M and Thibodeaux, S. (2005) Noble gas tracing of groundwater: coalbed methane interaction in the San Juan Basin, USA. *Geochim. Cosmochim. Acta* **69**: 5413–28.

SELECTED MAJOR REFERENCES

Benson, SM and Cole, DR (2008) CO_2 sequestration in deep sedimentary formations. *Elements* **4** (5), In: DR Cole and EH Oelkers (eds.). Carbon Dioxide Sequestration. pp. 305–10.

Benson, SM and Cook, P. (2005) Underground Geological Storage. In: *Carbon Dioxide Capture and Storage: Special Report of the Intergovernmental Panel on Climate Change (IPCC)*. Cambridge University Press, Interlachen, Switzerland, pp. 5–1 to 5–134.

Broecker, WS and Kunzig, R. (2008). CO_2, Fixing Climate, What Past Climate Changes Reveal about the Current Threat – and How to Counter it. Hill and Wang, New York, 253 pp.

Cole, DR and Oelkers, EH. (eds.), 2008, Carbon dioxide sequestration: Elements, v. 4 (5).

Hitchon, B (ed.). (1996) *Aquifer Disposal of Carbon Dioxide*. Geoscience Publishing Ltd., Sherwood Park, Alberta, Canada, 165 pp.

Intergovernmental panel on Climate Change (IPCC) (2005) *Special Report on Carbon Dioxide Capture and Storage. Prepared by the Working Group III of the Intergovernmental Panel on Climate Change* [Metz, B, O Davidson, HC de Coninck, M Loos and LA Meyer (eds.)]. Cambridge University Press, Cambridge, United Kingdom and New York, NY, USA, 442 pp.

Intergovernmental panel on Climate Change (IPCC), 2007, *Climate Change 2007: The Physical Science Basis. Contribution of Working Group 1 to the Fourth Assessment Report of the Intergovernmental Panel on Climate Change* [S Solomon, D Qin, M Manning, Z Chen, M Marquis, KB Averyt, M Tignor and HL Miller (eds.)]. Cambridge University Press, Cambridge, United Kingdom and New York, NY, USA, 996 pp.

Kharaka, YK, Thordsen, JJ, Hovorka, SD, Nance, HS, Cole, DR, Phelps, TJ, Knauss and KG. (2009). Potential environmental issues of CO_2 storage in deep saline aquifers: Geochemical results from the Frio-I Brine Pilot test, Texas, USA. *Applied Geochemistry* **24**(6): 1106–12.

McPherson, BJ and Sundquist, ET. (eds.) (2009) *Carbon Sequestration and Its Role in the Global Carbon*

Cycle. American Geophysical Union, Geophysical Monograph 183, Washington, D.C., 359 pp.

Pascale, P, Ménez, B and Noiriel, C. (2009) CO_2 geological storage: Integrating geochemical, hydrodynamical, mechanical and biological processes from the pore to the reservoir scale. *Chem. Geol.* **265**: 1–236.

9 Microbial Geochemistry: At the Intersection of Disciplines

PHILIP BENNETT AND CHRISTOPHER OMELON

The University of Texas at Austin, Department of Geological Sciences, Austin, TX, USA

ABSTRACT

Subsurface environments support active microbial communities that interact with the geological framework, carrying out redox reactions, assimilating nutrients, excreting waste and extracellular substances. These reactions perturb the extracellular environment, altering mineral-water equilibria, reaction rates and pathways, and are often the chemical basis for low-temperature mineral diagenesis. These complex biological-mineral interactions represent a coupled system where the microbe takes advantage of specific minerals, and the mineral-water reactions depend on a particular microbial population and the activities of the microbial community.

We review here the chemistry of microbial interactions with minerals and subsurface environments from the perspective of the microbial metabolism and subsurface microbial ecology. Using example microbe-mineral interactions we examine the driving forces for microbial geochemical interactions and outcomes, with a central theme of asking *why* a microbe would expend energy to perturb mineral-water equilibria and to drive geochemical reactions. In most cases we find that microorganisms drive geochemical reactions, but that they often gain something from the interaction; access to a nutrient, buffering of extreme habitat conditions, or perhaps protection from predation. Answering the question of why often leads to a richer understanding of how, how fast and how important the microbial geochemical interaction is.

Frontiers in Geochemistry: Contribution of Geochemistry to the Study of the Earth, First edition. Edited by Russell S. Harmon and Andrew Parker.
© 2011 Blackwell Publishing Ltd. Published 2011 by Blackwell Publishing Ltd.

INTRODUCTION

In virtually every subsurface environment investigated using appropriate techniques, to temperatures exceeding 120°C, we find evidence of active microbial communities, and interactions between microorganisms and the surrounding geological framework (Fredrickson and Onstott 1996; Ghiorse and Wilson 1988; Pedersen 1993; Phelps et al. 1989). Active microbial populations oxidize reduced chemical substrates for energy, take up macro and micronutrients, excrete reactive waste products, and exude extracellular proteins and polysaccharides (Chapelle 2000). These are necessary functions of life, but they also perturb the environment external to the cell, altering mineral-water equilibria, changing mineral surface conditions, and influencing reaction rates and pathways (Ehrlich 1996b). In many subsurface systems biology offers the principal source of reactive solutes that drive geochemical reactions and diagenesis at low temperature (Nealson 1997).

'Microbial geochemistry' falls generally within the field of 'Geobiology', or the study of biology in geological systems. Microbial geochemistry could encompass a broad array of processes and interactions, as essentially every microbial community outside of the laboratory interacts in some fashion with the geologic framework components (rock, air, water). This chapter, however, focuses on the study of the coupled microbe-mineral system from the perspective of geochemical reactions and outcomes that require consideration of geology, microbiology and chemistry. In this coupled system, microbes require minerals, as a source of nutrients or electron acceptors or attachment, while mineral reactions effectively require microbes to perturb the equilibrium or accelerate rate, and the link between microbe and the mineral is a set of chemical reactions and pathways (Ehrlich 1996a; Ehrlich 1996b).

Microbial geochemistry is often associated with the field of *geomicrobiology*. This broader field of research includes, in addition to the geochemical reactions between microbe and rock, the examination of the distribution and membership of microbial communities (microbial ecology), the genetic information contained in these communities (genomics), and the genetic specifics of the biochemical pathways responsible for geochemical reactions (proteomics). These biologic tools are useful for the microbial geochemist as well, but the primary goal is not to describe the community of organisms, or to elucidate the biomolecules being expressed, but rather what the community and the biomolecules do to the geologic framework, and what benefit the community gains from the minerals.

This chapter reviews the scope of inquiry of microbial geochemistry, some of the tools and approaches currently employed, and a selection of recent findings that demonstrate the importance of understanding the linked system of microbe and mineral.

MICROBIAL METABOLISM IN GEOLOGIC SYSTEMS

Microorganisms can influence geochemical systems in a variety of ways, ranging from isolating a mineral surface under a biofilm to nucleating mineral precipitates on cell surfaces or excreted biomolecules. But the primary basis for a microbial component in geochemistry is the metabolic processes carried out by viable microorganisms. From a narrow perspective, a microbe is a very small semi-permeable package carrying out a wide range of chemical reactions that are a consequence of life. Understanding the role of microbes in geology therefore often starts not with a phylogenetic description (a genotypic approach) but rather a description of what the organism does for a living (a phenotypic approach). From a simple mass-balance perspective (i.e. the greatest number of molecules participating in a reaction) the primary respiration pathways will have the greatest impact on the surrounding aquatic and mineral system.

Microbial respiration involves taking up chemically reduced (electron-rich) substrates[1] such as organic carbon compounds, molecular hydrogen or hydrogen sulphide, and oxidizing them (transfer electron(s) to an electron acceptor), conserving the energy released from the reaction in the form of ATP, the universal biochemical currency (Madigan et al. 2003; White 2007). This currency can be stored and used later to carry out metabolic functions such as protein synthesis or motility. The oxidation of substrates and the release of energy is a basic component of life, and is one of the distinctive attributes that distinguish living things from non-living but still-biological entities such as viruses or prions. This basic metabolic function, however, is also responsible for a substantial chemical perturbation of the surrounding environment as some solutes are consumed and others are produced as waste products.

Molecular oxygen as a terminal electron acceptor

In general, the microorganisms that can gain the greatest energy yield (the most ATP) from a particular substrate will outcompete those that utilize electron acceptors with lower redox potentials and less free energy (Chapelle 1993; Ehrlich and Newman 2008). The highest-energy yield for a generally available electron acceptor is derived from molecular oxygen (O_2 1.23 V). These reactions yield sufficient energy for the formation of several ATP molecules from ADP and orthophosphate, and a list of a few representative reactions and their standard state energy yields are found in Table 9.1. Where oxygen is present above a minimum threshold of about 1 mg/L, organisms that utilize O_2 as the terminal electron acceptor (TEA) will generally be more competitive than those using lower-potential TEAs (e.g. Atlas and Bartha 1986). This is a simplified view of competition between metabolic guilds, and other resources such as molecular hydrogen will also influence the succession of TEA processes (TEAPs) (Chapelle et al. 1996; Lovley and Goodwin 1988; Spear et al. 2005).

While the competition between guilds can be viewed in isolation as one involving the availability of TEAPs or other nutrients, in subsurface systems it also governs the uptake and excretion of reactive inorganic solutes that are central to defining the geochemistry of this environment. As organic carbon compounds are consumed, O_2 is depleted, limiting its ability to oxidize reduced carbon or other elements. The products of carbon oxidation are carbonic acid acidity and bicarbonate, with the exact proton balance dependent on the oxidation state of the carbon compounds (Table 9.1, Rxn 6 and 7). Ranging from fully reduced methane (C^{+4}), generating 0.5 moles of CO_2 per mole of O_2, to acetic acid (net oxidation number is C^0, representing aquifer organic matter) that generates 1 mole of CO_2 per mole of O_2, different substrates have significantly different geochemical outcomes in terms of reaction with solutes and minerals. The same organism, given different substrates, will have different effects, while the same substrate, given to different organisms or consortia, will result in different byproducts.

Organic carbon compounds oxidized against O_2 (Table 9.1) as well as other TEAs are a principal source for elevated partial pressures of carbon dioxide (PCO_2) in the subsurface and drive acid-base reactions. This includes dissolution and diagenesis of carbonate minerals and the accelerated hydrolysis of silicate minerals. For non-carbon substrates (e.g. H_2S, NH_3) there is typically substantial acidity generated by microbial oxidation (Table 9.1, Rxn 1–4), but in the absence of carbonate minerals, no CO_2. Both CO_2 and H^+ are obvious candidates for reactive components that would react with minerals, and the production of these reactants will perturb the local pore-scale equilibria, with the greatest perturbation closest to the source of reactants, the cell.

Alternative electron acceptors

It is tempting to think that oxidation is oxidation, and every time a reduced carbon substrate is oxidized via the Krebs cycle that reactive acidity or dissolved CO_2 are produced – the 'microbes dissolve rocks' generalization. This, however, is very much not the case, and the geochemical result of microbial metabolism is very closely tied to the metabolic pathway expressed by the dominant metabolic guild, and the eventual electron acceptor. When there is insufficient molecular oxygen to support efficient oxidation, then other 'alternative terminal electron acceptors' become more energetic, and organisms that can utilize them are more competitive. Examples are NO_3^- (where N^{+5} is the electron acceptor) or SO_4^{2-} (S^{+6} is the electron accepting species) and Fe^{+3} (Table 9.1, Rxn 9–11). While these reactions result in the oxidation of reduced carbon to a carbonate species, they do not generate substantial acidity, rather producing bicarbonate as the primary byproduct. This drives the system, via the common-ion effect, toward carbonate supersaturation, and potentially the precipitation of carbonate minerals, rather than dissolution. Figure 9.1 shows an example of dolomite precipitating on the cell wall

Table 9.1 Representative microbial respiration reactions and their reactive byproducts. DG0 is the standard state free energy of the reaction in kJ mol^{-1}. From Amend and Shock (2001).

#	Rxn	ΔG^0 25°C[*]	ΔG^0 100°C[*]	Reactive byproducts[**]
	Molecular oxygen	kJ mol^{-1}	kJ mol^{-1}	
1	$H_{2(aq)} + 0.5O_{2(aq)} \rightarrow H_2O_{(l)}$	-263.17	-258.44	None
2	$H_2S_{(aq)} + 2O'_{2(aq)} \rightarrow SO_4^= + 2H^+$	-749.62	-718.29	$+SO_4^=, +H^+$
3	$S_{(s)} + 1.5O_{2(aq)} + H_2O'_{(l)} \rightarrow SO_4^= + 2H^+$	-532.09	-508.10	$+SO_4^=, +H^+$
4	$2Fe^{2+} + 0.5O_{2(aq)} + 2H^+ \rightarrow 2Fe^{3+} + H_2O_{(l)}$	-96.92	-71.36	$+H^+$ with $Fe(OH)_{3(s)}$
5	$UO_{2(s)} + 0.5O_{2(aq)} + 2H^+ \rightarrow UO_2^{++} + H_2O_{(l)}$	-166.25	-153.70	$-H^+; +UO_2^{++}$
6	$CH_{4(aq)} + 2O_{2(aq)} \rightarrow CO_{2(aq)} + 2H_2O_{(l)}$	-858.97	-852.77	$+CO_2$
7	$CH_3COOH + 2O_{2(aq)} \rightarrow 2CO_{2(aq)} + 2H_2O_{(l)}$	-882.92	-881.20	$+CO_2$
8	$NH_{3(aq)} + 1.5O_{2(aq)} \rightarrow H^+ + NO_2^- + H_2O_{(l)}$	-267.50	-258.19	H^+
	Alternative electron acceptors			
9	$CH_3COOH + 8/5\ NO_3 + 8/5H^+ \rightarrow$ $4/5N_{2(aq)} + 2CO_{2(aq)} + 14/5\ H_2O_{(l)}$	-847.6	-860.814	$+CO2, -H^+$
10	$CH_3COOH + 2H^+ + SO_4^= \rightarrow 2CO_{2(aq)} + H_2S_{(aq)} + 2H_2O_{(l)}$	-133.30	-162.91	$-H^+, -SO_4^=; +H_2S, +CO_2$
11	$CH_3COOH + 8Fe^{3+} + 2H_2O(l) \rightarrow$ $8Fe^{2+} + 2CO_{2(aq)} + 8H+$	-495.25	-595.76	$-H^+$ (with $Fe(OH)_{3\ (s)}$) $+CO_2, +Fe^{2+}$
12	$CH_3COOH \rightarrow CO_{2(aq)} + CH_{4(aq)}$	-23.95	-28.43	$+CO_2,$
13	$CO_{2(aq)} + 4H_{2(aq)} \rightarrow CH_{4(aq)} + 2H_2O_{(l)}$	-193.73	-180.98	$-CO_2,$
	Autotrophy			
14	$CO_{2(aq)} + H_2O_{(l)} + hv \rightarrow CH_2O + O_{2(aq)}$	$+478.3$	$-\#-$	$-CO_2, +O_2$
15	$2H_2S + CO_2 + hv \rightarrow CH_2O + H_2O + S^0$		$-\#-$	$-CO_2$

* Note that the free energy values shown in Table 9.1 are at standard state, and do not reflect the actual energy conserved by a microorganism in nature – that is a function of the actual activities of the reactants.
** + is gain of a component, – is loss of a component. In all of the oxygen reactions there is a loss of molecular oxygen. #Photosynthesis does not occur at 100°C.

owing to local disequilibrium due to methanogenesis consuming CO_2 (Table 9.1, Rxn 13) (Roberts et al., 2004).

There is a wide variety of potential electron acceptors in nature – at the simplest level any compound with a higher oxidation potential than the electron donor substrate molecule (e.g. organic carbon) could potentially couple in a spontaneous redox reaction and act as the oxidizing species, the TEA. However, not every pairing of substrate

and TEA results in sufficient energy to be conserved by the organism (see below), not every energetic reaction occurs at a rate that is advantageous to an organism, and not every labile energetic pairing is actually utilized by a population for energy conservation (e.g. Gihring et al., 2001). It is not enough to have thermodynamics on your side, you also need a population that gains an advantage by carrying out that particular reaction – it must be able to utilize scarce resources or

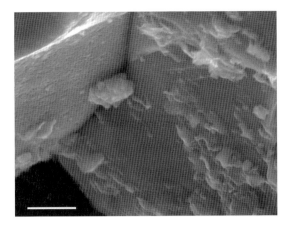

Fig. 9.1 Environmental SEM micrograph of a single cell attached to basalt (Columbia River) with nanocrystals of dolomite precipitating on the cell wall. Bar Scale = 2 µm (Roberts et al. 2004).

metabolize carbon in a TEA-limited environment better than another competing population.

Rock respiration

Many microorganisms are capable of reducing and mobilizing iron from mineral solid phases in order to assimilate the iron, to discharge excess reducing power, or to access adsorbed nutrients on the iron oxides, e.g. phosphate (Roden 1997). Rock respiration, however, refers to the utilization of oxidized metals in minerals and rocks, such as oxidized iron or manganese minerals, as terminal electron acceptors (Lovley et al. 1993; Lovley and Phillips 1988) (Table 9.1, Rxn 11). For many subsurface environments oxidized iron minerals such as goethite or ferrihydrite are highly abundant electron acceptors that offer substantial energy yield for organisms that are able to utilize it for energy conservation.

The geochemical consequences of rock respiration are initially the reductive dissolution of the iron oxide mineral and the mobilization of the metal as the soluble reduced species:

$$CH_3COOH + 8Fe(OH)_3 \leftrightarrow$$
$$2CO_2 + 6H_2O + 16OH^- + 8Fe^{2+} \quad (1)$$

The resultant ferrous iron sorbs to the oxides to produce magnetite or other mixed-valence iron phases. This metabolic pathway, however, also consumes substantial acidity, increasing alkalinity and shifting equilibria of pH-sensitive rock-water reactions. The increase in alkalinity can have a profound effect on carbonate mineral reactions, and the consequence of rock respiration in natural systems in many cases is carbonate-mineral precipitation.

Photosynthesis

A separate group of microorganisms, defined as phototrophs, harvest energy from sunlight in the process of photosynthesis, turning light energy into chemical energy. They are further divided into photoautotrophs, which assimilate carbon from CO_2 (Table 9.1, Rxn 14), and photoheterotrophs, acquiring carbon from organic sources. Phototrophic microorganisms include cyanobacteria and algae that use H_2O as an electron donor to produce oxygen (oxygenic photosynthesis, Rxn 14), in contrast to other bacteria such as purple sulphur bacteria that oxidize reduced forms of sulphur as electron donors but do not produce oxygen (anoxygenic photosynthesis).

Oxygenic photosynthesis is very likely the source of the rise of atmospheric O_2 in the early Precambrian (Bekker et al. 2004; Canfield 2005), with both modern and fossil evidence suggesting that organisms resembling modern cyanobacteria were present at least as far back as 3 Ga or more (Schopf 2006). The geochemical consequences of the Great Oxidation Event (~2.3 Ga) were profound as the world's oceans and atmosphere went from a slightly reducing condition to a strongly oxidizing one, precipitating some minerals while dissolving others, and completely changing the habitat for future life (Kasting and Siefert 2002). Cyanobacteria today are an important component of the larger photosynthetic community that utilizes CO_2, an important greenhouse gas.

Despite the fact that phototrophs receive energy from sunlight instead of redox-reactive elements such as Fe^{2+}, they are geochemically reactive with their surrounding environment. All photosynthetic pathways consume dissolved carbon dioxide, shifting equilibria toward super-saturation with respect to carbonate minerals. In addition, many species of cyanobacteria have evolved a carbon-concentrating mechanism (Badger and Price 2003; Kaplan and Reinhold 1999; Price et al. 2002) allowing them to take up HCO_3^- in addition to CO_2 in order to sustain photosynthesis efficiency. Uptake of HCO_3^-, however, leads to production of OH^- from the cell, resulting in the generation of high pH conditions in the surrounding environment:

$$HCO_3^- + H_2O + h\nu \rightarrow CH_2O + O_2 + OH^- \quad (2)$$

This process, known as bioalkalization (Büdel et al. 2004), may drive pH above 10 or even 11, producing a chemically aggressive environment that would increase the solubility and dissolution kinetics of both quartz (Bennett 1991) and aluminosilicate minerals (Helgeson et al. 1984; Holdren and Speyer 1986). The greatest impact is seen in terrestrial endolithic habitats where microorganisms inhabit the near subsurface environment of sedimentary rocks, most commonly sandstones (Friedmann and Ocampo 1976). High pH conditions generated by cyanobacteria in these environments leads to enhanced chemical weathering of the lithic habitat, and the ubiquity of cyanobacteria in these environments suggests that bioalkalization could play a fundamental role in the chemical weathering of silica minerals (Schwartzman and Volk 1989). Furthermore, alkalization has implications for porosity development, release of essential nutrients and mobilization of metals under alkaline conditions.

Bioenergetics

The ability of a microbial system to utilize a potential source of energy will be constrained by the free-energy yield available for metabolic processes (Schink 1997; Schink and Friedrich 1994), and the rate of the biocatalysed reaction (Jin 2007). Microbial bioenergetics investigates the available energy for a specific chemical reaction under the specific conditions present in a particular field or laboratory system, accounting for partial pressures, solute activity and temperature (Amend and Shock 2001).

Not every reaction that is technically spontaneous will be useful to microorganisms; there is a minimum energy yield that must be exceeded to allow the cell to conserve the released energy from biochemical redox reactions in a useable form, often referred to as the biological energy quantum, or BEQ (Schink and Friedrich 1994). This is an active area of research, and long-held dogma is being superseded by new theories as we gain further understanding of extreme environments at the thermodynamic edge of existence. The basic unit of biochemical currency that is at the centre of this is the energy required to phosphorylate one ADP molecule to form ATP, the energetic molecule used by all cells to drive metabolic processes. Depending on temperature (Amend and Shock 2001), and the ratio of ADP to orthophosphate, the required energy (including loss) for this reaction G_p is 50–70 kJ/mol P (Schink and Friedrich 1994; Thauer et al. 1977). For respiratory processes, phosphorylation of ADP occurs via proton translocation through the ATPase enzyme, with a number of H^+ molecules n required to pass through the ATPase to processes one ADP molecule to ATP (Madigan et al. 2003; Mahoney 1983; White 2007). The minimum energy therefore is the net energy required to translocate one proton to the outer membrane to support the chemiosmotic potential, and that energy is G_p/n. This minimum energy (the BEQ, Hoehler 2004) is the threshold energy yield below which, while the reaction might be spontaneous chemically, it cannot be conserved by the organism in the form of ATP.

While n was once thought to be a constant (~3), yielding an average BEQ of approximately 20 kJ mol^{-1}, there is mounting evidence that there could be many types of ATPase enzymes with different efficiencies, and that possibly a cell will utilize different pathways or modify internal biochemical conditions to adjust for changing energy availability (Jackson and McInerney 2002). Laboratory studies suggest that different carbon compounds oxidized against the same TEAP have different threshold energy values (Schink and Friedrich 1994). The value of n is now thought to vary from 0.2 to 0.5 (2–5 protons per ADP phosphorylated to ATP), and the BEQ could be as low as $10\,kJ\,mol^{-1}$ for methanogenic guilds, or as high as $30\,kJ\,mol^{-1}$ or more for nitrate-reducing bacteria (Hoehler 2004; Hoehler et al. 2001; Schink and Friedrich 1994). It is also increasingly clear that the energy requirement for growth is substantially greater than for maintenance, and that starved cells at the limit of survival may maintain viability at a very low-energy flux (reaction rate) with very high efficiency.

The BEQ, whatever the value is for the specific conditions or taxon, will govern to a certain extent whether a developing biogeochemical gradient can be utilized by a microbial population. The availability of a dissolved reduced solute, for example, may not support the growth of a population if the energy yield is less than the BEQ. Even reactions that support metabolism or growth under one set of conditions may not at higher temperature, or changing activities or proportions of reactants and products (Amend and Shock 2001; Spear et al. 2005).

Furthermore, not all microbial redox reactions of geochemical importance are the result of energy-conserving catabolic pathways. Some microbial geochemical reactions result in solute redox transformation that have large free-energy release but without apparently conserving energy. An example of this is the oxidation of ferrous iron by filamentous organisms in the *Betaproteobacteria*. While these organisms are ubiquitous in wetlands at redox interfaces, there is no evidence that *Leptothrix discophora* conserves energy from the reaction. Similar members of the *Betaproteobacteria* (*Gallionella ferruginea*) however, also oxidize iron while conserving energy from the reaction. Another important example is the redox transformation of arsenic. While a few organisms are capable of conserving energy from the reduction of arsenate as an electron acceptor (Ahmann et al. 1997), a great many others reduce arsenate as a detoxification pathway (see below). Similarly, there are both energy-conserving and non-conserving pathways for arsenite oxidation (Gihring and Banfield 2001; Gihring et al. 2001), even when it is clear that the free energy of reaction exceeds the biological energy quantum.

Example: Arsenic speciation

The redox transformation of arsenic is an example of a toxic result of life; for the organism the transformation of arsenic from one redox state to another is the chemical consequence of transporting electrons for the purpose of conserving energy or other, sometimes unknown purposes. For the other inhabitants, however, that activity that we define as life inevitably influences the mobility of key dissolved species, sometimes resulting in the mobilization of toxic elements. The example uppermost in the minds of millions in the Bengal basin is the mobilization of arsenic as the reduced arsenite species due to microbial activity:

$$2HAsO_4^{2-} + 4H^+\ CH_2O \rightarrow$$
$$2H_3AsO_3 + CO_2 + H_2O \tag{3}$$

This reaction is very slow in abiotic systems (Nordstrom and Archer 2003; Nordstrom et al. 2004), and arsenate would be expected to be stable for long periods of time, typically sorbed to the positively charged iron oxides ubiquitously found on mineral surfaces. Some microorganisms can utilize arsenate as a terminal electron acceptor coupled to the oxidation of carbon, or alternatively, they can reduce arsenate to arsenite as a detoxification mechanism to allow excretion of arsenic from the cell (Silver and Phung 2005). The reduced arsenite is relatively poorly associated with iron oxides, and in any case, the iron oxides are typically reduced as well by iron-respiring bacteria, resulting in the release of the more toxic form of arsenic to the environment (Smedley and Kinniburgh 2002).

GEOCHEMICAL REACTIONS FROM MICROBIAL ACTIVITY

Biofilms and the microbe-mineral microenvironment

The scale at which microbial geochemical reactions occur is a critical aspect in evaluating the significance of microorganisms in the broader context of geological processes. Microbes are of course tiny, and on the one hand the reaction zone of a single microorganism is equally tiny, and it is hard to imagine that a significantly reactive environment can be maintained to result in accelerated rock-water processes.

There are, however, several factors that work in favour of the organism that enhances their reactivity. One factor is sheer numbers; each microbe is tiny, but when there are 10^9 cells per millilitre of water there is the potential for substantial reaction. This results in a macroscopic alteration in the composition of the pore fluids, and the perturbation of geochemical reactions in a large volume of the subsurface compartment. This type of perturbation is often measurable by standard geochemical analytical procedures on a water or core sample, and the consequences of the reaction predicted by geochemical modelling tools (Bennett et al. 2000).

Probably more frequently, the abundance of microbes in subsurface environments is much lower, perhaps 10^2–10^5 cells per gram of sediment in a typical oligotrophic environment (Phelps et al. 1994; Phelps et al. 1989), and the summed ability to perturb geochemical processes is much less. Another factor, however, is the preference on the part of many microbes to attach to mineral surfaces as part of their lifestyle. This attachment can be in response to a variety of stimuli, including protection from predation, access to scarce nutrients, concentration of dilute organic compounds, or to buffer pH extremes. The net result, however, is that where the microbe is attached to the mineral surface, excretion of byproducts is concentrated at the cell-mineral interface, with potentially very high concentrations of reactants focused at the point of reaction. This results in a

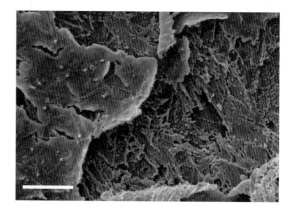

Fig. 9.2 SEM micrograph of Iceland Spar calcite with attached biofilm of filamentous sulfur oxidizing bacteria. Exposed mineral surface shows evidence of aggressive acid etching of the calcite under the biofilm. Bar scale = 20 µm.

zone that is greatly perturbed relative to the bulk solution and both the distance from equilibrium and the rate of reaction are much greater than the bulk-solution case (Bennett et al. 2000; Hiebert and Bennett 1992; Rogers and Bennett 2004; Rogers et al. 1998). This is illustrated in Fig. 9.2 and Fig. 9.3, where attached microbes have deeply etched calcite at the point of contact, but not distal to the filament, with the result that under a continuous biofilm there is extreme weathering of limestone.

The effect of a microenvironment is enhanced under conditions of a biofilm, where larger populations of microbes aggregate on a mineral substratum to form a continuous layer of biomass. Under this hydrophobic and diffusion-limiting layer, high concentrations of reactive byproducts build up in contact with the mineral, resulting in even greater perturbation and mineral reactivity. While this behaviour is not a universal trait of subsurface microbial communities (in fact we find biofilms to be relatively uncommon in deep subsurface environments), in highly productive near-surface environments at redox interfaces (boundaries between reduced anaerobic water and oxygenated surface water) where there is substan-

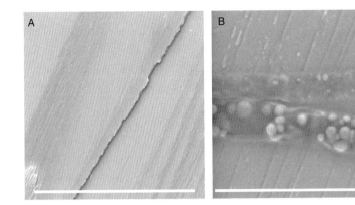
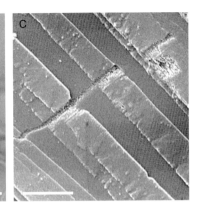

Fig. 9.3 Environmental SEM micrographs of calcite surfaces. A: Sterile calcite surface without attached biomass, Bar Scale = 20 μ; B: Calcite surface with attached microbial filaments, Bar Scale = 5 m; C: Calcite colonized by sulphur oxidizing bacterial filaments after biofilm has been cleaned off. Deep etch track follows trace of previously attached filament. Bar scale = 20 μm.

tial chemical energy available for growth, biofilms may be common.

Biofilm formation could have other influences in mineral-water reactions, depending on the microbial consortium and the mineral surface. With communities that involve carbon oxidation rather than sulphur, the sub-biofilm microenvironment is much less aggressive and the effect on mineral dissolution is much less. The extracellular polymeric substances associated with these biofilms may interfere with transport of reactants to the mineral surface (Wolfaardt et al. 1999), and potentially inhibit mineral dissolution reactions (Luttge and Conrad 2004).

Mineral dissolution

There is a rich literature on the enhanced dissolution of minerals by microorganisms, and a variety of genera have been implicated one way or the other in mineral corrosion. While initially this seems an obvious conclusion based on the chemical outcome of microbial metabolism, the evidence is actually somewhat conflicting, and different researchers reach different conclusions about the influence of even the same organism. We have found that the evidence for microbial enhancement of mineral corrosion is based on a wide spectrum of evidence, ranging from detailed laboratory experiments with single species and pure minerals, to observational evidence from field samples of biological material near weathered mineral surfaces. In interpreting this evidence we return to the basic question of *why*. Why would an organism dissolve a mineral, what is the advantage, and what is the consequence? If there is no such thing as a free lunch, and microbes don't do things randomly, then even conflicting evidence can often be understood when examined in an ecological context.

Carbonates

From a large-scale perspective, most carbonate weathering in the shallow subsurface is due to metabolic carbon dioxide. The classic equation:

$$CaCO_3 + H_2O + CO_2 \rightarrow Ca^{2+} + 2HCO_3^- \quad (4)$$

is hardly significant when the only source of CO_2 is the atmosphere, but becomes an overwhelmingly important reaction when CO_2 is constantly produced to very high partial pressure by life. In the very shallow critical zone a major source of CO_2 is via root respiration (Appelo and Postma 1993; Drever 1997), and the microbial populations associated with the root systems of photosynthetic plants. Deeper in the critical zone productive microbial populations oxidize soil organic carbon and produce substantial pressures of CO_2 that drives not just carbonate dissolution but also silicate weathering as well. PCO_2 values exceeding 0.1 bar (400x atmospheric partial pressure) are possible in groundwater and even soils with poor connection to the atmosphere, and unbuffered pure water in contact with this pore-scale atmosphere would be quite aggressive with an equilibrium pH of 4.5. This macroscale effect, the perturbation of the equilibrium PCO_2 resulting in carbonate dissolution, could occur over large volumes of an aquifer, driven by the aerobic oxidation of both natural and anthropogenic organic carbon compounds. This mechanism is widely applied in a variety of geoscience disciplines to explain weathering processes, and it contributes to the formation of karst features in carbonate rocks worldwide, in parallel with other abiotic mechanisms (e.g. mixing waters, change in temperature). In many caves, microbial communities are found at great depth and are associated with the remnant corrosion residues (Boston et al. 2001; Northup and Lavoie 2001).

Other metabolic pathways are more aggressive toward carbonate weathering and can result in the large-scale removal of carbonate minerals. Chemotrophic organisms that use inorganic chemicals for energy rather than organic carbon compounds often produce abundant acidity as a byproduct. These pathways (Table 9.1, Rxn 2–4, 8) include the oxidation of NH_3, Fe^{2+} and H_2S, and the associated acidity will react with, and typically dissolve, the labile carbonate minerals first. In a well-documented example the acidic runoff from sulphide ore waste generated by microbial pyrite oxidation drains into Pinal Creek, AZ resulting in the removal of carbonate minerals from the stream sediment (Brown and Glynn 2003).

Sulphuric acid from the biological oxidation of hydrogen sulphide has been linked to the formation of some of the largest cave systems so far explored, such as Carlsbad Caverns in New Mexico, USA (Egemeier 1981; Hill 1995; Hill

1990; Hose and Pisarowicz 1999; Thamdrup et al. 1993). This reaction is relatively straightforward in aqueous systems:

$$0.5H_2S + O_2 + CaCO_3 \rightarrow$$
$$Ca^{2+} + HCO_3^- + 0.5SO_4^{2-} \quad (5)$$

and has been documented in both ancient systems and a handful of modern examples.

At the microscale however, the evidence for microbial acceleration of carbonate dissolution, and the precise mechanisms potentially implicated, is more confusing. Some researchers have found that attached biofilms inhibit mineral surface reactions due possibly to the shielding or poisoning of the reactive mineral surface by EPS production (Luttge and Conrad 2004; Luttge et al. 2005; Welch and Vandevivere 1995). Others document direct evidence of corrosion associated with individual microbial filaments attached to the mineral surface (Engel et al. 2004a). The caution here is that any attempt to generalize about 'microbes' will almost certainly fail, as even the same organism grown under different nutrient or habitat conditions (Steinhauer et al. 2010; Vandevivere and Kirchman 1993; Welch and Vandevivere 1995) will alter its surrounding differently as it seeks a competitive advantage.

We have documented in field and laboratory observations and experiments that filamentous bacteria from several clades in the *Gammaproteobacteria* and *Epsilonproteobacteria* rapidly oxidize reduced sulphur compounds to produce sulphuric acid, and that when these organisms form biofilms on native carbonate mineral surfaces, the region under the biofilm develops an aggressive microenvironment that rapidly corrodes calcite (Fig. 9.2). In field experiments where clean chips of calcite were deployed in a cave system with mats of sulphur oxidizers the calcite under attached microbial filaments was deeply corroded while chips maintained in sterile solution were unaltered (Fig. 9.3). The rate of calcite dissolution in laboratory experiments with pure cultures of sulphur oxidizers as well as mixed environmental cultures rapidly increased to a maximum of seven times the abiotic rate, and the

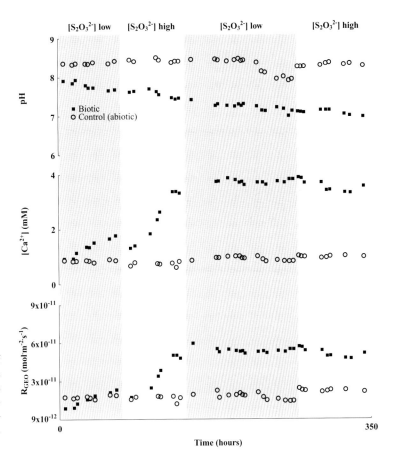

Fig. 9.4 Experimental results of calcite dissolution rate in abiotic and biotic flow-through reactors showing change in pH, calcium concentration, and zero-order dissolution rate expressed in moles $cm^{-2} s^{-1}$. Data from (Steinhauer et al. 2010).

magnitude of rate increase was independent of the thiosulphate concentration (Fig. 9.4).

In both the field and laboratory experiments, it was apparent that the environment under the biofilm is substantially out of equilibrium with the surrounding bulk solution, and both calcite dissolution and gypsum precipitation occur, even though the surrounding water chemistry would predict opposite behaviour (Engel et al. 2004b; Steinhauer et al. 2010). This is an important observation that can be extrapolated to biofilms in general, that the microenvironment can only be interrogated with difficulty, and that in a microbial system the composition of a water sampled from a well, for example, will not necessarily reflect the environment where reactions are occurring that change that water composition.

We have also found, however, that even under the most chemically aggressive chemotrophic pathways microbes are capable of turning on and off the metabolic pathways that result in carbonate dissolution (Steinhauer et al. 2010) on timescales of minutes to hours if it is to their advantage. When a mixed culture of filamentous sulphide-oxidizing bacteria dominated by *Thiothrix unzii* are grown on thiosulphate $(S_2O_3^{2-})$ as the sole energy substrate the culture produces substantial acidity that aggressively corrodes calcite (Fig. 9.2 and Fig. 9.4). The generalized reactions are:

$$S_2O_3^{2-} + 2O_2 + H_2O \leftrightarrow 2SO_4^{2-} + 2H^+$$
$$(\Delta G^\circ = -738.7 \text{ kJ} \cdot \text{mol}^{-1}) \tag{6}$$

However, if in the same culture experiment the medium is changed to one containing hydrogen sulphide as the sole electron source instead of thiosulphate, the production of acidity, and the concomitant dissolution of calcite, quickly (in a few hours) shuts down. The microbial community switches to a different metabolic pathway that produces elemental sulphur instead of sulphuric acid:

$$H_2S + 0.5O_2 \leftrightarrow S^0 + H_2O$$
$$(\Delta G° = -209.3 \text{ kJ} \cdot \text{mol}^{-1}) \qquad (7)$$

This reaction still yields substantial free energy for the organism, but does not produce acidity. For *T. Unzii*, this could be an advantage, since this organism is a neutrophilic sulphur oxidizer and requires circum-neutral pH conditions to survive, yet it produces acidity as a byproduct of some primary metabolic pathways, such as eqn. 6. This trait is shared by a wide variety of taxa such as several genera of *Thiobacillus* sp., as well as a group of *Epsilonproteobacteria* that we have isolated from a sulphidic cave environment (Engel et al. 2003; Engel et al. 2004a; Engel et al. 2004b). *T. Unzii* is capable of the partial oxidation pathway (Eqn. 7) with H_2S, but not with $S_2O_3^{2-}$, and for this organism the accelerated dissolution of calcite is entirely dependent on the substrate metabolized. For the *Epsilonproteobacteria*, our results to date suggest that it does not carry out this reaction to any significant extent, and produces acid from H_2S (Table 9.1, Rxn 2). In a mixed culture, therefore, the same substrate (H_2S) will result in different amounts of calcite dissolution depending on which organism is dominant, because that dictates which metabolic pathway dominates.

Silicates

There is a rich literature on the influence of microbes on silicate weathering reactions, and both bacteria and lichen have been implicated in feldspar and quartz dissolution in a variety of field and laboratory investigations. There appear to be three primary mechanisms: (i) generation of acidic environments; (ii) generation of alkaline environments; and (iii) the production of extracellular complexing ligands.

In acidic environments silicates will rapidly dissolve owing to the high mobility and solubility of aluminum, and metabolic pathways that produce either proton or CO_2 as byproducts (Table 9.1) would be expected to enhance feldspar weathering (Helgeson et al. 1984; Holdren and Speyer 1986; Huang and Longo 1992; Valsami-Jones et al. 1998; Welch and Ullman 1996). An example reaction coupled to iron oxidation would be:

$$Fe^{2+} + \tfrac{1}{4}O_2 + 2.5H_2O \rightarrow Fe(OH)_3 + 2H^+ \qquad (8)$$

$$2H^+ + KAlSi_3O_8 + 6H_2O \rightarrow$$
$$K^+ + Al(OH)_2^+ + 3H_4SiO_4 \qquad (9)$$

Similarly, at very high pH both feldspar and quartz are soluble owing to hydroxy complexation of aluminium and dissociation of dissolved silicic acid to silicate anion:

$$3OH^- + KAlSi_3O_8 + 5H_2O \rightarrow$$
$$K^+ + Al(OH)_4^- + 3H_3SiO_4^- \qquad (10)$$

This reaction becomes important in environments where pH exceeds approximately 9 (the pK_{a1} of silicic acid is ~9.9). While alkaline environments can be found in a variety of settings, they are not generally the result of a microbial metabolic pathway. Perhaps the most important mechanism is bioalkalization, driven by photosynthesis, described above. Another mechanism is the microbial degradation of urea via the urease mechanism to produce ammonia nitrogen (Karavaiko et al. 1985; Kutuzova 1969):

$$CO(NH_2)_2 + H_2O \rightarrow 2NH_4^+ + HCO_3^- + OH^- \qquad (11)$$

This can result in elevated pH and alkalinity, but would influence silicate weathering only in systems lacking dissolved Ca^{2+} as the precipitation of calcite (e.g. Fujita et al. 2008) would buffer pH and prevent the large excursion needed to destroy silicates.

There are numerous sources of reactive organic acids in the subsurface for feldspar weathering

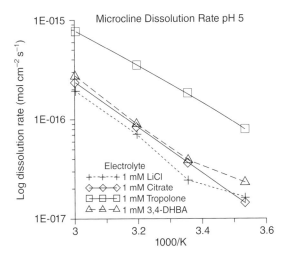

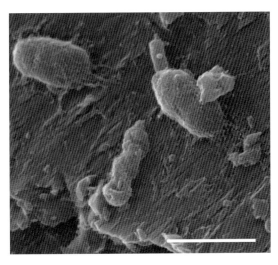

Fig. 9.5 Microcline dissolution rate at pH 5 in 1 mM LiCl and 1 mM select organic electrolyte solutions as a function of temperature, shown in an Arrhenius plot. Dissolution rate at lower temperature (right side of graph) is slightly higher in citrate and 3,4-dihydroxybenzoic acid, and significantly higher in tropolone solutions.

Fig. 9.6 Anorthoclase (Na/K feldspar) after 9 months exposure to groundwater in a petroleum-contaminated aquifer. Deep etching is visible only around attached microorganisms. Bar Scale = 3 μm.

(Horsfield et al. 2002; McMahon and Chapelle 1991), and there have been several studies on the effects of organic acids on feldspar dissolution kinetics (Franklin et al. 1994; Huang and Longo 1992; Petersen et al. 1992; Stillings et al. 1996; Valsami-Jones et al. 1998). Enhanced feldspar dissolution by organic ligands is most apparent at near neutral pH, where both the hydroxide and proton enhanced rate contribution is small (Vandevivere et al. 1994).

These reactions can result in greatly enhanced dissolution rates for silicates compared to the abiotic case (Fig. 9.5), but the influence of microbes has been largely ignored in the interpretation of early silicate diagenesis. In field experiments using well-characterized feldspar minerals as well as crystalline basalts we have found that silicate surfaces that are colonized are aggressively weathered, while surfaces of similar feldspars that are not colonized by the native microbial community are unaltered (Fig. 9.6) (Bennett and Casey 1994; Bennett et al. 2001; Hiebert and Bennett 1992;

Roberts et al. 2004; Rogers and Bennett 2004; Rogers et al. 1998). Even quartz is rapidly dissolved at near-neutral pH owing to the microbial degradation of petroleum (Bennett 1991; Bennett and Siegel 1987; Hiebert and Bennett 1992).

The effect of extracellular polymeric substances (aka EPS, glycocalyx) on feldspar and quartz weathering is controversial. Some research results suggest that EPS enhances the destruction of silicates via complexation mechanisms (see review in Ehrlich and Newman 2008). Recent studies however are more equivocal, and suggest that surface coverage by EPS may disrupt surface chemical reactions, separating aggressive microbial microenvironments from the mineral, and actually inhibiting dissolution (Ullman et al. 1996; Vandevivere and Kirchman 1993; Vandevivere et al. 1994; Welch and Vandevivere 1995).

Mineral precipitation

Carbonates

Many microbial metabolic pathways result in the precipitation of carbonates. Oxidation of organic carbon compounds with alternative TEAs results in the production of bicarbonate, increasing alkalinity and driving the equilibrium state with respect to calcite toward supersaturation (Table 9.1). Other reactions involving the degradation of proteins result in the production of ammonia with an increase in pH, also driving the geochemical system toward calcite precipitation (Warren et al. 2001).

Sulphate reduction coupled to organic carbon oxidation is carried out by a wide variety of microorganisms, many in the *Deltaproteobacteria* group. This reaction, using carbohydrate-type organic compounds, results in the production of bicarbonate without excess acidity, driving carbonate equilibria toward carbonate precipitation:

$$Ca^{2+} + 2CH_2O + SO_4^{2-} \rightarrow CaCO_3 + H_2CO_3 \qquad (12)$$

This is a well-studied mechanism for calcite precipitation (Baumgartner et al. 2006; Spiro 1977), and has been implicated in the precipitation of carbonates in stromatolites in both ancient and modern systems (Bosak and Newman 2003, 2005), and in both marine and lacustrine environments (Ferris et al. 1994; Morita 1980; Thompson and Ferris 1990). There is also evidence that both the sulphate-reducing metabolic pathway and the production of EPS by the SRB are important to carbonate precipitation (Braissant et al. 2007).

This general mechanism has recently been implicated with the microbial precipitation of dolomite (Vasconcelos and McKenzie 1997; Vasconcelos et al. 1995; Warthmann et al. 2000). In laboratory experiments the authors used water, initially supersaturated with respect to dolomite, but which is kinetically inhibited to form ordered dolomite (Land 1997). Sulphate-reducing bacteria were hypothesized to produce extreme supersaturated conditions around the cell itself, nucleating dolomite on the cell surface, with subsequent growth of micro-crystalline dolomite. For this mechanism both the microbial metabolism perturbing local equilibria, and the cell surface itself, are necessary to initiate mineral precipitation and growth.

It also important to consider that consumption of CO_2 by photoautotrophic or methanogenic microorganisms also results in an increasing SI (saturation index) with respect to carbonate minerals:

$$2HCO_3^- + Ca^{2+} + 4H_2 \rightarrow$$
$$CH_4 + CaCO_3 + 3H_2O \quad \text{(Methanogenesis)} \quad (13)$$

$$2HCO_3^- + Ca^{2+} + H_2O \rightarrow$$
$$CH_2O + CaCO_3 + 2O_2 \quad \text{(Photosynthesis)} \quad (14)$$

In both cases the result is an increase in SI with respect to carbonates, and the precipitation of calcite. This can occur on a microscale of individual microorganisms (Krumbein 1974; Obst et al. 2009; Thompson and Ferris 1990) up to the macroscale as whiting events in lakes (Thompson et al. 1997), or in the formation of freshwater microbialites (Brady et al. 2010). In particular the S-layer on some organisms have been specifically implicated in templating crystalline mineral nucleation and growth (Schultze-Lam and Beveridge 1994; Thompson and Ferris 1990).

An important observation is that the microenvironment around cyanobacteria is essentially diffusion-controlled (Beveridge 1988). This means that carbon-concentrating mechanisms and associated bioalkalization by cyanobacteria can create supersaturated conditions around the cell, leading to carbonate precipitation within the cyanobacterial sheath (Kah and Riding 2007; Merz 1992). Alternatively, reactive surface functional groups may be involved in carbonate precipitation owing to the adsorption of calcium, bicarbonate, or carbonate ions onto cell surfaces or associated EPS (Dittrich and Sibler 2005; Obst et al. 2009) in conjunction with carbon uptake (Martinez et al. 2008).

Silicates

Silicification of microorganisms has been observed primarily in terrestrial hot springs, which has been the focus of intensive study in areas such as Yellowstone Park, Chile, New Zealand and

Iceland (Ferris et al. 1986; McKenzie et al. 2001; Phoenix et al. 2006; Schultze-Lam et al. 1995). Discharging geothermal waters containing high concentrations of dissolved silica quickly equilibrate to ambient conditions at the Earth's surface, with cooling and evaporation leading to rapid oversaturation with respect to silica. Therefore, unlike precipitation of carbonates where bacteria can induce mineralization as described above, silica precipitation is predominantly an abiotic process whereby colloids passively attach to solid surfaces, including bacteria (Fig. 9.7). Laboratory work has shown that these microbes do not appear to catalyse silica precipitation (Benning et al. 2003; Fein et al. 2002). More recent work, however, reveals that bacteria can decrease the interfacial energy of silica nanoparticles to reduce the equilibrium solubility of silica, and enhance nucleation of silica nanoparticles on cell surfaces, leading to subsequent abiotic precipitation through hydrogen bonding, cation bridging, and electrostatic interactions (Benning et al. 2003; Ferris and Magalhaes 2008; Lalonde et al. 2005; Phoenix et al. 2002). This suggests that while non-biogenic processes dominate in the precipitation of silica in these environments, microorganisms can play a role in rapid preferential silicification (Fig. 9.8).

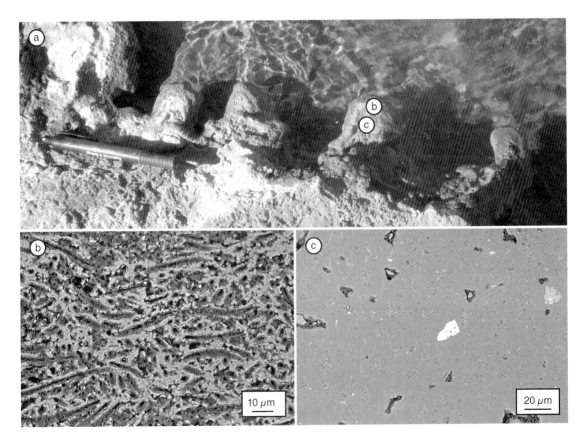

Fig. 9.7 Silicification of filamentous cyanobacteria at El Tatio Geyser Field, Chile. (a) Nodules forming at edge of main geyser pool, showing (b) moderate silicification of bacterial sheaths, and (c) complete mineralization of the microbial community and subsequent biosignature preservation. See Plate 9.7 for colour versions of these images.

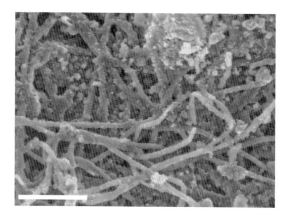

Fig. 9.8 SEM micrograph of silicified cyanobacteria filaments from soil crusts at the El Tatio Geyser Field, Chile. Bar Scale = 20 μm.

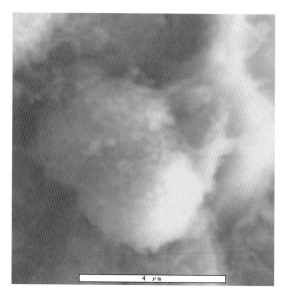

Fig. 9.9 Environmental SEM image of clay growing on cells attached to anorthoclase after 9 months exposure to the oil contaminated aquifer near Bemidji, MN. Bar Scale = 4 μm.

Clays

While clays are generally considered to form as a result of inorganic processes such as weathering of bedrock or soils, clay-like phases composed of (Fe, Al)-silicates coating microbes have been observed in freshwater environments such as lakes and rivers around the world (Ferris et al. 1987; Konhauser et al. 1998; Konhauser et al. 1993; Konhauser et al. 1994; Tazaki 1997). Early observations showed variability in crystallinity (Ferris et al. 1987), but similarities in grain size, structure, orientation and chemical makeup are evident, composed primarily of iron, silicon and aluminium (Konhauser and Urrutia 1999).

A model (Konhauser and Urrutia 1999) for natural clay-like authigenesis in these settings starts with microbial scavenging of iron by cell surface adsorption from the surrounding environment that can develop into aggregates, or the simple attachment of fine-grained ferric hydroxide minerals (Glasauer et al. 2001). Further growth of clay-like phases results from hydrogen bonding between dissolved silica or aluminosilicate complexes and the bound iron (Urrutia and Beveridge 1994, 1995) and can completely envelop the microbe (Konhauser and Urrutia 1999). The apparent ubiquitous nature of this form of mineral precipitation suggests that it is more common than often thought, and given the widespread distribution of microbes in freshwater habitats as biofilms and planktonic bacteria, this has important implications for the sequestration of dissolved metals and precipitation of minerals in aquatic environments.

We have found that clays precipitate on microbes colonizing feldspar surfaces in the oil-contaminated aquifer near Bemidji, MN (Roberts et al. 2004; Rogers and Bennett 2004). The clays are found predominantly on single cells attached to feldspars with evidence of associated surface etching (Fig. 9.9), but not basalt or quartz. The clay resembles hair standing up from the cell wall, and is not found directly on the mineral surface. The clay is only evident when examined by environmental SEM with a high humidity chamber environment, and when viewed by conventional SEM (Fig. 9.6) it appears only as a rough cell-surface texture. Circumstantial evidence (water chemistry, morphology) suggests that the clay is a smectite, and may be a mechanism for the cell to sequester aluminium into a non-toxic solid phase.

Ecology and geochemistry

It is not uncommon that, after even cursory examination, an environmental microbial community is found to influence geochemistry at a fundamental level. It is equally common to find in the literature that this interaction is viewed as the coincidental outcome of metabolism. Equal in importance to asking *how* a microbe influences rock-water reactions, however, is to ask *why* a microbe would do so; why would a microbe that seemingly needs every Joule of energy expend that energy precipitating a mineral, for example, that could impair its ability to absorb nutrients? We have found that microbes do almost nothing for free, that rarely does a microbial population do anything without gaining an advantage by that action. This can be summarized by the idea that evolution impacts geochemistry equally as directly as it does biochemistry. If there is a mutation in a population that results in the precipitation of clay on the cell surface, and a result is that predation is inhibited, that allows that population to grow faster, and that strategy will provide a competitive advantage over individuals lacking that adaptation. The result could then be that an otherwise obscure adaptation results in the widespread precipitation of clay, a geochemical outcome to evolutionary pressure. The precipitation is therefore not a coincidental outcome of metabolism but rather a specific adaptation that offers the organism an advantage. If we understand *why*, if we can discern the advantage, then we are more likely to understand the significance, the when, how and how much.

Here we present microbial geochemical interactions not as chemical pathways, but rather as ecological interactions expressed as factors in habitat, adaptation, competition, and survival, the *lingua franca* of the microbial ecologist.

Minerals as a source of nutrients

In the subsurface the mineral framework is more than simply a substratum upon which a biofilm can attach. It is an essential aspect of the habitat that can provide members of the microbial community with scarce or limiting nutrients, and the community members that can access those nutrients will have an advantage over those that can't. Microbes do not attach to mineral surfaces randomly; different minerals offer different habitats (Roberts et al. 2004; Rogers and Bennett 2004; Rogers et al. 1998; Scholl et al. 1990).

In a petroleum-contaminated aquifer near Bemidji, MN, the subsurface community has provided abundant organic carbon in the form of petroleum, but with limited access to electron acceptors and inorganic phosphate. In this environment some, but not all, of the native microorganisms will select silicate mineral surfaces that contain valuable inorganic P as phosphate, and will attach and dissolve those feldspars, while leaving virtually identical feldspars that lack P essentially barren of biomass (Rogers et al. 1998).

Rogers and Bennett (2004) found that adding feldspars containing minor apatite to laboratory microcosms as the sole source of P had three distinct effects: it accelerated the degradation of hydrocarbons, it increased the biomass and it increased the rate of weathering of the silicate mineral itself. The silicate weathering was enhanced over both sterile control experiments with the same mineral, as well as live experiments with similar feldspars but which contained no accessory apatite as a source of P (Rogers and Bennett 2004; Rogers et al. 1998) (Fig. 9.6). This suggests that a subsurface microbial population will expend energy to dissolve silicates in order to access a scarce nutrient. Further experiments using glasses with known compositions and P contents found that in P-limited media when a glass contains approximately 2% Al_2O_3 there was little colonization or dissolution, while identical glasses containing dispersed crystals of apatite were heavily colonized. Experiments with P-enriched media found no selection of glass by composition, and all surfaces supported a rich biofilm. This suggests that minerals containing potentially toxic elements (e.g. aluminium) may select against colonization that might result in mobilization of that element *unless* that mineral has a scarce resource to offer (e.g. P). In this scenario each mineral surface offers potential advantages (nutrients) and disadvantages (toxic metals) that represent forcing factors for colonization and biofilm development.

Minerals are also a major source of electron acceptors for dissimilatory microbial metabolism, such as Fe and Mn oxides. This normally occurs when in direct contact with the minerals, but when there is low availability of electron acceptors this extracellular electron transfer can occur through nanowires, 3–5 nm-wide pilus-like electrically conductive extracellular appendages (Gorby et al. 2006; Reguera et al. 2005). Nanowires can extend significant distances from bacteria, effectively extending their electron exchange ability far beyond the cell surface (El-Naggar et al. 2008). In addition to attaching to mineral surfaces, nanowires can develop between bacteria to form an electrically charged community where electrons are shared between microorganisms on an interspecies level, which may benefit the microbial community as a whole (Gorby et al. 2006). They may also play a structural role in the formation of biofilms (Reguera et al. 2007).

Buffering of extreme pH

Minerals offer the potential to buffer extremes of pH in a habitat, potentially to the advantage of some neutrophilic microorganisms, while other acidophilic organisms that thrive in very low pH might avoid highly buffered environments. We have characterized a sulphidic cave environment (Lower Kane Cave, WY, USA) that supports a rich population of sulphide-oxidizing bacteria downstream of springs that discharge sulphidic waters. The system pH is circum-neutral, and while the community produces substantial acidity that dissolves the host limestone rocks, the stream pH remains at neutral. Thick biofilms grow only on the carbonate rock surfaces (Fig. 9.2), and the carbonate mineral effectively buffers pH and maintains the habitat within the environmental limits of these bacteria.

In unbuffered systems associated with acid discharge from sulphide mineral oxidation much more acidic environments are commonly encountered, and in these extreme environments

acidophilic organisms dominate, such as *Acidothiobacillus* (Edwards et al. 2000; Edwards et al. 1999; Schrenk et al. 1998). In the Iron Mountain system of California, pH has been documented to be less than pH-2, and here the unique microorganisms that can survive and use the high hydrogen-ion activity are at a competitive advantage. The neutrophilic organisms found in carbonate rocks are completely excluded, and the surrounding host rock helps to define the nature of the microbial community.

Attenuation of extreme UV radiation

Exposure to prolonged or excessive UV radiation provides a challenge to surface microbial communities, especially phototrophic microorganisms that require sunlight for energy. UV radiation damages nucleic acids, proteins and photosynthetic pigments, most notably cyanobacterial phycobilins (Newton et al. 1979; Rozema et al. 2002; Sinha et al. 1998). Furthermore, this radiation can lead to the formation of reactive oxygen species both extracellularly and intracellularly, causing oxidative damage to biomolecules and leading to both structural alteration and impairment of physiological processes (Hansen et al. 2002; He and Häder 2002; Karsten 2008). In response to the damaging effects of UV radiation, cyanobacteria produce photo-protectants that quench free radicals or act as screening pigments such as scytonemin and mycosporine-like amino acids, which effectively absorb in the UV-A and UV-B regions of the electromagnetic spectrum, respectively. This allows for them to thrive in an environment that limits colonization by other microorganisms that are not equipped with similar protective strategies.

In high-altitude regions where UV radiation is extreme, phototrophic microorganisms can find refuge living under a protective mineral barrier that attenuates solar radiation (Figs. 9.7 and 9.8). At the El Tatio Geyser Field in northern Chile, these include both silicious sinters precipitating from cooling geyser discharge waters (Phoenix et al. 2006), as well as salt crusts forming on water-saturated soil surfaces as a result of evaporation (Omelon, personal observation). The microbes receive adequate solar radiation for photosynthesis, while being protected from high UV exposure by these minerals. It must be noted, however, that although some microorganisms do not appear to be hindered by mineral precipitation such as silicification (Phoenix et al. 2000) these microbes face constant pressure to maintain sufficient exposure to light by upward migration through the mineral matrix (phototaxis) to avoid burial and entombment.

OUTLOOK

Microbial geochemistry combines elements of microbial ecology, geochemistry and geology to examine the complex chemical interactions of microbes in the subsurface. We have investigated a wide variety of systems and repeatedly find that this interaction cannot be viewed in isolation as a microbe living in an inert mineral framework. Instead, it is always a coupled biogeochemical system, where the microorganisms are adapted to the mineral and the mineral-surface reaction is driven by metabolic processes. The important outcome of this interaction is the constant transfer of elements between the atmosphere, hydrosphere and lithosphere, acquiring energy for growth while precipitating and dissolving minerals both at the small scale (e.g. dissolving limestone under a biofilm by acid generation) to the regional scale (e.g. dissolving limestone in aquifers by creating high $P\text{CO}_2$ conditions in aquifers). Their effectiveness in changing redox states of metals for energy not only influences the state of minerals but also the toxicity of elements, which can have profound impacts on human populations.

The innumerable metabolisms of microbes makes them ubiquitous on Earth, inhabiting every conceivable environment, and this intersection between microorganisms and geochemistry means that they are more highly involved in catalysing geochemical processes that was once thought. Further, we have found that the geochemical processes that are the result of micro-

bial interaction with minerals are rarely coincidental – rather these reactions offer the organism an advantage in that habitat – nutrients, pH buffer capacity, protection, attachment or some other requirement. The population that can take advantage of these mineral attributes out competes those that do not, ultimately dominating the community. Understanding the question of *why* will increasingly guide our investigations of microbial geochemistry, and the relationship between the microbial community and low-temperature geochemical processes.

ACKNOWLEDGEMENTS

We wish to acknowledge the contributions and assistance of Annette Engel, Jennifer Roberts, Elspeth Steinhauer and Wan Joo Choi. This work was supported by the National Science Foundation, Grant # ARC-0909482; EAR-0617160; and EAR-0545336, The US Geological Survey Toxics program, and the Geology Foundation of the University of Texas at Austin.

FURTHER READING

Banfield, JF and Nealson, KH. (1997) Processes at minerals and surfaces with relevance to microorganisms and prebiotic synthesis, Reviews in Mineralogy Volume 35: Washington, DC, Mineralogical Society of America.

Beveridge, T. (1989) Role of cellular design in bacterial metal accumulation and mineralization. *Annu. Rev. Microbiol.* **43**: 147–71.

Davey, ME and O'Toole, GA. (2000) Microbial biofilms: from ecology to molecular genetics. *Microbiol Mol Biol Rev.* **64**: 847–67.

Ehrlich, HL and Newman, DK. (2008) Geomicrobiology. Boca Raton, CRC Press.

Konhauser, KO and Urrutia, MM. (1999) Bacterial clay authigenesis: a common biogeochemical process. *Chemical Geology* **161**: 399–413.

Riding, R. (2006) Cyanobacterial calcification, carbon dioxide concentrating mechanisms, and Proterozoic–Cambrian changes in atmospheric composition: *Geobiology* **4**: 299–316.

Xue, L, Zhang, Y, Zhang, T, An, L and Wang, X. (2005) Effects of Enhanced Ultraviolet-B Radiation on Algae and Cyanobacteria. *Critical Reviews in Microbiology* **31**: 79–89.

REFERENCES

Ahmann, D, Krumholz, LR, Hemond, HF, Lovley, DR and Morel, FMM. (1997) Microbial mobilization of arsenic from sediments of the Aberjona Watershed. *Environmental Science and Technology* **31**: 2923–30.

Amend, JP and Shock, EL. (2001) Energetics of overall metabolic reactions of thermophilic and hyperthermophilic Archaea and Bacteria. *FEMS Microbiology Reviews* **25**: 175–243.

Appelo, CAJ and Postma, D. (1993) Geochemistry, Groundwater, and Pollution. Rotterdam, AA Balkema, 536 pp.

Atlas, RM and Bartha, R. (1986) Microbial ecology: Fundamentals and applications. Menlo Park, Benjamin-Cummings, 533 pp.

Badger, MR, and Price, GD. (2003) CO_2 concentrating mechanisms in cyanobacteria: molecular components, their diversity and evolution. *J. of Experimental Botany* **54**: 609–22.

Baumgartner, LK, Reid, RP, Dupraz, C, Decho, AW, Buckley, DH, Spear, JR, Przekop, KM and Visscher, PT. (2006) Sulfate reducing bacteria in microbial mats: Changing paradigms, new discoveries. *Sedimentary Geology* **185**: 131–45.

Bekker, A, Holland, HD, Wang, P-L, Rumble, ID, Stein, HJ, Hannah, JL, Coetzee, LL and Beukes, NJ. (2004) Dating the rise of atmospheric oxygen. *Nature* **427**: 117–20.

Bennett, PC. (1991) Quartz dissolution in organic-rich aqueous systems. *Geochimica et Cosmochimica Acta* **55**: 1781–97.

Bennett, PC and Casey, WH. (1994) Organic acids and the dissolution of silicates, In: ED Pittman and M Lewan (eds.) The role of organic acids in geological processes, Springer-Verlag, p. 162–201.

Bennett, PC, Hiebert, FK and Rogers, JR. (2000) Microbial control of mineral-groundwater equilibria – macroscale to microscale. *Hydrogeology J.* **8**: 47–62.

Bennett, PC, Rogers, JR, Hiebert, FK and Choi, WJ. (2001) Silicates, silicate weathering, and microbial ecology. *Geomicrobiology J.* **18**: 3–19.

Bennett, PC and Siegel, DI. (1987) Increased solubility of quartz in water due to complexation by dissolved organic compounds. *Nature* **326**: 684–7.

Benning, LG, Phoenix, V, Yee, N and Konhauser, KO. (2003) The dynamics of cyanobacterial silicification: an infrared micro-spectroscopic investigation. *Geochimica et Cosmochimica Acta* **68**: 743–57.

Beveridge, TJ. (1988) The bacterial surface – general considerations towards design and function. *Canadian J. of Microbiology* **34**: 363–72.

Bosak, T and Newman, DK. (2003) Microbial nucleation of calcium carbonate in the Precambrian. *Geology* **31**: 577–80.

Bosak, T and Newman, DK. (2005) Microbial kinetic controls on calcite morphology in supersaturated solutions. *J. of Sedimentary Research* **75**: 190–9.

Boston, PJ Spilde, MN, Northup, DE, Melim, LA, Soroka, DS, Kleina, LG, Lavoie, KH, Hose, LD, Mallory, LM, Dahm, CN, Crossey, LJ and Schelble, RT. (2001) Cave biosignature suites: microbes, minerals, and Mars *Astrobiology* **1**: 25–55.

Brady, A, Slater, G, Omelon, CR, Southam, G, Druschel, G, Andersen, DT, Hawes, I, Laval, B and Lim, DSS, Photosynthetic isotope biosignatures in laminated micro-stromatolitic and non-laminated nodules associated with modern, freshwater microbialites in Pavilion Lake, BC. *Chemical Geology* **274**: 56–67.

Braissant, O, Decho, AW, Dupraz, C, Glunk, C, Przekop, KM and Visscher, PT. (2007) Exopolymeric substances of sulfate-reducing bacteria: Interactions with calcium at alkaline pH and implication for formation of carbonate minerals. *Geobiology* **5**: 401–411.

Brown, JG and Glynn, PD. (2003) Kinetic dissolution of carbonates and Mn oxides in acidic water: measurement of in situ field rates and reactive transport modeling. *Applied Geochemistry* **18**: 1225–39.

Büdel, B, Weber, B, Kühl, M, Pfanz, H, Sültemeyer, D and Wessels, D. (2004) Reshaping of sandstone surfaces by cryptoendolithic cyanobacteria: bioalkalization causes chemical weathering in arid landscapes. *Geobiology* **2**: 261–8.

Canfield, DE (2005) The early history of atmospheric oxygen. *Annu. Rev. Earth Planet. Sci.* **33**: 1–36.

Chapelle, FH. (1993) Ground-Water Microbiology and Geochemistry. New York, John Wiley & Sons, 424 pp.

Chapelle, FH. (2000) The significance of microbial processes in hydrogeology and geochemistry. *Hydrogeology J.* **8**: 41–6.

Chapelle, FH, Haack, SK, Adriaens, P, Henry, MA and Bradley, PM. (1996) Comparison of Eh and H_2 measurements for delineating redox processes in a contaminated aquifer. *Environmental Science and Technology* **30**: 3565–9.

Dittrich, M and Sibler, S. (2005) Cell surface groups of two picocyanobacteria strains studied by zeta potential investigations, potentiometric titration, and infrared spectroscopy. *J. of Colloid and Interface Science* **286**: 487–95.

Drever, JI. (1997) The Geochemistry of Natural Waters. New York, Prentice Hall, 436 pp.

Edwards, KJ, Bond, PL, Druschel, GK, McGuire, MM, Hamers, RJ and Banfield, JF. (2000) Geochemical and biological aspects of sulfide mineral dissolution: lessons from Iron Mountain, California. *Chemical Geology* **169**: 383–97.

Edwards, KJ, Goebel, BM, Rodgers, TM, Schrenk, MO, Gihring, TM, Cardona, MM, Hu, B, McGuire, MM, Hamers, RJ and Pace, NR. (1999) Geomicrobiology of Pyrite (FeS2) Dissolution: Case Study at Iron Mountain, California. *Geomicrobiology J.* **16**: 155–80.

Egemeier, SJ, (1981) Cavern development by thermal waters: *NSS Bull.* **43**: 31–51.

Ehrlich, HL. (1996a) Geomicrobiology. New York, Marcel Dekker, 719 pp.

Ehrlich, HL. (1996b) How microbes influence mineral growth and dissolution. *Chemical Geology.* **132**: 5–9.

Ehrlich, HL and Newman, DK. (2008) Geomicrobiology. Boca Raton, CRC Press.

El-Naggar, MY, Gorby, YA, Xia, W and Nealson, KH. (2008) The molecular density of states in bacterial nanowires. *Biophysical J.* **95**: L10–L12.

Engel, AS, Lee, N, Porter, ML, Stern, LA, Bennett, PC and Wagner, M. (2003) Filamentous 'Epsilonproteobacteria' dominate microbial mats from sulfidic cave springs. *Applied and Environmental Microbiology*: **69**: 5503–11.

Engel, AS, Porter, ML, Stern, LA, Quinlan, S and Bennett, PC. (2004a) Bacterial diversity and ecosystem function of filamentous microbial mats from aphotic (cave) sulfidic springs dominated by chemolithoautotrophic 'Epsilonproteobacteria'. *FEMS Microbiology Ecology* **51**: 31–.

Engel, AS, Stern, LA and Bennett, PC. (2004b) Microbial contributions to cave formation: new insight into sulfuric acid speleogenesis. *Geology* **32**: 369–72.

Fein, JB, Scott, S and Rivera, N. (2002) The effect of Fe and Si adsorption by *Bacillus subtilis* cell walls: insights into nonmetabolic bacterial precipitation of silicate minerals. *Chemical Geology* **182**: 265–73.

Ferris, FG, Beveridge, TJ and Fyfe, WS. (1986) Iron-silica crystallite nucleation by bacteria in a geothermal sediment. *Nature* **320**: 609–11.

Ferris, FG, Fyfe, WS and Beveridge, TJ. (1987) Bacteria as nucleation sites for authigenic minerals in a metal-contaminated lake sediment. *Chemical Geology* **63**: 225–32.

Ferris, FG and Magalhaes, E. (2008) Interfacial energetics of bacterial silicification: *Geomicrobiology J.* **25**: 333–7.

Ferris, FG, Wiese, RG and Fyfe, WS. (1994) Precipitation of carbonate minerals by microorganisms: Implications for silicate weathering and the global carbon dioxide budget. *Geomicrobiology* **12**: 1–13.

Franklin, SP, Hajash, A, Jr, Dewers, TA and Tieh, TT. (1994) The role of carboxylic acids in albite and quartz dissolution; an experimental study under diagenetic conditions. *Geochimica et Cosmochimica Acta* **58**: 4259–79.

Fredrickson, JK and Onstott, TC. (1996) Microbes deep inside the earth. *Scientific American* **275**: 68–73.

Friedmann, EI and Ocampo, R. (1976) Endolithic blue-green algae in the dry valleys: primary producers in the Antarctic desert ecosystem. *Science* **193**: 1274–9.

Fujita, Y, Taylor, JL, Gresham, TLT, Delwiche, ME, Colwell, FS, McLing, TL, Petzke, LM and Smith, RW. (2008) Stimulation Of Microbial Urea Hydrolysis In Groundwater To Enhance Calcite Precipitation. *Environmental Science & Technology* **42**: 3025–32.

Ghiorse, WC and Wilson, JL. (1988) Microbial ecology of the terrestrial subsurface: *Advances in Applied Microbiology* **33**: 107–72.

Gihring, TM and Banfield, JF. (2001) Arsenite oxidation and arsenate respiration by a new Thermus isolate. *Fems Microbiology Letters* **204**: 335–40.

Gihring, TM, Druschel, GK, McCleskey, RB, Hamers, RJ and Banfield, JF. (2001) Rapid arsenite oxidation by Thermus aquaticus and Thermus thermophilus: Field and laboratory investigations. *Environmental Science & Technology* **35**: 3857–62.

Glasauer, S, Langley, S and Beveridge, TJ. (2001) Sorption of Fe hydr(oxides) to the surface of *Shewanella putrefaciens*: cell-bound fine-grained minerals are not always formed de novo. *Applied and Environmental Microbiology* **67**: 5544–50.

Gorby, YA, Yanina, S, McLean, JS, Rosso, KM, Moyles, D, Dohnalkova, A, Beveridge, TJ, Chang, IS, Kim, BH, Kim, KS, Culley, DE, Reed, SB, Romine, MF, Saffarini, DA, Hill, EA, Shi, L, Elias, DA, Kennedy, DW, Pinchuk, G, Watanabe, K, Ishii, SI., Logan, B, Nealson, KH and Fredrickson, JK. (2006) Electrically conductive bacterial nanowires produced by Shewanella oneidensis strain MR-1 and other microorganisms.

Proceedings of the National Academy of Sciences
103: 11358–63.

Hansen, LJ, Whitehead, JA and Anderson, SL (2002)
Solar UV radiation enhances the toxicity of arsenic in
Ceriodaphnia dubia. *Ecotoxicology* **11**: 279–87.

He, Y-Y and Häder, D-P. (2002) UV-B-induced formation
of reactive oxygen species and oxidative damage of
the cyanobacterium Anabaena sp. protective effects
of ascorbic acid and N-acetyl–cysteine. *J. Photochemistry and Photobiology B: Biology* **66**: 115–24.

Helgeson, HC, Murphy, WM and Aagaard, P. (1984)
Thermodynamic and kinetic constraints on reaction
rates among minerals and aqueous solutions II. Rate
constants effective surface area, and hydrolysis of
feldspar. *Geochim. Cosmochim. Acta* **48**: 2405–32.

Hiebert, FK and Bennett, PC. (1992) Microbial control
of silicate weathering in organic-rich ground water.
Science **258**: 278–81.

Hill, C. (1995) Sulfur redox reactions – hydrocarbons,
native sulfur, Mississippi Valley-Type deposits, and
sulfuric-acid karst in the Delaware Basin, New-Mexico
and Texas. *Environmental Geology* **25**: 16–23.

Hill, CA. (1990) Sulfuric-acid speleogenesis of Carlsbad
Cavern and its relationship to hydrocarbons, Delaware
Basin, New-Mexico and Texas. *AAPG Bulletin* **74**:
1685–94.

Hoehler, TM. (2004) Biological energy requirements as
quantitative boundary conditions for life in the subsurface. *Geobiology* **2**: 205–15.

Hoehler, TM, Alperin, MJ, Albert, DB and Martens, CS.
(2001) Apparent minimum free energy requirements
for methanogenic Archaea and sulfate-reducing bacteria in an anoxic marine sediment. *FEMS Microbiology
Ecology* **38**: 33–41.

Holdren, GR and Speyer, PM. (1986) Stoichiometry of
alkali feldspar dissolution at room temperature and
various ph values. In: SM Colman and DP Dethier
(eds.) Rates of chemical weathering of rocks and minerals: New York, Academic Press, pp. 61–81.

Horsfield, B, Dieckmann, V, Mangelsdorf, K, Di Primio,
R, Wilkes, H, Schloemer, S, and Schenk, H. (2002)
Organic diagenesis; a potential provider of substrates
for deep microbial ecosystems. *Geochimica et
Cosmochimica Acta* **66**: 342.

Hose, L and Pisarowicz, J. (1999) Cueva de Villa Luz,
Tobasco, Mexico: Reconnaissance study of an active
sulfur spring cave and ecosystem. *J. of Cave and Karst
Studies.* **6**: 13–21.

Huang, WH and Longo, JM. (1992) The effect of organics
on feldspar dissolution and the development of secondary porosity. *Chemical Geology* **98**: 271–92.

Jackson, BE and McInerney, MJ. (2002) Anaerobic
microbial metabolism can proceed close to thermodynamic limits. *Nature* **415**: 454–6.

Jin, QaB, CM, (2007) The thermodynamics and kinetics
of microbial metabolism. *American J. of Science* **307**:
643–77.

Kah, C and Riding, R. (2007) Mesoproterozoic carbon
dioxide levels inferred from calcified cyanobacteria.
Geology **35**: 799–802.

Kaplan, A and Reinhold, L. (1999) CO_2 concentrating
mechanisms in photosynthetic microorganisms.
*Annual Review of Plant Physiology and Plant
Molecular Biology* **50**: 539–70.

Karavaiko, GI, Belkanova, NP, Eroshchev-Shak, VA and
Avakyan, ZA. (1985) Role of microorganisms and
some physicochemical factors of the medium in
quartz destruction. *Microbiologiya* **53**: 795–800.

Karsten, U. (2008) Defense strategies of algae and cyano-
bacteria against solar ultraviolet radiation. In: CD
Amsler (ed.) Algal Chemical Ecology: Berlin, Springer-
Verlag, p. 273–96.

Kasting, JF and Siefert, JL. (2002) Life and the evolution
of Earth's atmosphere. *Science* **296**: 1066–68.

Konhauser, KO, Fisher, QJ, Fyfe, WS, Longstaffe, FJ and
Powell, MA. (1998) Authigenic mineralization and
detrital clay binding by freshwater biofilms: the
Brahmani River, India. *Geomicrobiology J.* **15**:
209–222.

Konhauser, KO, Fyfe, WS, Ferris, FG and Beveridge, TJ.
(1993) Metal sorption and mineral precipitation by
bacteria in two Amazonian river systems: Rio
Solimões and Rio Negro, Brazil. *Geology* **21**:
1103–06.

Konhauser, KO, Schultze-Lam, S, Ferris, FG, Fyfe, WS,
Longstaffe, FJ and Beveridge, TJ. (1994) Mineral pre-
cipitation by epilithic biofilms in the Speed River,
Ontario, Canada. *Applied and Environmental
Microbiology* **60**: 549–553.

Konhauser, KO and Urrutia, MM. (1999) Bacterial clay
authigenesis: a common biogeochemical process.
Chemical Geology **161**: 399–413.

Krumbein, WE. (1974) On the precipitation of aragonite
on the surface of marine bacteria. *Naturwissenschaften*
61: 167.

Kutuzova, RS. (1969) Release of silica from minerals as
result of microbial activity. *Mikrobiologiya* **38**:
596–602.

Lalonde, S, Amskold, L, McDermott, T, Inskeep, BP and
Konhauser, KO. (2005) Chemical reactivity of microbe
and mineral surfaces in hydrous ferric oxide deposit-
ing hydrothermal springs. *Geobiology* **5**: 219–34.

Land, LS. (1997) Failure to precipitate dolomite at 25°C from dilute solution despite 1000-Fold oversaturation after 32 years. *Aquatic Geochemistry* **4**: 361–8.

Lovley, DR, Giovannoni, SJ, White, DC, Champine, JE, Phillips, EJP, Gorby, YA and Goodwin, S. (1993) Geobacter metallireducens gen. nov. sp. nov., a microorganism capable of coupling the complete oxidation of organic compounds to the reduction of iron and other metals. *Archives of Microbiology* **159**: 336–44.

Lovley, DR and Goodwin, S. (1988) Hydrogen concentrations as an indicator of the predominant terminal electron-accepting reactions in aquatic sediments: *Geochimica et Cosmochimica Acta* **52**: 2993–3003.

Lovley, DR and Phillips, JP. (1988) Novel mode of microbial energy metabolism: Organic carbon oxidation coupled to dissimilatory reduction of iron or manganese. *Applied and Environmental Microbiology* **54**: 1472–80.

Luttge, A and Conrad, PG. (2004) Direct observation of microbial inhibition of calcite dissolution. *Applied and Environmental Microbiology*, p. 1627–32.

Luttge, A, Zhang, L and Nealson, KH. (2005) Mineral surfaces and their implications for microbial attachment: Results from Monte Carlo simulations and direct surface observations. *American J. of Science* **305**: 766–90.

Madigan, MT, Martinko, JM and Parker, J. (2003) Brock Biology of Microorganisms. New York, Prentice Hall.

Mahoney, PC. (1983) Relationship between phosphorylation potential and electrochemical H+ gradient during glycolysis in Streptococcus lactis. *J. Bacteriol.* **153**: 1461–70.

Martinez, RE, Pokrovsky, OS, Schott, J and Oelkers, EH. (2008) Surface charge and zeta-potential of metabolically active and dead cyanobacteria. *J. of Colloid and Interface Science* **323**: 317–25.

McKenzie, EJ, Brown, KL, Cady, SL and Campbell, KA. (2001) Trace metal chemistry and silicification of microorganisms in geothermal sinter, Taupo volcanic zone, New Zealand: Geothermics, **30**: 483–502.

McMahon, PB and Chapelle, FH. (1991) Microbial production of organic acids in aquitard sediments and its role in aquifer geochemistry. *Nature* **349**: 233–35.

Merz, MUE. (1992) The biology of carbonate precipitation by cyanobacteria. *Facies* **26**: 81–102.

Morita, RM. (1980) Calcite precipitation by marine bacteria. *Geomicrobiology J.* **2**: 63–82.

Nealson, KH. (1997) Sediment bacteria: Who's there, what are they doing, and what's new? *Annual Reviews in Earth and Planetary Science* **25**: 403–34.

Newton, JW, Tyler, DD and Slodki, ME. (1979) Effect of ultraviolet-B (280–320nm) radiation on blue-green algae (Cyanobacteria), possible biologicla indicators of stratospheric ozone depletion. *Applied and Environmental Microbiology* **37**: 1137–41.

Nordstrom, DK and Archer, DG. (2003) Arsenic thermodynamic data and environmental geochemistry. In: Welch, AH and Stollenwerk, KG (eds.) Arsenic in Groundwater. Geochemistry and Occurrence. Boston, Kluwer Academic Publishers, pp. 1–25.

Nordstrom, DK, McCleskey, RB and Ball, JW. (2004) Processes governing arsenic geochemistry in thermal waters of Yellowstone National Park. *Geochimica et Cosmochimica Acta* **68**: A262–A262.

Northup, DE and Lavoie, KH. (2001) Geomicrobiology of Caves: A review. *Geomicrobiology J.* **18**: 199–222.

Obst, M, Dynes, JJ, Lawrence, JR, Swerhone, GDW, Benzerara, K, Karunakaran, C, Kaznatcheev, K, Tyliszczak, T and Hitchcock, AP. (2009) Precipitation of amorphous CaCO$_3$ (aragonite-like) by cyanobacteria: A STXM study of the influence of EPS on the nucleation process. *Geochimica et Cosmochimica Acta* **73**: 4180–98.

Pedersen, K. (1993) The deep subterranean biosphere. *Earth Science Reviews* **34**: 243–60.

Petersen, A, Matthess, G and Schenk, D. (1992) Experiments on the influence of organic ligands upon kinetics of feldspar weathering. In: Matthess, G, Frimmel, F-H, Hirsch, P, Schulz, H-D and Usdowski, HE (eds.) Progress in hydrogeochemistry; organics; carbonate systems; silicate systems; microbiology; models. Berlin, Springer-Verlag, pp. 86–92.

Phelps, TJ, Murphy, EM, Pfiffner, SM and White, DC. (1994) Comparison between geochemical and biological estimates of subsurface microbial activities. *Microbial Ecology* **28**: 335–49.

Phelps, TJ, Raione, EG, White, DC and Fliermans, CB. (1989) Microbial activities in deep subsurface environments. *Geomicrobiology J.* **7**: 79–92.

Phoenix, VR, Adams, DG and Konhauser, KO. (2000) Cyanobacterial viability during hydrothermal biomineralization. *Chemical Geology* **169**: 823–6.

Phoenix, VR, Bennett, PC, Engel, AS, Tyler, SW and Ferris, FG. (2006) Chilean high-altidude hot spring sinters: a model system for UV screening mechanisms by early Precambrian cyanobacteria. *Geobiology* **4**: 15–28.

Phoenix, VR, Martinez, RE, Konhauser, KO and Ferris, FG. (2002) Characterization and implictions of the cell surface reactivity of *Calothrix* sp. Strain KC97. *Applied and Environmental Microbiology* **68**: 4827–34.

Price, GD, Maeda, S-i, Omata, T and Badger, MA. (2002) Modes of active inorganic carbon uptake in the cyanobacterium, *Synechococcus* sp. PCC7942. *Functional Plant Biology* **29**: 131–49.

Reguera, G, McCarthy, KD, Mehta, T, Nicoll, JS, Tuominen, MT and Lovely, DR. (2005) Extracellular electron transfer via microbial nanowires. *Nature* **435**: 1098–101.

Reguera, G, Pollina, RB, Nicoll, JS and Lovely, DR. (2007) Possible nonconductive role of *Geobacter sulfurreducens* pilus nanowires in biofilm formation. *J. of Bacteriology* **189**: 2125–7.

Roberts, JA, Bennett, PC, Gonzalez, LA, Macpherson, GL and Milliken, KL. (2004) Microbial precipitation of dolomite in methanogenic groundwater. *Geology* **32**: 277–80.

Roden, EE. (1997) Phosphate mobilization in iron-rich anaerobic sediments: microbial Fe(III) oxide reduction versus iron-sulfide formation. *Archiv für Hydrobiologie* **139**: 347–78.

Rogers, JR and Bennett, PC. (2004) Mineral stimulation of subsurface microorganisms: release of limiting nutrients from silicates. *Chemical Geology* **203**: 91–108.

Rogers, JR, Bennett, PC and Choi, WJ. (1998) Feldspars as a source of nutrients for microorganisms. *American Mineralogist* **83**: 1532–40.

Rozema, J, Björnb, LO, Bornmanb, JF, Gaberikc, A, Häderd, D-P, Trotc, T, Germc, M, Klischd, M, Grönigerd, A, Sinhad, RP, Lebertd, M, Hed, Y-Y, Buffoni-Hallb, R, Bakkera, NVJd, Staaija, Jvd and Meijkampa, BB. (2002) The role of UV-B radiation in aquatic and terrestrial ecosystems: an experimental and functional analysis of the evolution of UV-absorbing compounds. *J. Photochemistry and Photobiology B: Biology* **66**: 2–12.

Schink, B. (1997) Energetics of syntrophic cooperation in methanogenic degradation. *Microbiol. Mol. Biol. Rev.* **61**: 262–80.

Schink, B and Friedrich, M. (1994) Energetics of syntrophic fatty acid oxidation. *FEMS Microbiology Reviews* **15**: 85–94.

Scholl, MA, Mills, AL, Herman, JS and Hornberger, GM. (1990) The influence of mineralogy and solution chemistry on the attachment of bacteria to representative aquifer minerals. *J. of Contaminant Hydrology* **6**.

Schopf, JW. (2006) Fossil evidence of Archaean life. *Phil. Trans. R. Soc. B* **361**: 869–85.

Schrenk, MO, Edwards, KJ, Goodman, RM, Hamers, RJ and Banfield, JF. (1998) Distribution of *Thiobacillus ferrooxidans* and *Leptospirillum ferrooxidans*: Implications for Generation of Acid Mine Drainage. *Science* **279**: 1519–22.

Schultze-Lam, S and Beveridge, TJ. (1994) Nucleation of celestite and strontianite on a cyanobacterial S-layer. *Applied and Environmental Microbiology* **60**: 447–53.

Schultze-Lam, S, Ferris, FG, Konhauser, KO and Wiese, RG. (1995) In situ silicification of an Icelandic hot spring microbial mat: implications for microfossil formation. *Canadian J. of Earth Sciences* **32**: 2021–26.

Schwartzman, DW and Volk, T. (1989) Biotic enhancement of weathering and the habitability of earth. *Nature* **340**: 457–9.

Silver, S and Phung, LT. (2005) Genes and enzymes involved in bacterial oxidation and reduction of inorganic arsenic. *Applied Environmental Microbiology* **71**: 599–608.

Sinha, RP, Klisch, M, Groeniger, A and Haeder, D-P. (1998) Ultraviolet-absorbing/screening substances in cyanobacteria, phytoplankton, and macroalgae. *J. Photochemistry and Photobiology. B.* **47**: 83–94.

Smedley, PL and Kinniburgh, DG. (2002) A review of the source, behaviour and distribution of arsenic in natural waters. *Applied Geochemistry* **17**. 517–68.

Spear, JR, Walker, JJ, McCollom, TM and Pace, NR. (2005) Hydrogen and bioenergetics in the Yellowstone geothermal ecosystem. *Proceedings of the National Academy of Sciences* **102**: 2555–60.

Spiro, B. (1977) Bacterial sulphate reduction and calcite precipitation in hypersaline deposition of bituminous shales. *Nature* **269**: 235–7.

Steinhauer, ES, Omelon, CR and Bennett, PC. Kinetics of microbial corrosion of limestone by sulfuric acid speleogenesis. *Geomicrobiology J.* **27**: 723–38.

Stillings, LL, Drever, JI, Brantley, SL, Sun, Y and Oxburgh, R. (1996) Rates of feldspar dissolution at pH 3–7 with 0–8 mM oxalic acid. *Chemical Geology* **132**: 79–90.

Tazaki, K. (1997) Biomineralization of layer silicates and hydrated Fe/Mn oxides in microbial mats: an electron microscopical study. *Clays and Clay Minerals* **45**: 203–12.

Thamdrup, B, Finster, K, Hansen, JW and Bak, F. (1993) Bacterial disproportionation of elemental sulfur coupled to chemical reduction of iron and manganese. *Applied and Environmental Microbiology* **59**: 101–8.

Thauer, RK, Jungermann, K and Decker, K. (1977) Energy conservation in chemotrophic anaerobic bacteria. *Bacteriol. Rev.* **41**: 100–118.

Thompson, JB and Ferris, FG. (1990) Cyanobacterial precipitation of gypsum, calcite, and magnesite from natural alkaline lake water. *Geology* **18**: 995–8.

Thompson, JB, Schultze-Lam, S, Beveridge, TJ and Des Marais, DJ. (1997) Whiting events: Biogenic origin due to the photosynthetic activity of cyanobacterial picoplankton. *Lirmol. Oceanogr.* **42**: 133–41.

Ullman, WJ, Kirchman, DL, Welch, SA and Vandevivere, P. (1996) Laboratory evidence for microbially mediated silicate mineral dissolution in nature. *Chemical Geology* **132**:.11–17.

Urrutia, MM and Beveridge, TJ. (1994) Formation of fine-grained metal and silicate precipitates on a bacterial surface (*Bacillus subtilis*). *Chemical Geology* **116**: 261–80.

Urrutia, MM and Beveridge, TJ. (1995) Formation of short-range ordered aluminosilicates in the presence of a bacterial surface (*Bacillus subtilis*) and organic ligands. *Geoderma* **65**: 149–165.

Valsami-Jones, E, McLean, J, McEldowney, S, Hinrichs, H and Pili, A. (1998) An experimental study of bacterially induced dissolution of K-feldspar. *Mineralogical Magazine* **62A**: 1563–4.

Vandevivere, P and Kirchman, DL. (1993) Attachment stimulates exopolysaccharide synthesis by a bacterium. *Applied and Environmental Microbiology* **59**: 3280–6.

Vandevivere, P, Welch, SA, Ullman, WJ and Kirchman, DL. (1994) Enhanced dissolution of silicate minerals by bacteria at near-neutral pH. *Microbial Ecology* **27**: 241–51.

Vasconcelos, CO and McKenzie, JA. (1997) Microbial mediation of modern dolomite precipitation and diagenesis under anoxic conditions (Lagoa Vermelha, Rio De Janeiro, Brazil). *J. of Sedimentary Research* **67**: 378–90.

Vasconcelos, CO, McKenzie, JA, Bernasconi, S, Grujic, D and Tien, AJ. (1995) Microbial mediation as a possible mechanism for natural dolomite formation at low temperatures: *Nature* **377**: 220–2.

Warren, LA, Maurice, PA, Parmer, N and Ferris, FG. (2001) Microbially Mediated Calcium Carbonate Precipitation: Implications for Interpreting Calcite Precipitation and for Solid-Phase Capture of Inorganic Contaminants. *Geomicrobiology J.* **18**: 93–115.

Warthmann, R, Lith, YV, Vasconcelos, C, McKenzie, JA and Karpoff, AM. (2000) Bacterially induced dolomite precipitation in anoxic culture experiments: *Geology* **28**: 1091–4.

Welch, SA and Ullman WJ. (1996) Feldspar dissolution in acidic and organic solutions: Compositional and pH dependence of dissolution rate. *Geochimica et Cosmochimica Acta* **60**: 2939–48.

Welch, SA and Vandevivere, P. (1995) Effect of microbial and other naturally occurring polymers on mineral dissolution: *Geomicrobiology J.* **12**: 227–38.

White, D. (2007) The Physiology and Biochemistry of Prokaryotes. New York, Oxford University Press, 628 pp.

Wolfaardt, G, Lawrence, J and Korber, D. (1999) Function of EPS. In: J Wingender, TR Neu and H-C Flemming (eds.) Microbial extracellular polymeric substances: characterization, structure, and function. New York, Springer, pp. 172–95.

NOTES

[1] We use the term substrates to denote the source of electrons for the specific respiratory pathway, and the term substratum for the mineral surface to which the cell is attached.

10 Nanogeochemistry: Nanostructures and Their Reactivity in Natural Systems

YIFENG WANG[1], HUIZHEN GAO[2] AND HUIFANG XU[3]

[1]Sandia National Laboratories, Albuquerque, NM, USA
[2]Sandia National Laboratories, Albuquerque, NM, USA
[3]University of Wisconsin, Madison, WI, USA

ABSTRACT

Nanophases and nanostructures are widely present in natural environments. As a newly emerging research area, nanogeochemistry studies the formation and the reactivity of these nanophases and nanostructures as well as their controls on geochemical reactions and mass transfer. Nanogeochemical study will provide a key linkage between the molecular-level understanding of geochemical processes and the macro-scale laboratory and field observations. The study will also possibly lead to the design of new materials and chemical processes for effective natural-resource extraction and environmental management. Nanogeochemistry is still in its infant stage. 'There's plenty of room' in this area for geochemists to explore.

INTRODUCTION

About half a century ago, physicist Richard Feynman gave an enlightening speech entitled

Frontiers in Geochemistry: Contribution of Geochemistry to the Study of the Earth, First edition. Edited by Russell S. Harmon and Andrew Parker. © 2011 Blackwell Publishing Ltd. Published 2011 by Blackwell Publishing Ltd.

'There's Plenty of Room at the Bottom' at the annual meeting of American Physical Society, in which he foresaw a new research area that would have a great potential for technological innovations:

I would like to describe a field, in which little has been done, but in which an enormous amount can be done in principle. This field is not quite the same as the others in that it will not tell us much of fundamental physics (in the sense of, 'What are the strange particles?') but it is more like solid-state physics in the sense that it might tell us much of great interest about the strange phenomena that occur in complex situations. Furthermore, a point that is most important is that it would have an enormous number of technical applications. (Ratner and Ratner 2004: p. 2)

Eventually, this research area started to emerge in early 1990s and today it is known as 'nanoscience' or 'nanotechnology'. Interest in this area largely stems from a general observation that novel physical or chemical properties of a material can emerge as the size of the material is reduced to nanometres (e.g. Roduner 2006). It has been found, for example, that the energy levels of both the conduction band and the valence band of a nanosized semiconductor material can vary with particle size, thus leading to a possibility for tuning the band gap of the material for a specific application (Lüning et al. 1999). Advances in nanoscience have greatly enhanced understanding of the control of nanostuctures on material properties. At the same time, progress in nanotechnology has enabled the purposeful manipulation of material structures at nanometre scales to obtain novel functional materials for various applications. Nanotechnology has been hailed as a key technological innovation that will potentially revolutionize industries in the 21st century (Ratner and Ratner 2003).

New concepts developed in nanoscience have a profound impact on almost every scientific discipline today. Nanogeochemistry came to exist about ten years ago by introducing these new concepts into existing geochemical studies (Banfield and Navrotsky 2001; Hochella 2002; Wang et al. 2003a). It loosely encompasses two general research areas:

(i) understanding geochemical reactions and mass transfers at nanometer scales, especially as regards the formation of nanostructures in geologic materials and their effects on geochemical processes;

(ii) using nanotechnology to design new materials and chemical processes for effective natural resource extraction and environmental management.

Modern geochemistry considers spatial scales that range from individual atoms (10^{-10} m) all the way up to planet Earth as a whole (10^7 m). Over this enormous spatial scale nanometre scales (1 nm = 10^{-9} m) play a vital role in mechanistic understanding of many important geochemical processes. For example, it is at this scale that molecular-level parts and bits (e.g., biomolecules, genes, etc.) are assembled inside microbial cells to give rise to amazingly diverse metabolic functionalities of the cells. Although the actual mechanism for this assembly process, to a large extent, still remains unknown, it is apparent that the metabolic behaviour of a cell must be determined by the functionalities of individual nanometer-scale compartmentalized sub-cell entities and the interactions among them (Gross 1999). It is known that ion transport in nanoscale channels across a cell membrane may directly regulate the rates of many biogeochemical processes

(Jin and Bethke 2002). Therefore, nanogeochemical studies will provide an important linkage between the molecular-level understanding and the macro-scale laboratory/field observations for many fundamental processes such as mineral-water interactions, microbial reactions, and subsurface contaminant migration, to name just a few.

This chapter provides a review of important studies in nanogeochemistry, as well as the recent progress in nanoscience which will potentially impact nanogeochemical studies in the future. This review focuses on nanostructures in geologic materials and their reactivity in natural environments. The term 'nanostructures' is used here to include nanophases, nanopores, nano-scale patterns in mineral structures, and nano-scale geochemical variations at solid-water or microbial cell-water interfaces. Specifically, the review will cover: (i) the occurrence of nanophases and nanostructures in natural systems, (ii) the size- and shape-dependent chemical reactivity and stability of mineral particles, (iii) the behaviours of water and ions in nanoconfinement and their geochemical implications, (iv) nanophases in biomineralization and nano-scale interactions between microbes and mineral surfaces, and (v) nanostructured environmental materials. By convention, the scale of nanophases and nanostructures ranges from 1 to 100 nanometres.

CHEMISTRY OF NANOPHASES

Origin and occurrence of nanophases

Nanophases, also called nanoparticles, are not completely new to geochemists who already know about colloids. By definition, nanoparticles (1–100 nm) are colloids (1–1000 nm). Natural nanoparticles may form through nucleation when a geologic fluid becomes supersaturated with a mineral phase. For a given saturation degree (Ω), the critical size of nuclei (r^*) is determined by (e.g. Steefel and Van Cappellen 1990):

$$r^\star = \frac{2\gamma V_m}{RT \ln \Omega} \tag{1}$$

where γ is the surface energy; V_m is the volume of a formula unit of the solid; R is the gas constant; and T is the absolute temperature. Clearly, for a given mineral phase, the saturation degree is an important factor controlling the size of precipitates; i.e. the higher the saturation degree, the smaller the particle size. Given the ubiquitous presence of colloid particles (e.g., Honeyman and Ranville 2002), precipitated nanophases may persist for quite a long time period in natural systems.

Nanophases are believed to play an important role in global biogeochemical cycling. Using ^{57}Fe Mossbauerove spectroscopy, van der Zee et al. (2003) have identified nanophase goethite (alpha-FeOOH) to be the major reactive iron oxyhydroxide phase in a large variety of lacustrine and marine environments. The predominant presence of this nanophase in such environments is not at all surprising because aquatic sediments are such a dynamic system that high supersaturation and intensive nucleation of iron oxyhydroxides are expected to occur at the water-sediment interface. The reactivity of nanophase goethite directly controls the bioavailability of this mineral for subsequent dissimilatory microbial iron reduction as well as the related biological cycling of phosphate and other trace elements during sediment early diagenesis (e.g. Wang and Van Cappellen 1996). Iron (oxyhydr)oxide nanoparticles are also found in suspended sediments from glacial meltwaters, supraglacial, and proglacial sediments, and sediments in basal ice, from Arctic, Alpine and Antarctic locations (Raiswell et al. 2006). Those particles, typically approximately 5 nm in diameter, are poorly crystalline and occur as single grains or aggregates that may be isolated or attached to sediment grains. Because of their high reactivity, such nanoparticles are potentially bioavailable. Iceberg delivery of sediment containing iron as (oxyhydr)oxide nanoparticles during the Last Glacial Maximum is estimated to be sufficient to fertilize the increase in oceanic productivity required to draw down atmospheric CO_2 to the levels observed in ice cores (Raiswell et al. 2006).

In many cases, nanoparticle formation in natural systems is closely coupled with microbial reactions. Microorganisms are able to precipitate nanoparticles, either intracellularly or extracellularly. Nanometer-size magnetite particles are known to form in certain microbial cells (Bazylinski and Moskowitz 1997). Roh et al. (2006) have found that Fe(III)-reducing bacteria such as *Theroanaerobacter ethanolicus* and *Shewanella sp.* have the ability to extracellularly synthesize magnetite and metal-substituted magnetite nanoparticles in aqueous media. Lee et al. (2008) have shown that a facultative dissimilatory metal-reducing bacterium, *Shewanella sp.* strain HN-41 can produce magnetite nanoparticles from a precursor, poorly crystalline iron-oxyhydroxide akaganeite (beta-FeOOH), by reducing Fe(III). The diameter of the biogenic magnetite nanoparticles ranges from 26–38 nm. It has been suggested that the formation of insoluble metal precipitates may be one of various mechanisms used by microorganisms to reduce the toxicity of a metal (e.g. Suzuki and Banfield 2004).

Nanophases can be formed mechanically in brittle faults, in which grain-size reduction and gouge formation are common at all scales (Wilson et al. 2005). It is shown that the ultrafine grains in gouge from both the San Andreas Fault in California and the rupture zone of a recent earthquake in a South African mine approach nanometer scales and the specific areas of gouge approach $80 \, m^2/g$. Wilson et al. (2005) have demonstrated that the formation of ultrafine gouge could significantly affect the energy balance of an earthquake. The gouge surface area of $80 \, m^2/g$ corresponds to a surface energy of $0.2–0.36 \, MJ/m^2$ of the fault surface for a gouge zone of 1 mm. The formation of such ultrafine material can consume more than 50% of the earthquake energy.

Gold nanoparticles are found in arsenian pyrite, a primary host-phase for Au in Carlin-type gold deposits. Gold nanoparticles constitute a significant fraction of invisible gold in those deposits (Palenik et al. 2004). Reich et al. (2006) have determined stability of gold nanoparticles as a function of temperature, particle size, and host material, using an in-situ transmission electron microscope (TEM) heating technique. They have found that the melting point of isolated Au nanoparticles decreases with decreasing particle size and, while incorporated in arsenian pyrite, these nanoparticles become unstable unless the nanoparticle size distribution coarsens by diffusion-driven, solid-state Ostwald ripening. This change in nanoparticle stability starts above 370°C, thus setting an upper temperature and size limit for the occurrence of nanoparticulate Au in refractory sulphides.

Interestingly, some seemingly exotic nano-phases can also be found in common geologic formations. For example, rare carbon forms, such as fullerenes and fullerite structures, were thought to form only at temperatures higher than 1500°C (Dresselhaus et al. 1996). However, it turns out that the very high temperatures required for fullerene synthesis in the laboratory do not restrict their occurrence in natural systems. These carbon nanophases, typically in form from barrel-like (practically fullerene) shapes to micrometer-size elongated shapes with an outer diameter of tens of nanometers, have been found in shungites, coals, marine carbonate–clayey sediments, chondrites, and stellar dust (Simakov et al. 2001). The origin of these nanophases is directly related to the reaction of hydrocarbon-containing fluids under metamorphic conditions at around 700°C and 500 MPa (Simakov et al. 2001).

Phase stability of nanophases

With the surface energy included, the total Gibbs free energy change of a solid (G) can be described by (e.g. Banfield and Zhang 2001):

$$dG = -SdT + VdP + \sum \mu_i dn_i + \sum \gamma_j dA_j \quad (2)$$

where S and V are the partial molar entropy and volume, respectively; P is the pressure; μ_i and n_i are the chemical potential and molar number of species i, respectively; and γ_j and A_j are the surface energy and the area of interface j, respectively. As the dimension of a phase is reduced to nanometers, the last term of Eqn. 2, representing the contribution of surface energy, becomes significant. As a rule of thumb, the stability of nanophases is size-dependent.

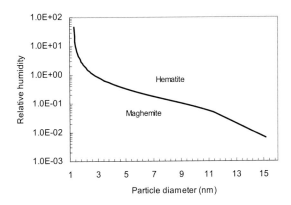

Fig. 10.1 Phase transition between haematite and maghemite as a function of particle size and relative humidity. Data are taken from Navrotsky et al. (2008) for phase stability and from Clarke and Hall (1991) for water sorption on ion oxides.

Consider the iron oxides, known to have a variety of polymorphs, as an example. Navrotsky et al. (2008) have convincingly demonstrated the importance of particle size and surface hydration in stabilizing nanophases of iron oxides. Figure 10.1 shows the phase boundary between haematite and maghemite, as controlled by particle size and relative humidity (or water activity). At a given relative humidity, maghemite becomes stable with respect to haematite as the particle size decreases, as observed in soils (Chen et al. 2005). For a very dry environment, for example, on the Mars surface, maghemite would become stable when the particle size is smaller than approximately 10 nm. Similarly, Barnard and Xu (2008) have demonstrated that the phase boundary between anatase and rutile nanocrystals depends on both the size and the degree of surface protonation. Therefore, the stability of nanophases may be a useful indicator for environmental conditions.

In a bulk solid phase, ideally assumed to have infinite extensions in all three directions, every individual atom or molecule would 'see' exactly the same forces exerted by its neighbours. However, if one cleaves the solid, the atoms near the cleavage experience an asymmetric interaction because the force on one side has disappeared. These atoms have to adjust their position to reach new equilibrium by lattice relaxation, i.e. lattice contraction (Jiang et al. 2001). As the size of the solid phase is on a nanometer scale, the lattice contraction starts to affect almost every atom in the particle, which is equivalent to exerting a confining pressure on the particle surface. This pressure (Δp) can be calculated by (Jiang et al. 2001; Roduner 2006) as:

$$\Delta p = \frac{2\gamma}{r} \tag{3}$$

where r is the radius of the particle and γ is the surface energy. Given a typical γ value of $1\,J/m2$ (Steefel and Van Cappellen 1990), the equivalent confining pressure can be as high as $2\,GPa$ for a particle with a radius of $1\,nm$. This equivalent pressure can potentially stabilize mineral polymorphs whose bulk phases can only be stable under high pressure conditions. This argument is supported by recent findings that diamonds and diamond-type nanophases have been observed to occur in various rocks of the Earth's crust (Simakov 2003). Simakov et al. (2008) have demonstrated diamond nanoparticles can form from carbon-containing hydrothermal fluids at a temperature of 500°C and a pressure of 1000 atm, much lower than the P-T conditions previously proposed for diamond formation in the upper mantle. They have shown that surface tension and particle size play an important role in the formation of these nanoparticles.

A similar argument can also be made for the effect of temperature. For example, zirconium dioxde (ZrO_2) exhibits an interesting dependence of phase transition on temperature and size (Barnard et al. 2006). ZrO_2 in a bulk phase can undergo successive phase transitions as the temperature increases: from a monoclinic form under ambient conditions, to a tetragonal one at approximately 1170°C, and then to a cubic structure at approximately 2370°C. Barnard et al. (2006) use a shape-dependent thermodynamic model to investigate the relationship between nanomorphology and phase stability. They show that as the particle size decreases to nanometers the high tempera-

ture tetragonal polymorph may become stable at low temperatures.

Surface chemistry of nanophases

Surface charge is a key factor determining the stability of a nanoparticle suspension as well as the ion sorption capability of the particles. The effect of particle size on surface charge has not been thoroughly understood, and traditional surface complexation models may not be applicable to particles of diameters smaller than 10 nm, because the planar surface assumption would be seriously violated due to the high curvature of the electric double layer (e.g. Abbas et al. 2008). Sonnefeld (1993) developed a model for calculating the surface charge of a spherical particle using the approximate solution of the Poisson-Boltzmann equation. The model predicts that the surface charge density of a particle increases with decreasing particle size. The charge increase becomes pronounced for a particle of diameter less than 10 nm. This effect has been attributed to the enhanced counterion-screening efficiency of the electrolyte solution around nanoparticles as compared with large particles (Lyklema 2000). More advanced modelling work has confirmed these results. Using Monte-Carlo simulations, Abbas et al. (2008) show that the surface charge density of nanoparticles is highly size-dependent (see Fig. 10.2). For instance, the surface charge calculated for 2 nm particles at an ionic strength of 0.1 M is close to the values for 100 nm particles at an ionic strength of 1 M.

The model predictions seem supported by experimental observations. Madden et al. (2006) measured zeta potentials on 7 nm and 25 nm hematite particles over a pH range of 4 to 11. Their measurements show that the finer particles systematically exhibit higher surface potentials than the coarser particles, although the point of zero charge (PZC) remains more or less the same. One explanation for this, as they proposed, is the presence of lower coordinated surface Fe ions in the finer particles. Interestingly, Madden et al. (2006) also observed the increased Cu^{2+} sorption capability for small particles. Their data indicate that the

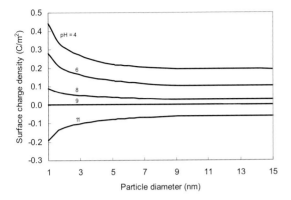

Fig. 10.2 Surface charge density predicted by Monte Carlo simulations for goethite nanoparticles as a function of particle size. Data are taken from Abbas et al. (2008).

position of the sorption edge, as measured at 50% of maximum sorption, shifts by approximately 0.6 pH units between experiments with 7 nm particles and all other experiments with larger (>25 nm) particles.

All these data suggest the necessity for revisiting some basic assumptions in the existing colloid-facilitated transport models. These models generally treat colloidal particles, regardless of particle size, in the same way as the corresponding bulk phase in terms of their surface properties and phase stabilities. The data obtained so far, however, suggest that small colloidal particles, especially with sizes less than 10 nm, may play a more significant role in contaminant transport in natural systems than previously thought (Banfield and Zhang 2001; Waychunas et al. 2005; Madden et al. 2006; Hochella et al. 2008).

However, caution should be exercised in generalizing the foregoing discussion on the effect of particle size on particle surface chemistry because contradicting data have also been reported. For example, Campos et al. (2007) determined the surface charge density of spinel-type ferrite nanoparticles using electrochemical measurements. They observed that, for large-size particles, the surface is fully ionized, whereas, for smaller particles, the saturation value of the superficial

charge density is significantly reduced. The contradicting results warrant further research.

Dissolution kinetics

According to a recent crystal dissolution theory (Lasaga and Lüttge 2003), the bulk dissolution rate is determined by the movement of step waves around etch pits developed on crystal surfaces. The spreading of step trains does not occur spontaneously until the pit size reaches a critical value (r^*). In analogue to crystal nucleation [eqn. 1], r^* for a two dimensional pit is described by:

$$r^* = -\frac{2\gamma V_m}{RT \ln \Omega} \tag{4}$$

The dissolution rate of a particle of radius r can be described by (Tang et al. 2004):

$$R(r) \approx R_0 \left(1 - \frac{r^*}{r}\right) \tag{5}$$

where R_0 is the dissolution rate of the bulk phase $(r \to \infty)$. Using the data provided by Tang et al. (2004), the functional dependence of the dissolution rate of apatite particles on saturation degree and particle size is illustrated in Fig. 10.3. As the degree of undersaturation increases, the critical

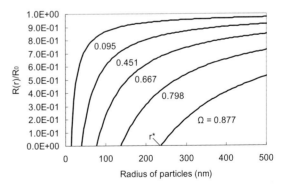

Fig. 10.3 Dissolution rate of nanoparticles as a function of particle size and undersaturation degree. Note that the critical pit size for dissolution decreases with increasing undersaturation degree.

pit size decreases, and, for a given initial particle size, the dissolution rate approaches the bulk phase value. By contrast, for a fixed saturation degree, as the particles become smaller due to dissolution, the dissolution rate decreases and eventually diminishes as the particle size approaches the critical pit size r^*. This self-inhibiting effect has been experimentally demonstrated by Tang et al. (2004) for the dissolution of apatite particles. For a given degree of undersaturation, the crystallites remaining at the end of each dissolution experiment practically have the same size distribution as predicted by Eqn. 4 regardless of initial particle sizes.

This self-inhibiting effect has two important geochemical implications. First, biological materials such as bones and shells are generally nanocomposite with nanoparticles as their building blocks (Alivisatos 2000). There is a good reason for organisms to select nanosized particles, because, according to Eqn. 5, these particles are resistant to demineralization even in undersaturated media, therefore rendering biomaterials more stable. Second, it is generally assumed that colloidal particles in a solution unsaturated with respect to their bulk phase will not be stable and therefore dissolve quickly. This assumption may not be valid. According to Eqn. 5, nanosized particles could be quite resistant to dissolution when their sizes approach a critical pit size r^*. This implies that colloidal particles, once formed, may persist for a long time period even in undersaturated geologic media. This is important for colloid-facilitated transport modelling. Consider as an example the release of the radionuclide element Pu from a nuclear waste disposal site. Close to the release source, the solution is highly saturated with PuO_2, and PuO_2 colloidal particles thus form through nucleation. These colloids are usually called 'true colloids' to attest to the fact that they are composed of only a contaminant-bearing mineral phase. As it moves away from the source, the contaminant plume becomes diluted owing to fluid mixing along the flow path, and eventually the solution becomes undersaturated with respect to the PuO_2 phase. Traditionally, it would be assumed that the 'true colloids' would

become insignificant and quickly disappear further down the flow path owing to dissolution. According to Eqn. 5, however, this assumption may no longer be true, and it may have led to underestimating the effect of 'true colloids' on contaminant transport.

CHEMISTRY OF NANOPORES

Occurrence of nanopores

Nanopores are ubiquitous in geologic media and constitute an integral part of the total porosity of geologic materials (Wang et al. 2003a). For example, TEM observations reveal that diatomaceous materials exhibit both micrometer-scale and nanometer-scale pores (Fig. 10.4a). The nanopores are regularly distributed with a pore size of approximately 3 nm. It has been proposed that these nanopores are formed in the template of biomolecules (probably proteins) (Ollver et al. 1995; Lobel et al. 1996), similar to a process used by material scientists to synthesize nanoporous materials (Kresge et al. 1992). Nanometer-scale channels are often seen along a grain boundary or a reaction front (Fig. 10.4b). Although their origin is still unknown, these channels are believed to provide necessary passages for mass transport during mineral reactions, such as polysomatic reactions (Veblen 1991). Nanopores are also commonly associated with iron oxyhydroxides in soils (Wang et al. 2003a).

The contribution of nanopores to the total surface area of geologic materials can be very significant. In B-horizon soils, the pores with diameters smaller than 100 nm account for 10–40% of total porosity of the soils (Görres et al. 2000). The porosity of Georgia kaolinite is dominated by pores smaller than 10-nm (Tardy and Nahon 1985). Suetsugu et al. (2004) have measured the pore size distributions of alluvial soils and volcanic soils from the B-horizon. They have found that nanopores with diameters less than 50 nm account for up to 20% of the total pore volumes. Since the specific area is inversely proportional to the pore diameter, the contribution of nanopores

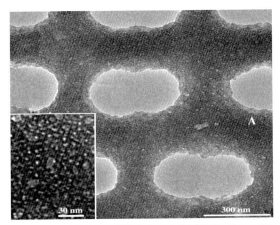

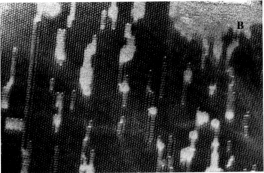

Fig. 10.4 Transmission electron microscope (TEM) images of nanopores in geologic materials (Wang et al., 2003a). (a) Diatomaceous materials displaying both micrometre-scale and nanometre-scale pore structures; (b) Partially alterated pyroxene crystal with the presence of nanometer channels at the tips of alteration front.

to the total surface area in those materials is expected to be very high, most likely greater than 90%.

Surface chemistry of nanopores

It is known that an electric double layer (EDL) develops on a solid surface when the surface contacts an aqueous solution (e.g. Stumm 1992). The thickness of EDL on a flat surface (*L*) can be estimated by:

$$L = \frac{1}{3.29 I^{1/2}} \qquad (6)$$

where *I* is the ionic strength. In dilute solutions, the EDL can extend up to approximately 100 nm. Now let us imagine rolling up the surface to make a cylindrical pore and shrinking the pore size to nanometres. It is reasonable to expect that the structure of the EDL in the nanopore, if it exists, would be very different from that on a flat surface; it has to be, at least, much more compressed. It has been shown the surface of nanopores behave quite differently from the corresponding unconfined surface in terms of surface acidity and ion sorption capability (Wang et al. 2002; Wang et al. 2003a,b).

Wang et al. (2002) performed acid-base titrations on two alumina materials: nanoporous materials and alumina particles. For a meaningful comparison, the two materials were characterized with TEM to ensure they had a similar mineral phase and crystallinity. From the titration data, the PZC is calculated to be 9.1 for nanoporous alumina and 8.7 for alumina particles, both within the range reported for aluminium oxides. Thus, the nanopore confinement seems to have little effect on the PZC of the pore surface, similar to the finding that the PZC of nanoparticles remains more or less the same regardless of particle sizes (see the discussion in the previous section). However, the titration results do suggest distinct behaviours in the surface charge density between the two materials. In Fig. 10.5(a), the surface charge density on the nanoporous alumina is plotted against that on the alumina particles for given ΔpH (= pH − PZC). If the surface charge difference between the two materials were mainly due to the specific surface area difference, all the experimental data points would fall on a 1:1 straight line. Apparently, this is not the case. The surface charge density on the nanoporous alumina is much higher than that on the alumina particles. This high surface charge density is postulated to be caused by the nanopore confinement. It is interesting to note in Fig. 10.5(a) that the surface charge on a confined surface is less sensitive to ionic strength changes than that on an

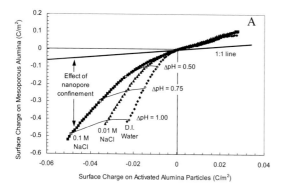

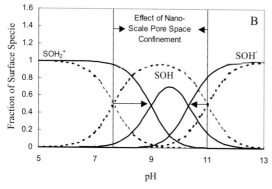

Fig. 10.5 Comparison of surface protonation and deprotonation between nanoporous alumina and alumina particles. Modified from Wang et al. (2002, 2003a).

nanotubes possess both confined (internal) and unconfined (external) surfaces whereas the nanorods only have the external surface. The surface acidity distributions, $f(pK)$, on the materials are calculated using a so-called single-pK model, which assumes that proton uptake on a single of population binding sites follows a Langmuir isotherm (Contescu et al. 1993):

$$Q = \int_{pK_{min}}^{pK_{max}} \frac{f(pK)}{1+10^{-pK+pH}} d(pK), \qquad (7)$$

where Q is the overall degree of protonation of material surfaces estimated from the pH titration results. As shown in Fig. 10.6(a), the two materials display distinct surface acidity distributions.

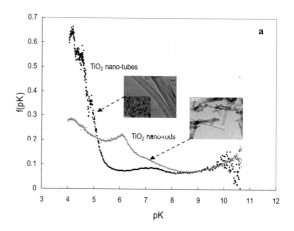

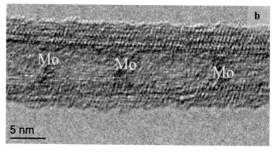

Fig. 10.6 Surface chemistry of TiO$_2$ nanomaterials. (a) Surface acidity distributions on TiO$_2$ nanotubes and nanorods; (b) Preferential precipitation of Mo inside a TiO$_2$ nanotube. Modified from Wang et al. (2003b; 2008).

unconfined surface. Wang et al. (2003a) have further estimated the surface acidity constants for the two materials: pK$_1$ = 9.0, pK$_2$ = 10.3 for nanoporous alumina, and pK$_1$ = 7.7, pK$_2$ = 11.0 for alumina particles. Therefore, because of the nanopore confinement, the separation between the two acidity constants is significantly reduced. Consequently, within the nanopores, the neutral surface species is suppressed, thus making the nanopore surface either more positively or more negatively charged (Fig. 10.5b).

The difference in surface acidity between confined and unconfined surfaces is further confirmed for a TiO$_2$ oxide system. Wang et al. (2008) conducted a similar titration study on TiO$_2$ nanotubes and nanorods. The two materials were synthesized under comparable hydrothermal conditions. TiO$_2$

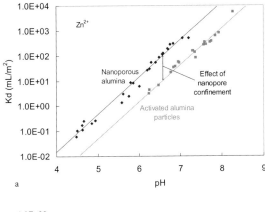

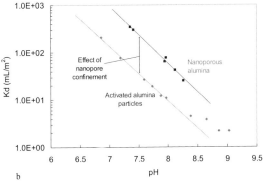

Fig. 10.7 Surface-normalized sorption coefficients (Kd) of Zn^{2+} and AsO_4^{3-} on nanoporous alumina and alumina particles.

Nanopore confinement also enhances the metal sorption capability of a solid surface. Wang et al. (2003a) measured the sorption coefficients (Kd) of Zn^{2+} and AsO_4^{3-} on both nanoporous alumina and alumina particles. They observed that, normalized to the surface area, the sorption coefficients for nanoporous alumina are systematically around 10 times higher than those for alumina particles for both the cation and the anion (Fig. 10.7). Two factors may contribute this enhanced sorption capability of nanopore surfaces. First, as discussed above, because of nanopore confinement, nanopore surfaces become either more positively or more negatively charged, thus enhancing both cation and anion sorption. Second, as discussed in more detail below, water inside nanopores is

more restrained and thus has a relatively low activity, which reduces the tendency for ion hydration and consequently increases the possibility for inner-sphere surface complexation of the adsorbates.

If one considers mineral precipitation as a dehydration process, the same mechanism can promote preferential mineral precipitation inside nanopores. To test this idea, Wang et al. (2003b) conducted a sorption experiment on TiO_2 nanotubes. Nanotubes provide an ideal system for this testing because both the inner surface (the confined surface) and the outer surface (the unconfined surface) of a tube almost have identical chemical compositions and crystal structures. Wang et al. (2003b) first suspended TiO_2 nanotubes in an aqueous solution spiked with MoO_4^{2-} and then imaged them, both chemically and structurally, using a high-resolution transmission electron microscope (HRTEM). The HRTEM images reveal clean outer surfaces and pillar-like Mo clusters precipitated inside the nanotubes (Fig. 10.6b), clearly indicating the preferential enrichment of trace metal in a confined space.

Caution should be exercised, however, in applying the nanopore confinement mechanism to organic molecules, especially to large organic molecules. In this case, the size-exclusion effect may become important and, as a consequence, large molecules may not be able to access nanopores. Thus, the sorption capability of a nanoporous material may decrease with a reduction in pore size. This is probably the reason for the observed reduction in $[Ni(en)_3]^{2+}$ sorption on alumina with decreasing pore sizes (Baca et al. 2008).

Water and ions in nanopores

The behaviour of water confined in nanopores is highly relevant to understanding many fundamental biological and geochemical processes. During the last decade, much research has been devoted to this topic. For example, using molecular dynamic simulations, Senapati and Chandra (2001) have calculated the dielectric constant of liquid water confined in a spherical nanocavity.

Their simulations suggest that the dielectric constant of water in the cavity is significantly smaller than that of bulk water. A nearly 50% decrease of the dielectric constant is estimated for water confined in a cavity with a diameter of 1.2 nm. Note that in those simulations the cavity surface is assumed to be uncharged and, therefore, there is no electrostatic interaction between the water molecules and the cavity surface. The calculated reduction of the dielectric constant thus originates purely from confinement. This reduction is expected to be even larger if the electrostatic interaction of water molecules with the charged sites of the cavity surface is taken into account. As far as surface complexation is concerned, the reduction of the dielectric constant (ε) inevitably leads to the decrease in the salvation energy of metal cation $(\Delta G_{s,M^{n+}})$, according to Born salvation equation:

$$\Delta G_{s,M^{n+}} = \omega_{M^{n+}}(1/\varepsilon - 1) \qquad (8)$$

where $\omega_{M^{n+}}$ is the Born salvation coefficient of cation M^{n+}. Therefore, inner sphere complexation is preferable owing to nanopore confinement.

Numerous studies have been conducted on the water phase transition under conditions of nanopore confinement. Teixeira et al. (1997) suggested that the behaviour of confined water is similar to that of supercooled water at low (~30°K) temperatures, implying a low water activity in nanopores. Using the perturbation theory of statistical mechanics, Keshavarzi et al. (2006) determined that the critical temperature of water increases with pore size and finally reaches its macroscopic value as the diameter of the pore approaches infinity. Based on extensive molecular dynamics simulations, Takaiwa et al. (2008) documented the existence of at least nine ice phases for water in single-walled carbon nanotubes at atmospheric pressure, with each having a structure that maximizes the number of hydrogen bonds under the cylindrical confinement. They then proposed a phase diagram which describes the ice structures and their melting points as a function of the tube diameter (Fig. 10.8). Furthermore, Cámara & Bresme (2004) have studied a Lennard-Jones liquid

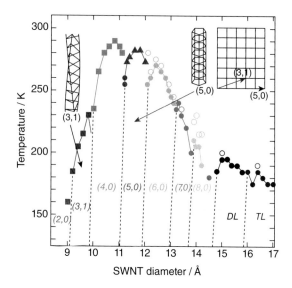

Fig. 10.8 Predicted phase diagram of water in single-walled carbon nanotubes at atmospheric pressure. Modifed from Takaiwa et al. (2008).

confined in a wedge-shaped nanopore and shown that small deviations from the parallel slit geometry result in non-uniform pressures and density profiles along the pore. Under conditions of high confinement and thermodynamic states close to the triple point, wedge-shaped pores can induce the formation of solid phases in specific regions within the pore.

Transport properties of water and ions in nanoscale confinement remain an active research area. Using molecular dynamics simulations, Liu et al. (2003) investigated the molecular distribution and diffusion of water confined in hydrophobic nanopores. They have found that these water molecules have some unusual behaviour as compared with those in the bulk phase. Density profiles are highly inhomogeneous in nanopores. The diffusivity of water in nanopores is much lower than that in the bulk phase and decreases as the pore size decreases. Interestingly, the diffusivity in the direction parallel to the channel is four to ten times higher than that in the perpendicular direction. Carrillo-Tripp et al. (2004) have shown that the energy cost of constraining a hydrated potassium ion inside a narrow nanopore is smaller than the energy cost of constraining the smaller hydrated sodium ion. The former allows for a greater distortion of its hydration shell and, therefore, can maintain a better coordination. In this way, the larger ion can go through narrow pores more easily than the smaller ions. This apparent contradiction is directly relevant to the molecular basis of ion selective nanopores. Since it does not depend on the molecular details of the pore, this mechanism could also operate in many different types of nanotubes, from biological to synthetic. Yang and Garde (2007) modelled the partitioning of cations into negatively charged nanopores from an aqueous solution. They demonstrated that, over a range of intermediate negative charge densities, nanopores display both thermodynamic as well as kinetic selectivity toward partitioning of the larger K^+ and Cs^+ ions into their interior over the smaller Na+ ions. Specifically, the driving force is in the order $K^+ > Cs^+ > Na^+$, and K^+ and Cs^+ ions enter the pore much more rapidly than Na^+ ions. At higher charge-densities, the driving force for partitioning increases for all cations – it is highest for K^+ ions – and becomes similar for Na^+ and Cs^+ ions. The variation of thermodynamic driving force and the average partitioning time with the pore-charge density together suggest the presence of free-energy barriers in the partitioning process.

Geochemical implications

The nanopore confinement mechanism discussed above offers a new understanding of many fundamental geochemical processes. First, since nanopores generally are too small to be accessible by microorganisms, the preferential enrichment of metals in nanopores can significantly reduce the bioavailability of metals in natural systems. Furthermore, being strongly bound on nanopore surfaces, the metals inside nanopores are expected to be less liable to release than those sorbed on large pore surfaces. This may explain why metal-ion desorption from geologic materials often displays a characteristic of two-stage release. One fraction desorbs rapidly, whereas the remainder desorbs rather slowly, and the proportion of slowly desorbing metal increases with sorption time (e.g. Glover et al. 2002; Wang et al. 2003a). The slow desorption may be attributed to strong sorption or even precipitation of metals in nanopores. Because of the effect of nanopore confinement, the surface sites on the same mineral surface may exhibit different sorption affinities.

Current evidence suggests that water confined in nanopores has a low activity and mobility (see the discussion in the previous section). This change in water behaviour has a profound impact on both ion sorption and mineral precipitation in nanopores. First, it reduces the tendency for ion hydration and, therefore, increases the possibility for inner-sphere complexation on nanopore surfaces. Second, it forces solutes to precipitate from solutions and therefore results in preferential precipitation of minerals in nanopores, which is similar to the non-electrical exclusion of ions in thin water-films (Zilberbrand 1997). Interestingly, the effect of nanopore confinement on ion hydration may provide a hint to solving the classic 'dolomite problem'. Dolomite is common in ancient platform carbonates but rare in Holocene sediments; there has been not much success in precipitating this mineral in the laboratory at Earth surface temperatures (e.g. Warren 2000). It has been proposed that dolomite formation is inhibited by strong Mg^{2+} hydration (de Leeuw and Parker 2001). The nanopore confinement may reduce Mg^{2+} hydration and thus enhance dolomite formation in natural systems.

The discussion above suggests that ion-sorption measurements on disaggregated geologic materials may not necessarily represent chemical/physical conditions in actual systems, because the disaggregation may destroy part of nanopore structures in the original material. Conca and Wright (1992) measured the apparent sorption coefficient of radionuclides in bentonite as a function of compaction density and observed that the sorption capability of the material first decreases with physical compaction and then increases at a high compaction density. This phenomenon can be easily explained by the nanopore confinement effect discussed above. The initial decrease in sorption can be attributed to the reduction of sorption site accessibility as the material is compacted. Then, further compaction probably creates more nanopores at the expense of large pores, thus enhancing radionuclide sorption.

The modification of surface chemistry and pore water activity by nanopore confinement may also affect mineral dissolution kinetics. It has been observed that diatomaceous materials display unique nonlinear dissolution kinetics (Van Cappellen and Qiu 1997), with dissolution rates nearly linear in the vicinity of the equilibrium point. With increasing distance from equilibrium, there is a pronounced transition in the functional dependence on the relative degree of understaturation. Beyond the transition, the dissolution rate rises much faster with increasing degree of solution undersaturation. As shown in Fig. 10.4(a), diatomaceous materials generally exhibit two levels of pore structures. The observed nonlinear dissolution kinetics are probably due to the presence of nanopores in these materials. Nanopores and large pores could display different chemical affinities for dissolution. A similar process can be envisioned to explain the large discrepancy between laboratory measurements and field observations in weathering rates (e.g. Brantley 1992), given the fact that nanopores in soils can account for more than 90% of their total surface area.

Another recently observed phenomenon related to nanopore confinement warrants mentioning. Anzalone et al. (2006) measured dissolution and deformation at grain-grain contacts of geologic materials under a high stress, using a surface-force apparatus (SFA). A thin water-film was observed to exist at grain contacts under the applied stress of 10 to 500 atm. The thickness of the film varies from 8 to 1 nm, as the stress increases. The film thickness also depends on solution chemistry. The diffusion of ions in the film was determined to be approximately 40 times slower than that in bulk water, but not sufficiently slow to prevent ion diffusion within the film. This study of pressure solution examined two cases – the symmetric case of mica in contact with mica and the asymmetric case of quartz in contact with mica. Interestingly, it was observed that the asymmetric contact situation promotes pressure solution. The actual mechanism for this is still not clear. It is probably due to the overlapping of two different electric double layers (EDLs) that creates an electric potential gradient across the water film. This is an area of research that deserved further exploration.

The foregoing discussion also suggests that existing surface-complexation models (e.g. Davis and Kent 1990; Dzombak and Morel 1990), which have been developed mostly for unconfined surfaces, probably are inadequate for modelling ion sorption on nanopore surfaces without taking into account the effect of nanopore confinement. Zhmud et al. (1997) have developed a charge-regulation model for the surface of porous matrices in which a pore is represented by a cylindrical cavity and the overlap of the electrical double-layer is taken into account by diminishing the radius of the cylinder. The model predicts a decrease in surface-charge density with decreasing pore size, which seems inconsistent with experimental results reported by Wang et al. (2002, 2003a). Therefore, further research is also needed in this area.

NANOSTRUCTURES AS ENVIRONMENTAL INDICATORS

As previously discussed, nanoparticles of iron (oxyhydr)oxides are sensitive to environmental changes near the Earth's surface (Banerjee 2006; Navrotsky et al. 2008), so they can be used as environmental indicators. Another example is shown here – the nanostructures in mixed-layer phyllosilicates. Interstratification in phyllosilicates is a common phenomenon, involving stacking of two different silicate layers (or silicate and hydroxide layers), either periodically or non-periodically along the crystallographic c-axis, leading to the formation of an amazingly rich collection of nanometre-scale layer structures (e.g. Veblen et al. 1990). Using electron diffraction and high-resolution TEM (HRTEM) analyses, Wang and Xu (2006) illustrated the occurrence of both

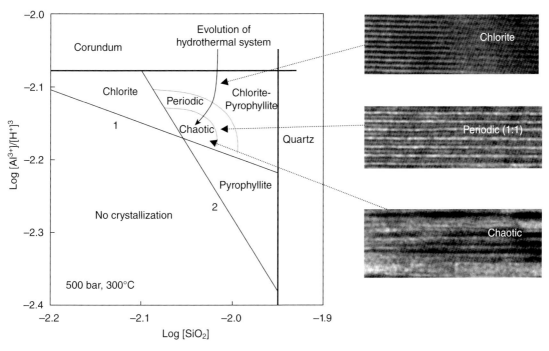

Fig. 10.9 Hydrothermal evolution of a chlorite-pyrophyllite system and the formation of related nanostructures. Modified from Wang and Xu (2006).

periodic and non-periodic layer structures in a chlorite-pyrophyllite sample from an ancient hydrothermal system. The non-periodic-structure domains were observed to be intergrown with the periodic-structure domains, as well as the domains dominated by either chlorite or pyrophyllite layers, with repeated transition from one interstratification mode to another. Selected-area electron diffraction (SAED) patterns indicate a typical layered stacking period of 2.34 nm along the c-axis direction, equal to the sum of d_{001} values for chlorite (1.42 nm) and pyrophyllite (0.92 nm).

Based on the chaos theory, Wang and Xu (2006) demonstrate that both periodic and non-periodic interstratification can autonomously arise from simple kinetics of mineral growth from solutions. In their model, the sequence of layer stacking is determined by two competing factors: (i) the affinity of each end-member structural component for attaching to the surface of the preceding layer: and (ii) the strain energy created by stacking

next to each other two silicate layers of different structural configuration. Chaotic (or non-periodic) interstratification emerges when the contacting solution becomes slightly supersaturated with respect to both structural components. Wang and Xu (2006) show that the transition between layer-stacking modes reflects a change in chemical environment during mineral crystallization. As the saturation of the solution with respect to the end-member phases decreases, the mineral precipitation changes from end-member layers to periodic layers and then to non-periodic layers (Fig. 10.9).

NANOPROBING AND NANOMATERIAL SYNTHESIS

In-situ probing of mineral–water interactions has been, and will remain to be, an active research area in nanogeochemistry. Various probing techniques, including the surface force apparatus (SFA) and the atomic force microscope (AFM), are tools available for this purpose. Work performed by Lower et al. (2001) on microbe–mineral interaction is particularly interesting. They have developed a force-sensing probe by linking a minute bacteria-coated bead to a silicon nitride cantilever (Lower et al. 2000). The probe is placed in a fluid cell used for force measurements and imaged with a scanning laser confocal microscope. Using this technique, Lower et al. (2001) have quantitatively measured the infinitesimal forces that characterize interactions between *Shewanella oneidensis* (a dissimilatory metal-reducing bacterium) and goethite. Force measurements with sub-nanonewton resolution are made in real time with living cells under aerobic and anaerobic solutions, as a function of the distance, in nanometers, between a cell and the mineral surface. The measurements show that the affinity between *S. oneidensis* and goethite rapidly increases by two to five times under anaerobic conditions. Such measurements combined with related modelling capability (e.g. Frink and van

Swol 1996) will provide a powerful tool for understanding many interfacial phenomena.

Geologic systems are notoriously known for being 'messy'. In order to probe a specific nanoscale phenomenon, it is often critical to synthesize 'model' nanomaterials with controllable structures and dimensions. Thanks to the development of nanotechnology, it is now possible to synthesize these materials in a routine manner. The materials of interest include two general types: nanoparticles and nanoporous materials. Many methods have been developed for synthesis of metal (hydr)oxide nanoparticles. The methods include hydrothermal synthesis, reversed micelles, and chemical precipitation (see a comprehensive review by Burda et al. (2005)). Fig. 10.10(a) shows the magnetite nanoparticles synthesized through chemical precipitation. Nanoporous materials are usually synthesized by a supramolecular templating process (Kresge et al. 1992). Typically, this process employs organic molecules of low symmetry that can be organized into a well-defined supramolecular assembly in a solution, most commonly amphiphilic surfactant

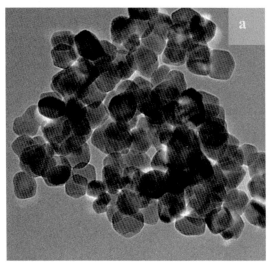 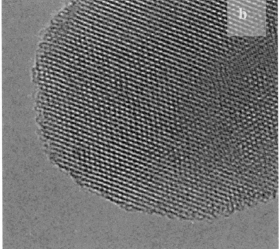

Fig. 10.10 Transmission electron microscope (TEM) images of synthetic nanomaterials. (a) Magnetite nanoparticles synthesized through chemical precipitation. The particle size is approximately 6 nm. (b) Nanoporous silica synthesized with a supramolecular templating process. The pore size is approximately 3 nm.

molecules or polymers composed of hydrophobic and hydrophilic parts. Above the critical concentration, surfactants in an aqueous solution assemble into micelles, spherical or cylindrical structures that maintain the hydrophilic parts of the surfactant in contact with water while shielding the hydrophobic parts within the micellar interior. Further increases in surfactant concentration result in the self-organization of micelles into periodic hexagonal, cubic or lamellar nanophases. Inorganic precursors then precipitate on the surfactant structural template. Inorganic nanoporous materials are obtained through removal of surfactants by chemical methods or calcination. Fig. 10.10(b) shows nanoporous silica synthesized with a supramolecular templating process.

NANOSTRUCTURED ENVIRONMENTAL MATERIALS

The current energy-environment dilemma poses a significant challenge. The development of nanotechnology may hold part of the solution to this problem. Possible environmental application of nanomaterials has been explored over the past two decades since the beginning of the nanotechnology era. Most studies to date have been focused on using nanomaterials as catalysts to degrade specific contaminants or as adsorbents to remove/ immobilize contaminants. For example, TiO_2 oxide nanoparticles have been used for photocatalytic purification of water and air (e.g. Ruan et al. 2001). Xu et al. (2006) suggest that the titanium oxide nanotube could be a potential catalyst for photocatalytic oxidation of low concentrations of acetaldehyde at room temperature, because the commercial P25 nanoparticles usually deactivate quickly and nanotubes are more reactive than P25 titania over a long time period. They postulate that the bending and curving of the Ti–O octahedral sheets of a nanotube result in electrical polarity being developed on outer and inner surfaces, thus enhancing the electron-hole separation and the observed photo-oxidation by the nanotubes. Furthermore, the idea of using bimetallic iron

nanoparticles to reduce halogenated organic solvents has been proposed for subsurface remediation (e.g. Zhang et al. 1998; Xu and Zhang 2000). Bimetallic nanoparticles are expected to be much more reactive than traditional zero-valent materials, owing to their large specific areas and the rapid reaction rates enhanced by noble metals.

Owing to their flexibility for surface modification, and also because of their large specific surface areas, nanoporous materials have been widely proposed as highly effective adsorbents. Nanoporous silica, with a monolayer of thiol (-SH) group grafted on its pore surface, displays a high sorption capacity for removing mercury from aqueous solutions (Feng et al. 1997; Chen et al. 1999). Uncalcined nanoporous silicate materials synthesized with hexadecyl-trimethylammonium bromide as a template are able to remove significant amounts of trichlorethylene and tetrachloroethylene from water (Zhao et al. 2000), while calcined nanoporous silicates or titanosilicates are found to have a capability for removing copper, lead and uranyl ions from aqueous solutions (Xu et al. 1999; Shin et al. 1999; Jung et al. 2001). An ordered nanoporous anion-exchange inorganic/ organic hybrid resin has been suggested for radionuclide separation (Ju et al. 2000). In addition, nanostructured materials have been explored for their potential use as ion getters or waste forms to immobilize radionuclides in nuclear waste management (Wang et al. 2003b; Wang et al., 2007).

PERSPECTIVES

Nanogeochemistry will remain to be an active research area for decades to come. For such a young and dynamic discipline, it is difficult, if not impossible, to offer any personal perspective on its future development. It is certain, however, that nanogeochemistry will continue adsorbing new ideas and importing new techniques from nanoscience and nanotechnology studies. There are several research areas in nanoscience and nanotechnology that merit attention. First, water behaviour in nanopores will continue to be an important topic in nanoscience research.

Nanofluidic studies have potential to enhance understanding of fluid movement in nanoconfinement (e.g. Surani and Qiao 2006; Qiao and Aluru 2005; Gong et al. 2007), which will have important implications to many geochemical processes, specifically including the movement of subsurface fluids and the related mass transfers. Second, ion transport in biological ion channels will continue to attract the attentions of both biologists and material scientists (e.g. Carrillo-Tripp et al. 2004). Any breakthrough in this area will have a huge impact, not only on understanding of the metabolic behaviour of biological cells but also on the ability to mimic biomembranes for highly selective chemical separations for water purification and desalination. Finally, progress in nanotechnology can potentially lead to the development of chemical devices able to detect a single molecule (Dekker 2007), which could potentially revolutionize chemical and geochemical analyses. To conclude, nanogeochemistry is still in its infant stage, and 'there's plenty of room' in this area for geochemists to explore.

ACKNOWLEDGEMENTS

Sandia is a multi-program laboratory operated by Sandia Corporation, a Lockheed Martin Company for the United States Department of Energy's National Nuclear Security Administration under contract DE-AC04-94AL85000. This work is supported by US DOE Laboratory Directed Research & Development (LDRD) Program, Advanced Fuel Cycle Initiative (AFCI), NASA Astrobiology Institute (N07-5489), and National Science Foundation (EAR-0810150).

FURTHER READING

Banfield, JF and Navrotsky, A. (eds.) (2001) Nanoparticles and the Environment. *Mineralogical Society of America*, v. 23, Washington, DC.
Burda, C, Chen, XB, Narayanan, R and El-Sayed, MA. (2005) Chemistry and properties of nanocrystals of different shapes. *Chemical Reviews* 105, 1025–102.

Gross, M. (1999) *Travels to the Nanoworld: Miniature Machinery in Nature and Technology.* Perseus Publishing, Cambridge.
Roduner, E. (2006) *Nanoscopic Materials: Size-Dependent Phenomena.* RSC Publishing, Cambridge.
Zhang, J, Wang, ZL, Liu, J, Chen, S and Liu, GY. (2002) *Self-Assembled Nanostructures.* Kluwer Academic Publishers, New York.

REFERENCES

Abbas, Z, Labbez, C, Nordholm, S and Ahlberg, E. (2008) Size-dependent surface charging of nanoparticles. *Journal of Physical Chemistry C* 112: 5715–23.
Alivisatos, AP. (2000) Enhanced: Naturally aligned nanocrystals. *Science* 289: 736–7.
Anzalone, A, Boles, J, Greene, G, Young, K, Israelachvili, J and Alcantar, N. (2006) Confined fluids and their role in pressure solution. *Chemical Geology* 230: 220–31.
Baca, M, Carrier, X and Blanchard, J. (2008) Confinement in nanopores at the oxide/water interface: Modification of alumina adsorption properties. *Chemistry-A European Journal* 14: 6142–8.
Banfield, JF and Navrotsky, A. (2001) Nanoparticles and the Environment. *Mineralogical Society of America*, v. 23, Washington, DC.
Banfield, JF and Zhang HZ. (2001) Nanoparticles in the environment. *Reviews In Mineralogy & Geochemistry* 44: 1–58.
Banerjee, SK. (2006) Environmental magnetism of nanophase iron minerals: Testing the biomineralization pathway. *Physics of the Earth and Planetary Interiors* 154: 210–21.
Barnard, AS and Xu, H. (2008) An environmentally sensitive phase map of titania nanocrystals. *ACS NANO* 2(11): 2237–42.
Barnard, AS, Yeredla, RR and Xu, HF. (2006) Modelling the effect of particle shape on the phase stability of ZrO$_2$ nanoparticles. *Nanotechnology* 17: 3039–47.
Bazylinski, DA and Moskowitz, BM. (1997) Microbial biomineralization of magnetic iron minerals: Microbiology, magnetism and environmental significance. *Reviews in Mineralogy* 35: 181–223.
Brantley, SL. (1992) Kinetic dissolution and precipitation – Experimental and field results. In: Y Kharaka (ed.) *Water-Rock Interaction*, v. 1. Balkema, Rotterdam, Park City, pp. 3–6.
Burda, C, Chen, XB, Narayanan, R and El-Sayed, MA. (2005) Chemistry and properties of nanocrystals

of different shapes. *Chemical Reviews* **105**: 1025–1102.

Cámara, LG and Bresme, F. (2004) Liquids confined in wedge shaped pores: nonuniform pressure induced by pore geometry. *Journal of Chemical Physics* **120**: 11355–8.

Campos, AFC, Tourinho, FA, Aquino, R and Depeyrot, J. (2007) Probing interface and finite size effects in magnetic ferrite nanoparticles by electrochemical measurements. *Journal of Magnetism and Magnetic Materials* **310**: 2847–9.

Carrillo-Tripp, M, Saint-Martin, H and Ortega-Blake, I. (2004) Minimalist molecular model for nanopore selectivity. *Physical Review Letters* **93**: 168104/1–4.

Chen, T, Xu, H, Xie, Q, Chen, J, Ji, J and Lu, H. (2005) Characteristics and genesis of maghemite in China loess and paleosols: Mechanism for magnetic susceptibility enhancement in paleosols. *Earth and Planetary Science Letters* **240**: 790–802.

Chen, X, Feng, X, Liu, J, Fryxell, GE and Gong, M. (1999) Mercury separation and immobilization using self-assembled monolayers on mesoporous supports (SAMMS). *Separation Science and Technology* **34**: 1121–31.

Clarke, NS and Hall, PG. (1991) Adsorption of water vapor by iron oxides. 2. Water isotherms and X-ray photoelectron spectroscopy. *Langmuir* **7**: 678–82.

Conca, JL and Wright, J. (1992) Diffusion and flow in gravel, soil, and whole rock. *Applied Hydrogeology* **1**: 5–24.

Contescu, C, Jagiello, J and Schwarz, JA. (1993) Heterogeneity of proton binding-sites at the oxide solution interface. *Langmuir* **9**: 1754–65.

Davis, JA and Kent, DB. (1990) Surface complexation modeling in aqueous geochemistry. *Reviews in Mineralogy & Geochemistry* **23**: 177–260.

Dekker, C. (2007) Solid-state nanopores. *Nature Nanotechnology* **2**: 209–15.

de Leeuw, NH and Parker, SC. (2001) Surface-water interactions in the dolomite problem. *Physical Chemistry and Chemical Physics* **3**: 3217–21.

Dresselhaus, MS, Dresselhaus, G and Eklund, PC. (1996) *The Science of Fullerenes and Carbon Nanotubes: Their Properties and Applications*. Academic Press, San Diego.

Dzombak, DA and Morel, FMM. (1990) *Surface Complexation Modeling: Hydrous Ferric Oxide*. Wiley-Interscience, New York.

Feng, X, Fryxell, GE, Wang, LQ, Kim, AY, Liu, J and Kemner, KM. (1997) Functionalized monolayers on ordered mesoporous supports. *Science* **276**: 923–6.

Frink, LJD and van Swol, F. (1996) Oscillatory surface forces: A test of the superposition approximation. *Journal of Chemical Physics* **105**: 2884–90.

Gong, XJ, Li, JY, Lu, HJ, Wan, RZ, Li, JC, Hu, J and Fang, HP. (2007) A charge-driven molecular water pump. *Nature Nanotechnology* **2**: 709–12.

Görres, JH, Stolt, MA, Amador, JA, Schulthess CP and Johnson P. (2000) Soil pore manipulations to increase bioaccessible pore volume. In: O Sililo (ed.) *Groundwater: Past Achievement and Future Challenge*, AA Balkema, Rotterdam, Park City, pp.755–6.

Glover II, LJ, Eick, MJ and Brady, PV. (2002) Desorption kinetics of cadmium^{2+} and lead^{2+} from goethite: Influence of time and organic acids. *Soil Science Society of America Journal* **66**: 797–804.

Gross, M. (1999) *Travels to the Nanoworld: Miniature Machinery in Nature and Technology*. Perseus Publishing, Cambridge.

Hochella, MF. (2002) There's plenty of room at the bottom: Nanoscience in geochemistry. *Geochimica et Cosmochimica Acta* **66**: 735–43.

Hochella, MF, Jr, Lower, SK, Maurice, PA, Penn, RL, Sahai, N, Sparks, DL and Twining, BS. (2008) Nanominerals, mineral nanoparticles and Earth systems. *Science* **319**: 1631–5.

Honeyman, BB and Ranville, JF. (2002) Colloid properties and their effects on radionuclide transport through soils and groundwaters. In: P-C Zhang and PV Brady (eds.) Geochemistry of Soil Radionuclides. Madison, Wisconsin, *Soil Science Society of America* **59**: 131–63.

Jiang, Q, Lang, LH and Zhao, DS. (2001) Lattice contraction and surface stress of fcc nanocrystals. *Journal of Physical Chemistry B* **105**: 6275–7.

Jin, Q and Bethke, CM. (2002) Kinetics of electron transfer through the respiratory chain. *Biophysical Journal* **83**: 1797–808.

Ju, YH, Webb, OF, Dai, S, Lin, JS and Barnes, CE. (2000) Synthesis and characterization of ordered mesoporous anion-exchange inorganic/organic hybrid resins for radionuclide separation. *Industrial & Engineering Chemistry Research* **39**: 550–3.

Jung, J, Kim, JA, Suh, JK, Lee, JM and Ryu, SK. (2001) Microscopic and macroscopic approaches of Cu(II) removal by FSM-16. *Water Research* **35**: 937–42.

Keshavarzi, TE, Sohrabi, R and Mansoori, GA. (2006) An analytic model for nano confined fluids phase-transition: Applications for confined fluids in nanotube and nanoslit. *Journal of Computational and Theoretical Nanoscience* **3**: 134–41.

Kresge, CT, Leonowicz, ME, Roth, WJ, Vartuli, JC and Beck, JS. (1992) Ordered mesoporous molecular-sieves synthesized by a liquid-crystal template mechanism. *Nature* **359**: 710–12.

Lasaga, AC and Lüttge, A. (2003) A model for crystal dissolution. *European Journal of Mineralogy* **15**: 603–15.

Lee, JH, Roh, Y and Hur, HG. (2008) Microbial production and characterization of superparamagnetic magnetite nanoparticles by Shewanella sp HN-41. *Journal of Microbiology and Biotechnology* **18**: 1572–7.

Liu, YC, Wang, Q and Lu, LH. (2003) Water confined in nanopores: its molecular distribution and diffusion at lower density. *Chemical Physics Letters* **381**: 210–15.

Lobel, KD, West, JK and Hench, LL. (1996) Computational model for protein-mediated biomineralization of the diatom frustule. *Marine Biology* **126**: 353–60.

Lower, SK, Hochella, MF and Beveridge, TJ. (2001) Bacterial recognition of mineral surfaces: Nanoscale interactions between Shewanella and alpha-FeOOH. *Science* **292**: 1360–3.

Lower, SK, Tadanier, CJ and Hochella, MF. (2000) Measuring interfacial and adhesion forces between bacteria and mineral surfaces with biological force microscopy. *Geochimica et Cosmochimica Acta* **64**: 3133–9.

Lüning J, Rockenberger, J, Eisebitt, S, Rubensson, J-E, Karl, A, Kornowski, A, Weller, H and Eberhardt, W. (1999) Soft X-ray spectroscopy of single sized CdS nanocrystals: Size confinement and electronic structure. *Solid State Communications* **112**: 5–9.

Lyklema, J. (2000) *Fundamentals of Interface and Colloid Science*. Academic Press, New York.

Madden, AS, Hochella, MF and Luxton, TP. (2006) Insights for size-dependent reactivity of hematite nanomineral surfaces through Cu^{2+} sorption. *Geochimica et Cosmochimica Acta* **70**: 4095–104.

Navrotsky, A, Mazeina, L and Majzlan, J. (2008) Size-driven structural and thermodynamic complexity in iron oxides. *Science* **319**: 1635–8.

Ollver, S, Kuperman, A, Coombs, N, Lough, A and Ozln, GA. (1995) Lamellar aluminophosphates with surface patterns than mimic diatom and radiolarian microskeletons. *Nature* **378**: 47–50.

Palenik, CS, Utsunomiya, S, Reich, M, Kesler, SE, Wang, LM and Ewing, RC. (2004) 'Invisible' gold revealed: Direct imaging of gold nanoparticles in a Carlin-type deposit. *American Mineralogist* **89**: 1359–66.

Qiao, R and Aluru, NR. (2005) Surface-charge-induced asymmetric electrokinetic transport in confined silicon nanochannels. *Applied Physics Letters* **86**(14): 143105.

Raiswell, R, Tranter, M, Benning, LG, Siegert, M, De'ath, R, Huybrechts, P and Payne, T. (2006) Contributions from glacially derived sediment to the global iron (oxyhydr)oxide cycle: Implications for iron delivery to the oceans. *Geochimica et Cosmochimica Acta* **70**: 2765–80.

Ratner, D and Ratner, M. (2004) *Nanotechnology and Homeland Security*. Pearson Education, Inc., New Jersey.

Ratner, M and Ratner, D. (2003) *Nanotechnology: A General Introduction to the Next Big Idea*. Prentice Hall, New Jersey.

Reich, M, Utsunomiya, S, Kesler, SE, Wang, LM, Ewing, RC and Becker, U. (2006) Thermal behaviour of metal nanoparticles in geologic materials. *Geology* **34**: 1033–6.

Roduner, E. (2006) *Nanoscopic Materials: Size-Dependent Phenomena*. RSC Publishing, Cambridge.

Roh, Y, Vali, H, Phelps, TJ and Moon, JW. (2006) Extracellular synthesis of magnetite and metal-substituted magnetite nanoparticies. *Journal of Nanoscience and Nanotechnology* **6**: 3517–20.

Ruan, SP, Wu, FQ, Zhang, T, Gao, W, Xu, BK and Zhao, MY. (2001) Surface state studies of TiO_2 nanoparticles and photocatalytic degradation of methyl orange in aqueous TiO_2 dispersions. *Materials Chemistry and Physics* **69**: 7–9.

Senapati, S and Chandra, A. (2001) Dielectric constant of water confined in a nanocavity. *Journal of Physical Chemistry B* **105**: 5106–9.

Shin, YS, Burleigh, MC, Dai, S, Barnes, CE and Xue, ZL. (1999) Investigation of uranyl adsorption on mesoporous titanium-based sorbents. *Radiochimica Acta* **84**: 37–42.

Simakov, SK. (2003) *Physico-Chemical Aspects of Diamond-Bearing Eclogites Formation in the Upper Mantle and Earth Crust Rocks*. NESC, Magadan.

Simakov, SK, Dubinchuk VT, Novikov, MP and Drozdova, IA. (2008) Formation of diamond and diamond-type phases from carbon-bearing fluid at PT parameters corresponding to processes in the earth's crust. *Doklady Earth Sciences* **421**: 835–7.

Simakov, SK, Grafchikov, AA, Sirotkin, AK, Drozdova, IA, Lapshin, AE and Grebenshchikova, EA. (2001) Synthesis of carbon nanotubes and fullerite structures at *PT* parameters corresponding to natural mineral formation. *Doklady Earth Sciences* **376**: 87–9.

Sonnefeld J. (1993) An analytic expression for the particle size dependence of surface acidity of colloidal silica. *Journal of Colloid and interfacial science* **155**: 191–9.

Steefel, CI & Van Cappellen, P. (1990) A new kinetic approach to modeling water-rock interaction – The role of nucleation, precursors, and Ostwald ripening. *Geochimica et Cosmochimica Acta* **54**: 2657–77.

Stumm, W. (1992) *Chemistry of the Solid-Water Interface*. Wiley Interscience, New York.

Suetsugu, A, Imoto, H, Mizoguchi, M and Miyazaki, T. (2004) Effects of nanoscale pores in soils on carbon evolution under extremely dry conditions. *Soil Science and Plant Nutrition* **50**: 891–7.

Surani, FB and Qiao Y. (2006) Infiltration and defiltration of an electrolyte solution in nanopores. *Journal of Applied Physics* **100**: 34311-1-4.

Suzuki, Y and Banfield, JF. (2004) Resistance to and accumulation of uranium by bacteria from a uranium-contaminated site. *Geomicrobiology Journal* **21**: 113–21.

Takaiwa, D, Hatano, I, Koga, K and Tanaka, H. (2008) Phase diagram of water in carbon nanotubes. *Proceedings of the National Academy of Sciences of the United States of America* **105**: 39–43.

Tang, RK, Wang, LJ and Nancollas, GH. (2004) Size-effects in the dissolution of hydroxyapatite: an understanding of biological demineralization. *Journal of Materials Chemistry* **14**: 2341–6.

Tardy, Y and Nahon, D. (1985) Geochemistry of laterites, stability of al-goethite, al-hematite, and Fe^{3+}-kaolinite in bauxites and ferricretes: an approach to the mechanism of concretion formation. *American Journal of Science* **285**: 865–903.

Teixeira, J, Zanotti, JM, Bellissent-Funel, MC and Chen, SH. (1997) Water in confined geometries. *Physica B* **234**: 370–4.

Van Cappellen, P and Qiu, L. (1997) Biogenic silica dissolution in sediments of the Southern Ocean. II. Kinetics. *Deep-Sea Research II* **44**: 1129–49.

van der Zee, C, Roberts, DR, Rancourt, DG and Slomp, CP. (2003) Nanogoethite is the dominant reactive oxyhydroxide phase in lake and marine sediments. *Geology* **31**: 993–6.

Veblen, DR. (1991) Polysomatism and polysomatic series: A review and applications. *American Mineralogist* **76**: 801–26.

Veblen, DR, Guthrie, GD, Jr, Livi, KJT and Reynolds, RC, Jr. (1990) High resolution transmission electron microscopy and electron diffraction of mixed-layer illite/smectite: Experimental results. *Clays and Clay Minerals* **38**: 1–13.

Wang, Y, Bryan, C, Xu, H, Pohl, P, Yang, Y and Brinker, CJ. (2002) Interface chemistry of nanostructured materials: Ion adsorption on mesoporous alumina. *Journal of Colloid and Interface Science* **254**: 23–30.

Wang, YF, Gao, H, Xu, H, Siegel, M and Konishi, H. (2008) Surface chemistry and stability of nanostructured materials in natural aquatic environments. *Geochimica et Cosmochimica Acta* **72**: A1002.

Wang, Y, Bryan, C, Xu, H and Gao, H. (2003a) Nanogeochemistry: Geochemical reactions and mass transfers in nanopores. *Geology* **31**: 387–90.

Wang, Y, Bryan C, Gao, H, Pohl, P, Brinker, CJ, Yu, K, Xu, H, Yang, Y, Braterman, PS and Xu, Z. (2003b) Potential Applications of Nanostructured Materials in Nuclear Waste Management. Sandia National Laboratories, Albuquerque, NM, SAND2003-3313.

Wang, Y, Gao, H, Yeredla, R, Xu, H and Abrecht, M. (2007) Control of surface functional groups on pertechnetate sorption on activated carbon. *Journal of Colloid and Interface Science* **305**: 209–17.

Wang, Y and Van Cappellen, P. (1996) A multicomponent reactive transport model of early diagenesis: Application to redox cycling in coastal marine sediments. *Geochimica et Cosmochimica Acta* **60**: 2993–3014.

Wang, Y and Xu, H. (2006) Geochemical chaos: Periodic and nonperiodic growth of mixed-layer phyllosilicates. *Geochimica et Cosmochimica Acta* **70**: 1995–2005.

Warren, J. (2000) Dolomite: Occurrence, evolution and economically important associations. *Earth-Science Reviews* **52**: 1–81.

Wilson, B, Dewers, T, Reches, Z and Brune, J. (2005) Particle size and energetics of gouge from earthquake rupture zones. *Nature* **434**: 749–52.

Waychunas, GA, Kim, CS and Banfield, JF. (2005) Nanoparticulate iron oxide minerals in soils and sediments: unique properties and contaminant scavenging mechanisms. *Journal of Nanoparticle Research* **7**: 409–33.

Xu, Y-M, Wang, R-S and Wu, F. (1999) Surface characterization and adsorption behavior of Pb(II) onto a mesoporous titanosilicate molecular sieve. *Journal of Colloid and Interface Science* **209**: 380–5.

Xu, HF, Vanamu, G, Nie, ZM, Konishi, H, Yeredla, R, Phillips, J and Wang, Y. (2006) Photocatalytic oxidation of a volatile organic component of acetaldehyde using titanium oxide nanotubes. *Journal of Nanomaterials spec. iss.* **2**: 78902.

Xu, Y and Zhang, WX. (2000) Subcolloidal Fe/Ag particles for reductive dehalogenation of chlorinated benzenes. *Industrial & Engineering Chemistry Research* **39**: 2238–44.

Yang L and Garde, S. (2007) Modeling the selective partitioning of cations into negatively charged nanopores in water. *Journal of Chemical Physics* **126**: 84706-1-8.

Zhang, WX, Wang, CB and Lien, HL. (1998) Treatment of chlorinated organic contaminants with nanoscale bimetallic particles. *Catalysis Today* **40**: 387–95.

Zhao, H, Nagy, KL, Waples, JS and Vance, GF. (2000) Surfactant-templated mesoporous silicate materials as sorbents for organic pollutants in water. *Environmental Science and Technology* **34**: 4822–7.

Zhmud, BV, Sonnefeld, J and House, WA. (1997) Role of ion hydration in the charge regulation at the surface of porous silica gel. *Journal of the Chemical Society-Faraday Transactions* **93**: 3129–36.

Zilberbrand, M. (1997) A nonelectric mechanism of ion exclusion in thin water films in finely dispersed media. *Journal of Colloid and Interface Science* **192**: 471–4.

11 Urban Geochemistry

MORTEN JARTUN AND ROLF TORE OTTESEN

Geological Survey of Norway, Trondheim, Norway

ABSTRACT

Urban geochemistry is a fairly new field of geochemical studies, comprising an amalgam of sciences such as environmental chemistry,
toxicology, food sciences, sociology, history and
engineering. Multidisciplinary studies of the
environmental condition of soils and sediments
within anthropogenic areas have been performed
all over the world, and have led to advances in
science as to how we might relate to the soil that
we actually live upon, such as clean playgrounds
for our children, city awareness maps and health
risk assessments. This chapter on urban geochemistry takes you down the road of an important field of research that is so applied and relevant
that it cannot, or at least should not, be performed
without close collaboration with local city
authorities, historians and medical expertise. The
authors of this chapter humbly recognize the tremendous work of fellow ambassadors of urban
geochemistry all over the world, and realize that
we are not able to cover all the important
approaches made in different studies, different
strategies and analytical methodologies. However,
we have tried to explain what we think are important aspects and challenges of urban geochemistry
studies today. With the establishment of an Urban
Geochemistry group within the International
Association of GeoChemistry (IAGC), we hope in
the future to create a united approach as to how
we should perform soil studies in urban areas and
the important results that may follow such
surveys. In this chapter we challenge the geosci-
entists around the world in (at least) one particular area: the introduction of organic substances,
such as polychlorinated biphenyls (PCBs), polycyclic aromatic hydrocarbons (PAHs), dioxins and
brominated flame retardants (BFR) in studies of
geochemistry. Consequently, a substantial part of
this chapter will be devoted to the hunt for local,
active sources of pollutants in the urban environment and their significance upon environmental
load, human health and environmental management within the cities and towns around the
world, big or small.

INTRODUCTION

We are walking down a busy street in a major city
somewhere in northern Europe. A truck loaded
with $10–15\,m^3$ of excavated materials is driving
out from a large construction-demolition site.
Inside the fence of the site lies a building torn
down to small pieces. Concrete, painted wood,

bricks, pipes, sand and soil constitute one big mess, and will soon be replaced by an underground parking garage with a business establishment on top. The driver of the truck does not look back. All these materials must be removed. Time is of the essence, and operational costs must be held low. His truck is heavy. Ten minutes after the first truck has left, another one leaves with the same content. The street is about to change. A new façade is being built. We recognize the obvious change and modernization, but we tend to forget some of the dynamics of the urban environment. 'Where do these trucks go?' 'What happens to these materials?' 'What was the chemical content of that waste?'

On the other side of the city, a new public day-care centre for small children is being built. The landscape architect has recommended a change in the outdoor area, and has ordered several truck loads of soil and sand to make small hills and heaps for the active children. The trucks arrive with 'new' materials, constructing a new face for the playground area of the day-care centre. We can ask again: 'What is the chemical content of these materials?' 'May the levels of chemical substances in soil induce any negative health effects for the small, fast-growing children playing here?'

Unfortunately, there are not many people who are familiar with, willing, or in the position to ask, these questions. Every day, excavated materials, with a high probability of being contaminated with a wide range of substances, are being transported uncritically from one site to another within every single city of the world (Fig. 11.1). Statistics show that we can account for only a few per cent of the tonnage of soil within the urban environment. In Oslo, the capital city of Norway with about 550,000 inhabitants, 1.6 million metric tonnes of soil is being excavated each year. We know by the records of landfill facilities that only about 5 per cent of these materials arrive at a proper landfill or treatment plant. We also know from detailed geochemical mapping of soil within the cities that the soil in older, central parts of any city is contaminated with heavy metals and organic pollutants. Where does

Fig. 11.1   The most important dispersion mechanism for urban soil, and consequently its chemical content, is by truck.

the rest end up? Inside the new day-care centre (Fig. 11.2). In the city harbour? At the next construction site? These are some of the issues rising from the research field of urban geochemistry, a multidisciplinary field consisting of environmental geochemists, biologists, medical expertise, social workers, historians, city officials and engineers. But most importantly, it concerns you and me who live and breathe the urban environment every day.

THE URBAN ENVIRONMENT

The twentieth century was, par excellence, the century of urbanization. Around the world the supremacy of rural populations over urban ones was reversed and cities experienced an accelerated growth, often beyond the desirable. They have been through unthinkable transformations, which left a fantastic array of challenges and possibilities as a legacy.

If the last century was the century of urbanization, the twenty-first will be the century of cities. It is in the cities that decisive battles for the quality of life will be fought, and their outcomes will have a defining effect on the planet's environment and on human relations. (The honorable Jaime Lerner, 'Foreword' in Starke (ed.) 2007)

Fig. 11.2 Excavated urban soil from constructions sites often end up in the wrong places, such as day-care centres for small, growing children.

By the end of 2008 more than half of the World's population was living in urban areas; towns and cities (UNFPA 2007). Urbanization has been stabilized in America and Europe, with three out of four people living in urban areas. In Africa and Asia the number of urban dwellers is rapidly increasing, and will exceed 50 per cent within a few years. Although most cities constitute the foundation of innovation and wealth, urban local governments are faced with complicated challenges such as slum development, unemployment and the risk of terror. Sustainable advances in water and food supply, waste treatment and sewer systems are needed in every expanding city. In addition, there are important challenges regarding air, water and soil pollution, and the associated health implications (Genske 2003; Starke 2007).

The urban soils in these cities are important recipients of centuries of contamination. Typically urban soils are moved, turned and recycled several times. Still, with depth, a chemical fingerprint from each generation can be found. The concentrations of many of the contaminants often reach levels of concern for human health. Already in 1980, in a report to the National Academy of

Sciences, geochemist Clair C. Patterson wrote a sentence that caught immediate attention: 'Sometime in the near future it probably will be shown that the older urban areas of the United States have been rendered more or less uninhabitable by the millions of tons of poisonous industrial lead residues that have accumulated in cities during the past century' (Patterson 1980: p. 271; Mielke 2005).

Empirical evaluations have demonstrated a strong association between the amount of Pb in soil and children's Pb exposure (Pb in blood), from which serious health effects have been documented (Mielke et al. 2007; Miranda et al. 2007).

Several Geological Surveys throughout Europe have carried out urban geochemical studies (e.g. Birke and Rauch 2000; Ottesen et al. 2008). In Norway, systematic and directly comparable geochemical mapping based on sampling and analysis of surface soils (0–2 cm) has been carried out in several cities during the last 15 years. Main results indicate that the soils in the oldest parts of the cities are typically contaminated with metals, especially lead (see example in Fig. 11.3),

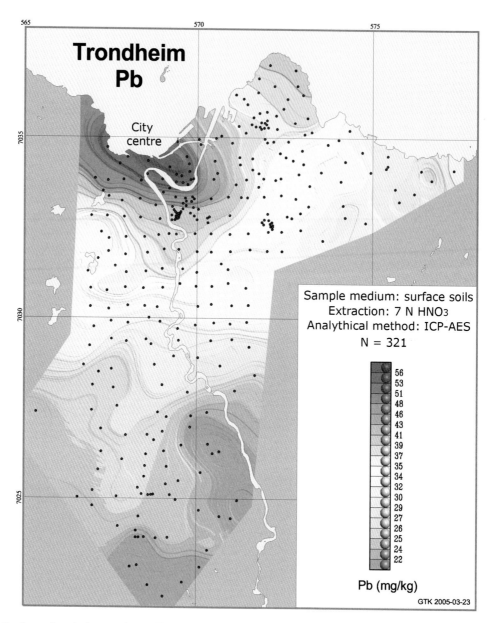

Fig. 11.3 A geochemical map of Trondheim, the third largest city in Norway, showing the concentration of lead (Pb) in urban surface soil. The highest concentrations are found in the inner and older parts of the city. The same pattern is observed for most cities.

Fig. 11.4 The urban environment – where we live and breathe – where products are brought in, but don't leave – most of the activities of man end up in the soil layer of a city.

in addition to polycyclic aromatic hydrocarbons (PAHs). Surface soils in the younger suburban parts of the cities, more specifically from the time period between 1950 and 1980, normally show lower concentrations of metals and PAH. However, polychlorinated biphenyls (PCBs) often occur in elevated concentrations here. The importance of specific sources for different pollutants in an urban environment are being studied all over the world, and the Norwegian studies have especially pointed to the importance of local sources such as contaminated building materials, city fires and traffic as the most important ones (Ottesen and Langedal 2001; Andersson et al. 2004; Jartun et al. 2009). Local urban pollution sources may contribute to a majority of the environmental load of different pollutants in surface

soil, marine sediments and urban runoff sediments compared to, e.g., long-range transport.

The urban environment (Fig. 11.4) is affected by a wide variety of local, anthropogenic activities, and this contamination is a continuous and diffuse process. Road networks, housing and industrial activities will tend to increase the content of heavy metals and organic pollutants in urban soils (Kelly et al. 1996; Mielke et al. 1999; Ottesen and Langedal 2001; Madrid et al. 2002; Mielke et al. 2004; Möller et al. 2005). Various types of soil have been found within the urban environment, ranging from relatively undisturbed soils, similar in some respects to their rural counterparts (Bridges 1991; Hollis 1991), to completely man-made soils (Bockheim 1974; Ottesen and Langedal 2001). An early definition of urban soils was put forward by Bockheim (1974): 'A soil material having a non-agricultural, man-made surface layer more than 50 cm thick, that has been produced by mixing, filling, or by contamination of land surfaces in urban and suburban areas.' Average thickness of what we may describe as the 'urban soil layer' in a typical Norwegian city is about 2 m in depth (Ottesen et al. 2000; Jartun et al. 2003). This layer is also called 'the cultural soil layer' in some cases, and reflects the influence that anthropogenic activities have had upon the soil during the entire history of the city. In this layer, we have left traces of our lifestyle for decades and centuries. Small pieces of bricks from old buildings, household waste and sewage are all mixed together in a soil cocktail with various amounts of underlying natural clay, soil or rocks. We actually live on a historical landfill. On the other hand, 'historical' may be wrong, because the urban soil layer is a very dynamic medium as described throughout this chapter.

The most significant impact of urbanization on soil is the complete loss or burial as a result of construction activities. Assuming that an average proportion of between 50–100% of the urban area has been built upon, this represents a considerable volume of buried or displaced soil in addition to an important change in the hydrological distribution of urban surface water. Impervious areas such as roofs, asphalt roads and concrete will allow only a small amount of stormwater to be infiltrated, evaporated, detained or retained by vegetation within the urban environment (Mays 2001). This will cause high-volume runoff of stormwater followed by challenges with the hydraulic efficiency of old sewer systems, and consequently a direct discharge to a downstream recipient such as a local lake or the marine environment. The impervious areas of a typical urban environment will often constitute 60–100% of the total area (Lindholm 2004; Lu and Weng 2006), which means that the study of urban geochemistry is not exclusively dedicated to soil, but will also have to include stormwater, runoff sediments, and consequently the identification of contamination sources.

At the European scale, sampling and analytical methods being used differ so widely that the results thus far are hardly comparable, though they all may demonstrate serious soil contamination levels in parts of our cities. The newly formed working group called 'Urban Geochemistry' within the International Association of Geochemistry (IAGC) strongly calls upon fellow geochemists to use a harmonized sampling strategy and identical analytical methods, to obtain directly comparable values for soil contaminants in the urban environment at the European (continental) scale. A joint research program for a selection of European cities was initiated in 2009. Such a universal scientific approach will promote the establishment of directly comparable concentrations of pollutants in urban soils from different parts of Europe.

OBJECTIVES OF URBAN GEOCHEMISTRY

There is an increasing interest in urban geochemistry globally. Individual projects are carried out by many different organizations (geological surveys, universities and city administrations). There is an obvious need for a common arena for exchange of scientific results, and to demonstrate the practical use of geochemical data for city planning and for environmental health. When IAGC established the Urban Geochemistry working group in 2008, we were moving one step closer to a wider understanding of the challenges connected to contaminated urban soil.

Some of the documented benefits of urban geochemical mapping include:
• Raised awareness that soil pollution often occurs in city centres, independent of industrial history. Thus more appropriate legislation, e.g. in terms of dispersion control, has to be developed.
• Development of health-based guidelines for different land uses in urban areas.
• Improved understanding of dispersion of urban pollution, and the documentation of proper handling, e.g. remediation, or delivery of contaminated soil to an approved landfill facility.

• Only directly comparable data at the continental, and finally global, scale allow the comparison of contamination status between countries (e.g. different contamination sources; different traditions in the use of chemicals in the urban environment).

The study of urban geochemistry must include close collaboration between city or national environmental authorities, politicians and scientists, working for one or several overall objectives. A central part of the establishment of urban soil pollution studies in, e.g., Norway, has been to make national and local politicians and authorities aware of the environmental hazard that urban soils may represent to the public, more specifically the health of small children. As a result, detailed studies of each and all public and private day-care centres (Fig. 11.2) within the ten largest cities and five most important industrial areas in Norway were initiated in 2008. These are examples of joint environmental initiatives supported by the Norwegian government in an effort to provide clean and safe outdoor play environments for small children (Ottesen et al. 2008).

In 2009, the Geological Survey of Norway together with EuroGeoSurveys initiated a European network project on urban geochemical mapping, using a common sampling protocol and identical analytical methods to study the urban soil pollution in ten selected cities. Important in this project was to design and publish a field manual detailing proper sampling methods for urban geochemical mapping. As discussed later in this chapter, different methodology may be a challenge we have to overcome in terms of comparable data. In addition, to provide a sampling strategy, the design and publication of an analytical manual detailing specific methods for both inorganic and organic analyses of urban (soil) samples will be a central point.

METHODS IN URBAN STUDIES

Sampling

The term urban ge<u>o</u>chemistry implies that some geological material should be a major subject of interest, such as soil (topsoil, C-horizon) or local bedrock. However, working with environmental studies there is no doubt that other sample media play a significant role. The collection of air samples will tell us something about the possibility for different pollutants to evaporate and accumulate in urban air compared to other areas, in addition to the relevance of pollution dispersion from local sources. Sampling of road dust will also be an appropriate sample medium, indicating contamination from, e.g., asphalt, automobile wear-and-tear (rubber tyres, breaks) and the sanding of icy roads. Rain water, groundwater, and stormwater sampling could tell us something about the ability of different pollutants to be dispersed in solution or attached to particles in the water phase. Other sample media more related to the direct exposure to plants, humans or other animals could include house dust, sand/soil/dust attached to leaves/stem/hands/skin, or direct biological samples (e.g. plants/blood/liver/urine). Environmental studies may in many cases be too focused on giving an exact description of the current situation regarding contamination levels, and may often forget to include suggestions on how the (pollution) situation got this way (description of sources), or what could be done to perform suitable environmental management to improve the situation in the future (remediation). A key word here is 'applied'. Applied urban geochemistry is a composite field of science, and as mentioned above, the environmental load in an urban area may include an undefined network of contamination sources, materials, fluxes and uncontrollable anthropogenic activities such as soil dispersion by trucks instead of water, for instance (Fig. 11.1). It is all connected. How do we manage these challenges? What is our ultimate goal?

Soil sampling

If we allow ourselves to focus on the collection of soil samples, and let them represent the environmental load of a metropolitan area, there are still many ways to perform such a sampling. In Norway and in New Orleans, a large number of single samples of topsoil (0–2 cm) are most often

Fig. 11.5 Surface soil sampling. Urban geochemistry surveys often use surface soils from parks, gardens, playgrounds and ditches. Vegetation is usually removed to capture only soil.

Fig. 11.6 Sampling of products likely to be significant sources of contamination in the urban environment, in this case flaking paint from a building façade.

collected using simple tools, such as a garden shovel/spade (e.g. Mielke et al. 1999; Ottesen and Volden 1999; Ottesen and Langedal 2001). Other studies have defined the term 'topsoil' to cover the top 10 or top 20 cm of the soil layer (Birke and Rauch 2000; Madrid et al. 2002; Möller et al. 2005). Soil samples from both top layer and deeper parts of urban soil have also been collected in some studies (Ottesen et al. 2000; Angelone et al. 2002). The choice of sampling depth is somewhat related to what one wants to study. Based on health criteria and the risk of exposure to pollutants in the soil, the immediate top layer of urban soil would be the appropriate sample medium. Often, urban soil is collected from city parks, back yards, trenches and other open areas where children may stay and play (Figs. 11.5 and 11.6). During regular play activities, a child will come in close contact with the top 0–2 cm of soil as it will adsorb to toys and hands, and whirled-up dust will be easily transported to the respiratory passage. In addition, there are always small soil-eating children out there who will be directly exposed to the soil and its various substances. We will discuss this further later in the chapter.

Given that you have decided on a proper sample medium, how do you collect the samples, and how many samples do you have to collect to provide representative and reproducible results?

There are many different practices in this area, too, which makes it difficult to perform direct comparisons from one study to another. However, there are two different methods that are worth mentioning. The first is **composite sampling**, which again could be divided into two main categories. The first category involves a collection of several subsamples that are bulked together in the field to form a composite (Ren et al. 2007). In some cases these subsamples are homogenized to get a large representative sample from one given area. The second category involves a single-sample collection, followed by mixing in the laboratory (Chen et al. 2005). If a high concentration of a given pollutant is determined in the composite sample, available single samples from the study are profitable, because of the possibility to go back and define the extent of the contamination.

The second method involves the collection of as many samples as possible, more or less independent of the size of the study area, followed by analysis of all single samples (Ottesen and Langedal 2001; Mielke et al. 2004; Ottesen et al. 2008). One of the advantages gained by this method is that you cannot dilute or hide a high concentration from a single sample. Duplicate sampling in Norwegian urban-soil studies have shown that the concentration of a given substance may vary within short distances (Jartun et al. 2002; Ottesen et al. 2008).

Regarding the number of samples that should be collected in each study, the strategies vary from study to study and from country to country. Urban studies from Italian cities ranged from 15 to 48 soil samples per city (Angelone et al. 2002). A general idea, proposed at the 33rd International Geological Congress in 2008 (IGC33), where the Urban Geochemistry working group of IAGC was established, was that four samples per km^2 should be appropriate. This density will be cost-effective for any city, and will describe the variance of most anthropogenic trace elements in urban soil. More samples may be required, however, to delineate the pollution 'hot spots' at polluted sites. On the other hand, in a study of soil pollution in (all) day-care centres from the city of Oslo, 7000 samples were collected from a total area of $454\,km^2$ or 15 samples per km^2 (Ottesen et al. 2008). The reproducibility may be illustrated in a figure such as Fig. 11.7. In this figure, the actual

number of collected soil samples is indicated on the x-axis, and a general, unspecified value of error in reproducibility is indicated on the y-axis. The figure is drawn based on large empirical datasets from various studies of urban soil pollution in Norway, and the key question is: 'How many samples do you need to collect from an area in order to be able to reproduce the same results a second time (or even better: from a second sampling team)?' The reproducibility may, for instance, be quantified in terms of the median concentration of a given substance within the sampling area. To do this, we have designed a small script that extracts a specific number of random results from one or several large datasets. If you apply this procedure repeatedly, you will soon experience that you will need at least 20 to 25 samples to approach a suitable reproducibility. Furthermore, we have used the same approach to study how many samples you need to be able to reproduce the same geochemical map over and over again, and the number of samples in that case will be closer to 300 in each case, (see example in Fig. 11.3). The key message is: **collect and analyse as many samples as you can to get the most reliable result.**

Analytical issues

Once you have collected a suitable number of samples of a proper sample medium, you need to process them in the laboratory, for which within the area of urban geochemistry there are no general guidelines (yet). More-or-less standard preparation of soil samples for heavy metal analysis is to let the sample dry in 30–40°C for about one week or until the sample is completely dry. Furthermore, the larger than 2 mm grain-size fraction should be used, which means that the soil sample should be sieved through a 2 mm nylon screen (Zhai et al. 2003; Mielke et al. 2004; Ottesen et al. 2008). The next step of the sample preparation before the instrumental analysis is perhaps of greatest importance, and the source of several discrepancies between studies from different countries. The first concern is the **extraction**. Here, there are several methods to choose from

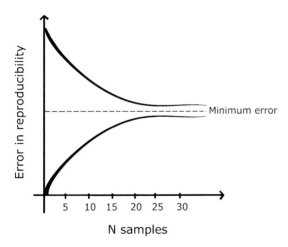

Fig. 11.7 Figure showing the reproducibility of a soil survey. The fewer samples you collect, the larger the error. Empirical data show that a certain number of samples need to be collected, independent of survey area size, in order for a second survey to provide the same general results.

(unfortunately). It seems to be difficult to agree on one single practice. In Norway, the Norwegian Standard (NS 4770) states that the proper extraction for heavy metals is performed using a boiling 7N nitric acid (HNO_3) solution, which provides a strong acid attack on the soil particles (Haugland et al. 2008; Jartun et al. 2008; Ottesen et al. 2008). Other institutions have a different approach, such as Howard Mielke's urban soil group in New Orleans. They have performed urban soil studies in New Orleans and other American cities for several years, and use a much weaker acid attack for extraction: a 1M HNO_3 solution in room temperature, which may represent the acidity in your stomach more closely, consequently resembling the possible abdominal/intestinal uptake of heavy metals from soil particles (Mielke et al. 1999; Mielke et al. 2000). Surprising or not, the results from a non-published study showed that the concentrations of most heavy metals were the same for the two methods.

Another study of urban soils from playgrounds in Oslo and New Orleans, focusing on different techniques of extraction, will be initiated in 2009. An even stronger attack than the 7N HNO_3 may be applied by using Aqua regia extraction, which resembles a 'total' extraction of elements (e.g. Möller et al. 2005). Other possibilities include hydrofluoric acid (HF), perchloric acid ($HClO_4$), and 'total' determination by X-ray fluorescence (XRF) (Birke and Rauch 2000; Zhai et al. 2003).

Traditionally, heavy metals and other elements have received most attention in the field of urban geochemistry. However, recent studies include some of the organic substances that may cause harm to the environment or human health, such as polychlorinated biphenyls (PCBs), PAHs, dioxins and brominated flame retardants (BFRs). The methods to bring these compounds into solution from a soil sample may be even more complex than for the inorganic elements. Solvents include, amongst others hexane, acetone and dichloromethane (DCM) (Ren et al. 2007; Jartun et al. 2008). The subsequent cleaning procedures of the extract before analysis may involve several chemicals, which again gives room for a large number of laboratory protocols and difficulties for com-

parison. Even though standards, blanks and field duplicates are analysed regularly, it is often very difficult to be able to obtain the same result twice. The margin of error may exceed 100% in several cases. One of the possible explanations for this is that you often select a minor part of your soil sample for analysis (in the order of 0.4–10 g.). Trace elements and organic substances often occur in very low concentrations, down to only some ng/g or even lower. If, for instance, a small flake of brittle paint should find its way to your selected soil portion and into the extraction, this will influence the final result significantly. It is therefore often suggested to use quite large portions of soil sample for extraction (maybe as much as 50 g or more), or use a finer grid for sieving (e.g. 0.63 μm). This may reduce the large variation in concentration that may be obtained from the same soil sample if it is analysed several times. Then again, this is often a matter of economy.

Element determination is often performed with Inductively Coupled Plasma analysis (ICP) with atomic emission spectrometry (AES) or mass spectrometry (MS) or atomic absorption spectrometry (AAS). Mercury, however, is often determined by another method called cold-vapour atomic absorption (CVAAS). Organic substances, such as PCBs and PAHs, may be determined using a gas chromatograph with an electron capture detector, or a mass spectrometer (GC-ECD/MS).

POLLUTANTS – OR NATURAL OCCURRING SUBSTANCES

Natural variation

So far, we have just mentioned that the urban soil may contain various substances. Which substances? And what makes them interesting to an environmental (urban) geochemist? These questions, of course, may have as many answers as there are scientists working on urban geochemistry. One of the conspicuous characteristics of a geochemist may be the light in her/his eyes and glow on her/his face when stumbling upon what seems like a higher concentration of any substance than expected. This may not be a suitable feature for an environmental geochemist, because it evidently may represent a risk of harmful exposure to the environment or people. But this feeling that she/he experiences may be a heritage from the time when geochemists were often occupied with searching for ores. In Norway today, several of the experts on urban geochemistry gained most of their experience collecting stream sediments or floodplain sediments hunting for elevated concentrations of various metals in nature. The bedrock and soil do not contain a uniform 'background' concentration of chemical elements; some places have a high natural concentration of one or several elements, and these areas may be susceptible for mining, as we see examples from all over the world.

Thoughts on toxicology

Some of the ore-seeking geochemists ultimately came back to civilization, bringing with them a highly qualified competence in the composition of our soils and sediments. The focus, however, shifted from ore prospecting to environmental issues and health risk when the soils within areas heavily affected by anthropogenic activities were to be studied. When the first large urban geochemical mapping projects were initiated in Norway in the 1990s, the focus was still on chemical elements, especially arsenic and the heavy metals. The laboratory procedures were more or less routinely continued, but the attention as to which elements to consider important changed. One of the metals that were important in Norwegian prospecting for some time was molybdenum. This element, however, does not receive the same attention when we are discussing the health risk to children or the environmental effects in marine sediments, for instance. This has something to do with the definition of pollutant. All elements, at least with respect to geochemistry, may be found in nature, so what makes us define some substances as pollutants? Walker et al. (2001) defines pollutants as 'chemicals which exist at levels judged to be above those that would normally occur in any particular com-

ponent of the environment'. This raises several follow-up questions, such as, What is normal? What is background? Heavy metals, together with several of the organic substances that also are recognized as 'pollutants' (e.g. PAHs, PCBs, dioxins, methyl mercury), are naturally occurring and were present in nature long before man. With geochemical and geological processes not discussed here, variation in their concentration on the surface of the Earth varies both in place and time, and therefore a globally accepted normal concentration range for any chemical is meaningless.

A general toxicological principle is that toxicity, and consequently the term **pollutant** are closely related to dose. The Swiss physician, journeyman miner, and philosopher, born Philip von Hohenheim, later called Philippus Theophrastus Aureolus Bombastus von Hohenheim, but mostly known as Paracelsus (1493–1541), is often called the father of toxicology. He suggested that 'All things are poison and nothing is without poison, only the dose permits something not to be poisonous' ('Septum Defensiones' 1538) Which, consequently, means that a harmless, ordinary substance may become harmful if the dose is high enough, and conversely that a substance known to be toxic may be beneficial in low doses. A jar of simple table salt may actually kill you (Smith and Palevsky 1990). The relationship between any performance (e.g. growth, survival) and the concentration of a substance may be summarized considering essential and non-essential substances. Essential substances, such as copper, iron and zinc will always have a 'window of essentiality', which means that both a deficiency and high concentrations may be harmful. The dose determines whether or not the substance becomes a poison. A non-essential substance, such as lead, cadmium or mercury, may be tolerated up to a certain point where the substance becomes toxic.

Organic compounds

Organic compounds are chemical structures that contain carbon, and have the property of forming a wide range of complex substances such as hydrocarbons, proteins and nucleic acids. A common property is that they have low water-solubility, which makes them susceptible to bio-accumulation and biomagnification in food chains, resulting in high concentrations in top predators such as the polar bear (Lie et al. 2003; AMAP 2004). Organic pollutants are often described as predominantly anthropogenic, which means that they are believed to have appeared in nature during the last, industrialized century. This is, however, not necessarily true for all compounds. Dioxins and furans (PCDDs/Fs), for instance, have been shown to accidentally form as by-products inside various kinds of incinerating processes (Alcock and Jones 1996; Lee et al. 2007; Tejima et al. 2007) leading to emissions to air, and consequently accumulation in surface soils. Studies have also shown that PCBs, which exhibit very similar chemical structures to dioxins, may be formed *de novo* in such processes (Schoonenboom et al. 1995; Ishikawa et al. 2007).

POPs and PCBs

The Stockholm Convention on Persistent Organic Pollutants (POPs) calls on the acknowledging parties to recognize the toxic and persistent properties of compounds such as PCBs, DDT, HCB, furans and dioxins, with the objective of protecting human and environmental health (UNEP 2001). Obligations to reduce or eliminate emissions of POPs include restriction or prohibition of intentional production, and use in addition to regulations on import/export. Polychlorinated biphenyls (PCBs) are an important group of organic pollutants, causing environmental challenges on a global scale. Many studies have been dedicated to the long-range atmospheric transport from medium latitudes to the Arctic in a distillation process (Wania and Mackay 1995). However, PCBs all over the world were added to a wide range of common products and applications, including hydraulic oils, electrical transformers and capacitors, double-glazed windows, concrete constructions, sealants and paint (Sundahl et al. 1999; Hellman and Puhakka 2001; Poland et al. 2001; Andersson et al. 2004; Herrick et al. 2004; Kohler et al. 2005; Shin and Kim 2006). Studies in Norway have shown that building materials may be the most significant sources of PCBs in the urban environment, see Fig. 11.8 (Andersson et al. 2004; Jartun et al. 2009). Emissions of PCBs

Fig. 11.8 Renovation and demolition of houses built or renovated between 1940 and 1980 may be considerable sources of polychlorinated biphenyls (PCBs) and other pollutants in the urban environment. PCBs are often found in paint, sealants, electrical installations or window frames.

to air from primary sources, such as the manufacturing and intentional use of PCBs, have been reduced after the ban of PCBs in various products executed during the 1970s and 1980s (Breivik et al. 2002a,b). However, PCBs are susceptible to volatilization from secondary source compartments such as soil, vegetation, water, atmospheric particles and products containing PCBs with temperature as one of the factors controlling the dispersion of PCBs in the environment (Halsall et al. 1995; Wania et al. 1998; Halsall et al. 1999; Breivik et al. 2004). A number of studies of PCBs in the environment have been dedicated to study levels on a regional or global scale (Breivik et al., 2002a,b; AMAP 2002) followed by studies of PCBs in specific environmental media, such as soil, sediments and water (Meijer et al. 2003; Rossi et al. 2004). Particle-bound PCBs, transported from unknown active sources to urban soil and runoff sediments, however, appear to be a major challenge in urban areas (Jartun et al. 2008).

A Norwegian project called 'Urban Risk' is dedicated to study specific contaminants such as the PCBs, PAHs and heavy metals in the urban environment and their significance to a downstream recipient, consequently the marine sediments (Jartun 2008). Dispersion of pollutants from urban areas may cause active contamination of marine sediments, and specific advice against the consumption of seafood exist in more than 30 fjords and harbours in Norway alone (Økland 2005). High concentrations of pollutants in sediments, both as bottom and suspended sediments, will subsequently lead to a biomagnification through food chains, depending on specific properties of a given pollutant such as lipophilicity and resistance to degradation (Ruus et al. 2005). The presence of pollutants such as PCBs in air, water, soil or sediment becomes especially important to address in Arctic areas, where the accumulation and biomagnification is enhanced by a number of factors such as several trophic levels and a general high fat content in animals. Consequently, as mentioned previously, the levels of pollutants in Arctic areas are causing high concentrations in top predators such as the polar bear (Skaare et al. 2000; Haave et al. 2003).

Dispersion and mobilization of pollutants within environmental compartments may give rise to implications on both human health and the environment. The main exposure of PCBs to humans is by ingesting contaminated fish or other seafood. In addition, occupational exposure may be significant, e.g. during management of old electrical transformers or capacitors, and renovation of contaminated buildings. Children may be exposed to PCBs prenatally and from breast milk since the PCBs are stored in fat tissue in the mother's body and subsequently released during pregnancy, cross the placenta, and finally enter foetal tissues. Human health effects associated with PCB exposure may include liver, thyroid, dermal and ocular alterations, immunological alterations, neurodevelopmental changes, reduced birth weight, reproductive toxicity and cancer (ATSDR 2000).

OUTLOOK – RESULTS OF URBAN GEOCHEMISTRY STUDIES

Health implications – clean playgrounds for our children

Empirical evaluations have demonstrated a strong association between children's Pb exposure and the amount of Pb in soil at a community scale in New Orleans. The children are particularly vulnerable to environmental toxins in their early developmental years (ages of 12–24 months) and are especially susceptible to, e.g., neurotoxic effects from metals such as Pb (Mielke et al. 1983; Mielke 1999; Mielke et al. 2007). Urban geochemistry studies, covering dense populated areas, form the basis of such health-risk studies, providing essential information on pollutant levels, geographical distribution, dispersion mechanisms and potential exposure. No such correlation between urban soil pollution and human health has so far been documented in Norway, but Norway has been especially active in involving the national authorities to bring the subject of clean outdoor play environments for our children to attention. In that respect, soil samples from each and all day-care centres in Oslo, the capital and largest city of Norway (N = 700) were collected and analysed in 2006–2007 (Ottesen et al. 2008). An important approach is to take action in those playgrounds or areas that actually are polluted, replace the contaminated soil, and remove the possible exposure sources before a health problem occurs. It is crucial that the soil within play areas does not represent any unnecessary exposure of harmful substances to developing young minds.

Awareness maps – preventing further dispersion of pollutants

Another important connection to urban geochemistry is to create an awareness that soil pollution often occurs in city centres, independent of the industrial history. Thus more appropriate legislation must be developed to prevent contaminated excavation materials to be transported from one location to another (maybe more vulnerable) area. Today, there is more or less a 'free flow' of soil, sand and sediment without anyone knowing the content of possible harmful pollutants. This logistic challenge, in addition to suitable landfill facilities or remediation processes, must be included in the environmental management of any major city. In Norway, this will, in some cases, be done by considering the innermost part of the city as one large contaminated location, and that all excavation, including the movement of soil and building refuse, must be chemically documented.

Locating sources

Even though urban soil has proven to be a suitable sample medium to establish the environmental impact of an urban area, the specific sources of contamination may remain unidentified for a given location. First of all, urban soil is a dynamic sample medium, meaning that it has been turned and used over and over again following the beat of the city since people first sat their foot in that particular area. Excavation, relocation and burial through several decades have generally made

urban soils into inhomogeneous signals of anthropogenic activities. For this reason, urban soil may be unsuitable for the concept of detecting ongoing contamination to, e.g., the marine environment. Based on the knowledge from several studies of urban soil pollution, the need for a suitable and practical method appropriate for studying the role of *active* sources of contamination to, e.g., the marine environment, has become evident. Urban runoff sediments entrapped in small stormwater units (traps) suited most of the intentions for this purpose. The method is described in detail in Jartun (2008). The main motive behind choosing stormwater sediments as a sample medium for ongoing contamination in an urban environment is that these materials are 'young', given that they are removed more or less regularly each year by the proper authorities dependent on where the traps are located. In most cases the removal responsibilities apply to the Norwegian Public Roads Administration or the Municipality. Furthermore, each stormwater trap has a limited catchment area from which it receives materials. If a high concentration of any given pollutant is found in the runoff sediments, the source of contamination is confined to a highly specific area of no more than $5–600\,m^2$.

FURTHER READING

This chapter has scratched the surface of the research area of urban geochemistry. Several references have been made in the text, leading interested readers to go deeper into this interesting field of science. The authors humbly recognize the work of fellow colleagues around the world, and realize that we were not able to include all aspects and approaches. There are several important publications in the field of urban geochemistry, but not yet any dominating large works that covers all the aspects mentioned in this chapter. Let the reference list be a starting point to explore the exciting and important field of urban geochemistry.

REFERENCES

Alcock, RE and Jones, KC. (1996). Dioxins in the environment: A review of trend data. *Environmental Science and Technology* **30**(11): 3133–43.

AMAP (2002). Arctic Pollution 2002. *Persistent organic pollutants, heavy metals, radiactivity, human health, changing pathways.* Arctic Monitoring and Assessment Programme (AMAP), Oslo, Norway.

AMAP (2004). *AMAP assessment 2002: Persistent organic pollutants in the Arctic.* Arctic Monitoring and Assessment Porgramme (AMAP), Oslo, Norway.

Andersson, M, Ottesen, RT and Volden, T. (2004). Building materials as a source of PCB pollution in Bergen, Norway. *Science of the Total Environment* **325**: 139–44.

Angelone, M, Armiento, G, Cinti, D, Somma, R and Trocciola, A. (2002). Platinum and heavy metal concentration levels in urban soils of Naples (Italy). *Fresenius Environmental Bulletin* **11**(8): 432–6.

ATSDR (2000). *Toxicological profile for polychlorinated biphenyls (PCBs).* U.S. Department of health and human services, Public health service, Agency for toxic substances and disease registry (ATSDR), Atlanta, Georgia, US.

Birke, M and Rauch, U. (2000). Urban Geochemistry: Investigations in the Berlin metropolitan area. *Environmental Geochemistry and Health* **22**: 233–48.

Bockheim, JG. (1974). Nature and properties of highly disturbed urban soils, Philadelphia, Pennsylvania. Paper presented before Div. S-5, Soil Science Society of America, Chicago Illinois.

Breivik, K, Sweetman, A, Pacyna, JM and Jones, KC. (2002a). Towards a global historical emission inventory for selected PCB congeners – a mass balance approach. 1. Global production and consumption. *Science of the Total Environment* **290**(1–3): 181–98.

Breivik, K, Sweetman, A, Pacyna, JM and Jones, KC. (2002b). Towards a global historical emission inventory for selected PCB congeners – a mass balance approach. 2. *Emissions. Science of the Total Environment* **290**(1–3): 199–224.

Breivik, K, Alcock, R, Li, YF, Bailey, RE, Fiedler, H and Pacyna, JM. (2004). Primary sources of selected POPs: regional and global scale emission inventories. *Environmental Pollution* **128**: 3–16.

Bridges, EM. (1991). Waste materials in urban soils. In: P Bullock and PJ Gregory (eds.). *Soils in the urban environment*, Blackwell Scientific Publications, Oxford, 28–36.

Chen, T, Zheng, Y, Lei, M, Huang, Z, Wu, H, Chen, H, Fan, K, Yu, K, Wu, X and Tian, Q. (2005). Assessment of heavy metal pollution in surface soils of urban parks in Beijing, China. *Chemosphere* **60**: 542–51.

Genske, DD. (2003). *Urban Land – degradation, investigation, remediation.* Spreinger-Verlag Berlin, Germany.

Haave, M, Ropstad, E, Derocher, AE, Lie, E, Dahl, E, Wiig, Ø, Skaare, JU and Jenssen, BM. (2003). Polychlorinated biphenyls and reproductive hormones in female polar bears at Svalbard. *Environmental Health Perspectives* **111**: 431–6.

Halsall, CJ, Lee, RGM, Coleman, PJ, Burnett, V, Harding-Jones, P and Jones, KC. (1995). PCBs in UK urban air. *Environmental Science and Technology* **29**(9): 2368–76.

Halsall, CJ, Gevao, B, Howsam, M, Lee, RGM, Ockenden, WA and Jones, KC. (1999). Temperature dependence of PCBs in the UK atmosphere. *Atmospheric Environment* **33**: 541–52.

Haugland, T, Ottesen, RT and Volden, T. (2008). Lead and polycyclic aromatic hydrocarbons (PAHs) in surface soil from day care centers in the city of Bergen, Norway. *Environmental Pollution* **153**: 266–72.

Hellman, SJ and Puhakka, JA. (2001). Polychlorinated biphenyl (PCB) contamination of apartment building and its surroundings by construction block sealants. Geological Survey of Finland, Special Paper 32, 123–7.

Herrick, RF, McClean, MD, Meeker, JD, Baxter, LK and Weymouth, GA. (2004). An unrecognized source of PCB contamination in schools and other buildings. *Environmental Health Perspectives* **112**(10): 1051–3.

Hollis, JM. (1991). The classification of soils in urban areas. In: P Bullock, PJ Gregory (eds.) *Soils in the Urban Environment*, Blackwell Scientific Publications, Chap. 2: 5–27.

Ishikawa, Y, Noma, Y, Yamamoto, T, Mori, Y and Sakai, S. (2007). PCB decomposition and formation in thermal treatment plant equipment. *Chemosphere* **67**(7): 1383–93.

Jartun, M. (2008). Active sources and dispersion mechanisms for pollutants, especially polychlorinated biphenyls (PCBs), in the urban environment. PhD-thesis 2008:229, Norwegian University of Science and Technology (NTNU), Department of Chemistry, ISBN-978-82-471-1148-2.

Jartun, M, Ottesen, RT and Steinnes, E. (2003). Urban soil pollution and the playfields of small children. *Journal de Physique IV* **107**: 671–4.

Jartun, M, Ottesen, RT, and og Volden, T. (2002). Jordforurensning i Tromsø. *Urban soil pollution in Tromsø.* NGU-report 2002.041, 44 pp. *(in Norwegian).*

Jartun, M, Ottesen, RT, Steinnes, E and Volden, T. (2008). Runoff of particle bound pollutants from urban impervious surfaces studied by analysis of sediments from stormwater traps. *Science of the Total Environment* 396: 147–63.

Jartun, M, Ottesen, RT, Steinnes, E and Volden, T. (2009). Painted surfaces – important sources of polychlorinated biphenyls (PCBs) contamination to the urban and marine environment. *Environmental Pollution* 157: 295–302.

Kelly, J, Thornton, I and Simpson, PR. (1996). Urban geochemistry: A study of the influence of anthropogenic activity on the heavy metal content in soils in traditionally industrial and non-industrial areas of Britain. *Applied Geochemistry* 11: 363–70.

Kohler, M, Tremp, J, Zennegg, M, Seiler, C, Minder-Kohler, S, Beck, M, Lienemann, P, Wegmann, L and Schmid, P. (2005). Joint sealants: an overlooked diffuse source of polychlorinated biphenyls in buildings. *Environmental Science and Technology* 39(7): 1967–73.

Lee, S-J, Choi, S-D, Jin, G-Z, Oh, J-E, Chang, Y-S and Shin and SK. (2007). Assessment of PCDD/F risk after implementation of emission reduction at a MSWI. *Chemosphere* 68: 856–63.

Lie, E, Bernhoft, A, Riget, F, Belikov, SE, Boltunov, AN, Derocher, AE, Garner, GW, Wiig, Ø and Skaare, JU. (2003). Geographical distribution of organochlorine pesticides (OCPs) in polar bears (Ursus maritimus) in the Norwegian and Russian Arctic. *Science of the Total Environment* 306: 159–70.

Lindholm, O. (2004). Miljøgifter I overvann fra tette flater. *(Pollutants in stormwater runoff from impervious surfaces).* NIVA-report no. 4775-2004 (in Norwegian).

Lu, D and Weng, Q. (2006). Use of impervious surface in urban land-use classification. *Remote Sensing of Environment* 102: 146–60.

Madrid, L, Díaz-Barrientos, E and Madrid, F. (2002). Distribution of heavy metal contents of urban soils in parks of Seville. *Chemosphere* 49: 1301–8.

Mays, LW. (2001). Introduction to Storm Drainage (Chapter 1). In: LW Mays (ed.) *Stormwater collection systems design handbook*, McGraw-Hill, NY, USA. ISBN 0-07-135471-9.

Meijer, SN, Ockenden, WA, Sweetman, A, Breivik, K, Grimalt, JO and Jones, KC. (2003). Global distribution and budget of PCBs and HCB in background surface soils: Implications for sources and environmental processes. *Environmental Science and Technology* 37(4): 667–72.

Mielke, HW. (1999). Lead in the inner-cities. *American Scientist* 87: 62–73.

Mielke, HW. (2005). Lead's toxic urban legacy and children's health. *Geotimes* 50(5): 22–6.

Mielke, HW, Anderson, JC, Berry, KJ, Mielke, PW and Chaney, RL. (1983). Lead concentrations in inner city soils as a factor in the child lead problem. *American Journal of Public Health* 73: 1366–9.

Mielke, HW, Smith, MK, Gonzales, CR and Mielke Jr, PW. (1999). The urban environment and children's health: Soils as an integrator of lead, zinc and cadmium in New Orleans, Louisiana, USA. *Environmental Research Section A* 81: 117–29.

Mielke, HW, Gonzales, CR, Smith, MK and Mielke, PW. (2000). Quantities and associations of Lead, Zinc, Cadmium, Manganese, Chromium, Nickel, Vanadium, and Copper in fresh Mississippi Delta alluvium and New Orleans alluvial soils. *Science of the Total Environment* 246: 249–59.

Mielke, HW, Wang, G, Gonzales, CR, Powell, ET, Le, B and Nancy Quach, V. (2004). PAHs and metals in the soils of inner-city and suburban New Orleans, Louisiana, USA. *Environmental Toxicology and Pharmacology* 18: 243–7.

Mielke, HW, Gonzales, CR, Powell, ET, Jartun, M and Mielke, PW. (2007).

Nonlinear association between soil lead and blood lead of children in metropolitan New Orleans, Louisana: 2000–2005. *Science of the Total Environment* 388: 43–53.

Miranda, M, Kim, D, Galeano, AO, Paul, CJ, Hull, AP and Morgan, P. (2007). The relationship between early childhood blood levels and performance on end-grade tests. *Environmental Health Perspectives* 115: 1242–7.

Möller, A, Müller, HW, Abdullah, A, Abdelgawad, G and Utermann, J. (2005). Urban soil pollution in Damascus, Syria: concentrations and patterns of heavy metals in the soils of the Damascus Ghouta. *Geoderma* 124: 63–71.

Ottesen, RT, Langedal, M, Cramer, J, Elvebakk, H, Finne, TE, Haugland, T, Jæger, Ø, Longva, O, Storstad, TM and Volden, T. (2000). Forurenset grunn og sedimenter i Trondheim kommune: Datarapport. *Contaminated soils and sediments in the municipality of Trondheim: Data report.* NGU-report 2000.115, 119 pp. (in Norwegian).

Ottesen, RT and Langedal, M. (2001). Urban geochemistry in Trondheim, Norway. *NGU Bulletin* 438: 63–9.

Ottesen, RT and Volden, T. (1999). Jordforurensning i Bergen. *Urban soil pollution in Bergen.* NGU-report 99.022: 27 pp (in Norwegian).

Ottesen, RT, Alexander, J, Langedal, M, Haugland, T and Høygaard, E. (2008). Soil Pollution in Day-Care Centers and Playgrounds in Norway – National Action Plan for Mapping and Remediation. *Geochemistry and Environmental Health* **30**: 623–37.

Patterson, CC. (1980). An alternative perspective – lead pollution in the human environment: origin, extent, and significance. Lead in the human environment. National Research Council, National Academy of Sciences. Nation Academy Press, Washington, DC; pp. 265–349.

Poland, JS, Mitchell, S and Rutter, A. (2001). Remediation of former military bases in the Canadian Arctic. *Cold Regions Science and Technology* **32**: 93105.

Ren, N, Que, M, Li, Y, Liu, Y, Wan, X, Xu, D, Sverko, E and Ma, J. (2007). Polychlorinated biphenyls in Chinese surface soils. *Environmental Science and Technology* **41**: 3871–6.

Rossi, L, de Alencastro, L, Kupper, T and Tarradellas, J. (2004). Urban stormwater contamination by polychlorinated biphenyls (PCBs) and its importance for urban water systems in Switzerland. *Science of the Total Environment* **322**: 179–89.

Ruus, A, Schaanning, M, Øxnevad, S and Hylland, K. (2005). Experimental results on bioaccumulation of metals and organic contaminants from marine sediments. *Aquatic Toxicology* **72**: 273–92.

Schoonenboom, MH, Tromp, PC and Olie, K. (1995). The formation of coplanar PCBs, PCDDs and PCDFs in a fly ash model system. *Chemosphere* **30**(7): 1341–9.

Shin, SK and Kim, TS. (2006). Levels of polychlorinated biphenyls (PCBs) in transformer oils from Korea. *Journal of Hazardous Materials B* **137**: 1514–22.

Skaare, JU, Bernhoft, A, Derocher, A, Gabrielsen, GW, Goksoyr, A, Henriksen, E, Larsen, HJ, Lie, E and Wiig, Ø. (2000). Organochlorines in top predators at Svalbard – occurrence, levels and effects. *Toxicology Letters* **112–113**: 103–9.

Smith, EJ and Palevsky, S. (1990). Salt poisoning in a two-year old child. *The American Journal of Emergency Medicine* **8**(6): 571–2 (with references).

Starke, L. (ed.) (2007). *State of the world 2007 – Our urban future.* Worldwatch Institute. W.W. Norton & Company, Inc, New York, USA.

Sundahl, M, Sikander, E, Ek-Olausson, B, Hjorthage, A, Rosell, L and Tornevall, M. (1999). Determination of PCB within a project to develop cleanup methods for PCB-containing elastic sealant used in outdoor joints between concrete blocks in buildings. *Journal of Environmental Monitoring* **1**: 383–7.

Tejima, H, Nishigaki, M, Fujita, Y, Matsumoto, A, Takeda, N and Takaoka, M. (2007). Characteristics of dioxin emissions at startup and shutdown of MSW incinerators. *Chemosphere* **66**: 1123–30.

UNEP (2001). United Nations Environmental Programme. *Stockholm Convention on Persistent Organic Pollutants (POPs).* Convention text. Stockholm, Sweden.

UNFPA (2007). *The state of world population 2007 – Unleashing the potential of urban growth.* United Nations Population Fund, New York, USA, E/31,000/2007.

Walker, CH, Hopkin, SP, Sibly, RM and og Peakall, DB. (2001). Principles of Ecotoxicology. 2nd edition, Taylor and Francis.

Wania, F and Mackay, D. (1995). A global distribution model for persistent organic chemicals. *Science of the Total Environment* **160/161**: 211–32.

Wania, F, Haugen, J-E, Lei, YD and Mackay, D. (1998). Temperature dependence of atmospheric concentrations of semivolatile organic compounds. *Environmental Science and Technology* **32**(8): 1013–21.

Zhai, M, Kampunzu, HAB, Modisi, MP and Totolo, O. (2003). Distribution of heavy metals in Gaborone urban soils (Botswana) and its relationship to soil pollution and bedrock composition. *Environmental Geology* **45**: 171–80.

Økland, Te. (2005). Kostholdsråd i norske havner og fjorder. *(Dietary advice from Norwegian harbors and fjords).* Report from Bergfald & Co as. in co-operation with Norwegian Food Safety Authority, Norwegian Scientific Committee for Food Safety and Norwegian Pollution Control Authority (in Norwegian).

12 Archaeological and Anthropological Applications of Isotopic and Elemental Geochemistry

HENRY P. SCHWARCZ

McMaster University, Hamilton, Ontario, Canada

ABSTRACT

The study of the past history of the human species has greatly benefited from the application of various chemical and physical analytical methods to materials recovered from sites, and to human skeletal remains. The age of archaeological sites and fossil humans have been determined using radioisotopes such as ^{14}C and ^{230}Th, as well as measurement of trapped charges in crystals using electron spin resonance (ESR) and luminescence. Stable isotopes are used to measure palaeoclimate and past human diet as well as human migration. Trace elements can be used to track movement as place of manufacture of artifacts.

INTRODUCTION

Archaeology and anthropology are concerned with the study of the past condition of the human species. Archaeologists study artifacts and sites where humans lived, while physical anthropologists study skeletal remains of our ancient ancestors as well as more recent human populations.

Frontiers in Geochemistry: Contribution of Geochemistry to the Study of the Earth, First edition. Edited by Russell S. Harmon and Andrew Parker.
© 2011 Blackwell Publishing Ltd. Published 2011 by Blackwell Publishing Ltd.

Broadly defined, the goals of all of these studies are to reconstruct the history and prehistory of our species and the cultures and civilizations of the past.

Over the last few decades both archaeology and anthropology have greatly benefited from the work of geochemists collaborating with them or retrospectively analysing materials from older excavations and discoveries. Initially, the reports of analytical chemists were consigned to appendices in the end of archaeological site reports, as matters of possible fleeting interest. Gradually it has become apparent to archaeologists that detailed analysis of materials from sites obtained even in real time during the excavations can contribute essential information to a study.

The impact of geochemistry on archaeology and anthropology has been strongest in a few specific areas of specialization, as follows:

Chronology

Our species evolved through the Quaternary era; determining the chronology of human evolution has made use of a wide range of geochemically based dating methods which have been applied both to human remains themselves, and more commonly to sedimentary materials enclosing archaeological deposits. The methods applied include argon-argon dating of volcanic rocks, uranium series dating of speleothems, ESR dating of teeth, thermoluminescence (TL) dating of heated artifacts, and optically stimulated luminescence (OSL) dating of quartz. Radiocarbon dating has been applied to sedimentary deposits as well as to human remains.

Palaeoclimate

The evolution of our species took place largely within the period of Earth history marked by glacial/interglacial transitions in the temperate and polar latitudes, while other climatic transitions in equatorial regions have also had a profound impact on human prehistory. The magnitudes of these climatic shifts have been studied using isotopic and other geochemical characteristics of shells, faunal bones and teeth, speleothems, and soils. As well, seasonality of site occupation has been studied using stable isotopic analyses of teeth and shells.

Palaeodiet

Subsistence and nutrition are always important issues in understanding the living conditions of past populations. Human teeth and bones have been analysed for stable isotopes and trace elements. Food residues adhering to ancient ceramics have been studied for their lipid residues and isotopes.

Migration

Humans are the most widely dispersed animal species on the planet. Our migration has been traced by analysis of bones and teeth for stable and radiogenic isotopes.

Provenance

Artifacts can be made from stone, bone, wood, shell and metal or shaped from clay. These raw materials can be traced back to their sources by analyses of major and trace elements, stable and radiogenic isotopes. These analyses also shed light on human migration and trade routes.

Site analyses

Humans alter their environment, as we are painfully aware today. The traces of their ancient impact on soils and sediments can be detected by chemical analyses of these deposits as well as lacustrine and marine deposits containing materials that were eroded from the land due to anthropogenic deforestation.

While these are the principal areas of impact of geochemistry on studies of the human past, there are other less well-developed topics that are beyond the scope of this review, such as geochemical impacts on past human health, on soil fertility, etc. In general, the impact of geochemistry in archaeology has been so significant that we might consider the definition of a new field called archaeological geochemistry.

CHRONOLOGY

Radiocarbon

The discovery of natural ^{14}C and the consequent revolution in the chronology of human history set the stage for a major onslaught by geochemists on the problem of dating archaeological materials (Taylor 1987). Willard Libby won the Nobel Prize in 1960 for his discovery, and was in his later years a professor and director of the Institute of Geophysics and Planetary Physics at UCLA. The first demonstrations of the power of the method were in its application to unravelling the timescale of ancient human history and prehistory (Libby 1961), and the significance of this discovery was immediately appreciated by archaeologists.

The refinement of the method and especially the calibration using tree rings and other independently datable materials has largely been carried out by geochemists working in tandem with physicists and dendrochronologists. Dating was initially based on the measurement of the radioactivity of large, purified samples of carbon. In the 1970s , physicists showed that, thanks to the physical instability of negative ions of nitrogen, it was possible to detect ^{14}C ions using a mass spectrometer with a particle accelerator acting as ion source, without significant interference from [14]N (Nelson et al. 1977; Bennett et al.1978). Accelerator mass-spectrometric (AMS) dating can be carried out on μg-sized samples of pure C, and allows geochemists to analyse single biogenic molecular species from an ancient sample, to the exclusion of most interfering species produced by diagenesis and decay of the sample (Stafford et al. 1988). Analyses on individual amino-acids separated from and unique to collagen can be used to eliminate interference from bacterial, fungal and other post-mortem biogenic contaminants (van Klinken et al. 1994).

Radiocarbon dating is typically carried out on organic materials associated with an archaeological site: collagen extracted from human or animal bone, seeds, charcoal and wood. Other less-well characterized materials such as soil humic matter and food residues charred onto ceramics are potential targets as well, but it is much more difficult to identify a population of C atoms that are unequivocally coeval with the archaeological site.

Potassium-Argon, $^{40}Ar/^{39}Ar$

Measurement of the decay of ^{40}K to ^{40}Ar opened up another window in the early history of humans. Fortunately for archaeologists, our most ancient ancestors lived close to volcanoes in East Africa, many of which erupted widespread layers of K-rich tephra ideal for the application of this method. Initially, following Reynolds' development of a high-sensitivity Ar mass spectrometer, dates were obtained from the ratio of ^{40}Ar to K as determined by chemical analysis. (Dalrymple and Lanphere 1969). By this means Curtis and Evernden were able to sketch the outlines of the multi-million-year history of Australopithecus and early Homo (Howell 1962).

Merrihue and Turner (1966) showed that ^{40}Ar and K could be simultaneously determined by neutron activation, leading to the development of the $^{40}Ar/^{39}Ar$ dating method. This greatly improved the precision and range of the method, while the advent of single-crystal laser fusion (SCLF) analysis of sanidine and other K-rich min-

erals further reduced (but did not eliminate) contamination by atmospheric ^{40}Ar (Walter 1997).

$^{40}Ar/^{39}Ar$ dating can now be applied to rocks as young as 30,000 years (Calvert et al. 2005); its upper limit clearly extends beyond the range of human prehistory. The assumption of closed-system behaviour (no Ar loss) which is problematic for older geological materials (greater than $10^8 y$) is less critical in the archaeological time range and opens up the possibility of dating feldspars, pyroxenes and other igneous minerals in younger (less than 4 Ma) rocks, principally by SCLF. The principal applications of this method have continued to be in eastern Africa, from the Red Sea to the southern end of the East African Rift Zone (McDougall and Brown 2008: Fig. 12.1), although striking insights have been obtained at other sites in Europe and Asia. For example, at Dmanisi in south Georgia, tephra $^{40}Ar/^{39}Ar$ dated to 1.8 Ma was associated with Homo erectus (DeLumley et al. 2002), while $^{40}Ar/^{39}Ar$ dating has demonstrated the presence of humans in Indonesia at 1.8 Ma and later (Swisher et al. 1994).

Uranium series

The decays of ^{238}U and ^{235}U to their respective daughters ^{230}Th and ^{231}Pa provide a potential geochronometer for the time range from a few thousand years to c. 0.6 Ma (Schwarcz et al., 1997; Pike and Pettitt, 2003). Materials suitable for dating must be formed initially free of these daughters. In the archaeological context, the most commonly dated materials have been speleothems (stalagmitic layers and stalagmites) and travertines found associated with archaeological sites. Their dates have allowed us to define the chronology of the early occurrences of Neanderthals in Europe and Asia. For example, at the site of La Chaise de Vouthon in the Charente district of France, Neanderthal remains were found interstratified with a series of flowstones which were U-series dated by α-spectrometry (Blackwell et al. 1983) showing that Neanderthals occupied the cave from Oxygen Isotope Stages 7 to 5a (Fig. 12.2).

Some speleothems and travertines have been contaminated with detritus at the time of

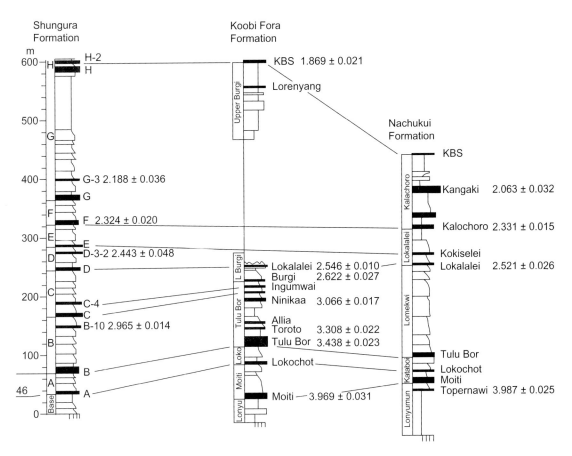

Fig. 12.1 $^{40}Ar/\,^{39}Ar$ dates for volcanic units in the Turkana Basin, obtained by single-crystal laser fusion of alkali feldspars. From McDougall and Brown (2008), with permission.

deposition, In order to correct for the resultant non-zero initial activities of ^{230}Th, isochrons can be constructed using analyses of multiple samples of the same calcite deposit, which are presumed to contain variable amounts of the same contaminant and chemically precipitated calcite. Isochrons can be constructed by plotting $^{230}Th/^{232}$Th versus $^{234}U/^{232}$Th. The slope of the resultant line gave the $^{230}Th/^{234}U$ ratio of the initial chemically precipitated calcite (Schwarcz and Latham 1989; Bischoff and Fitzpatrick 1991). Various approaches to this method have been tried, including using leachates of the detritus, or total dissolution.

The use of thermal ionization mass spectrometry (TIMS) and, more recently, multi-collector inductively-coupled plasma mass spectrometry (MC-ICP-MS) have allowed us to obtain dates with a precision of better than 1%, and extended the dating range to c. 0.6 Ma. (Li et al.1989). Relatively few applications of these new methods to archaeological dating of speleothems or travertines have yet been attempted. Falguéres et al. (2004) restudied the French site of Caune de l'Arago, and found that the age of the hominid remains found at this site was more than 350 ka. Smith et al. (2004) used TIMS dates to determine

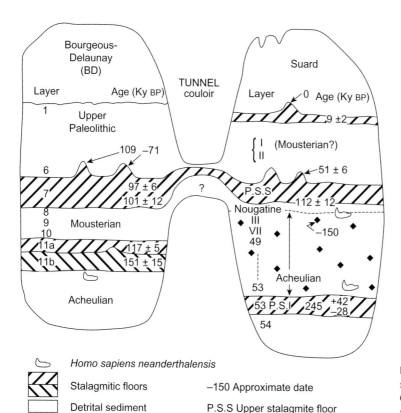

Stalagmitic floors   −150 Approximate date

Detrital sediment   P.S.S Upper stalagmite floor

Calcite cement   P.S.I Lower stalagmitic floor

Fig. 12.2 U-series dates for stalagmitic layers from the cave of La Chaise de Vouthon, Charente district, France. From Blackwell et al. (1983), with permission.

times of occupation of Palaeolithic sites in the Western Desert of Egypt.

Trapped-charge dating

Three closely related methods of dating are all based on the same principle: trapping of electronic charges in crystals as a result of bombardment by ionizing radiation. Electron spin resonance (ESR) dating, thermoluminescence (TL) and optically stimulated luminescence (OSL). In each case the age is given by the ratio of the apparent dose to the environmental dose-rate, although in some applications, a modelled time-dependent increase in dose rate is also assumed. The radiation dose is mainly derived from the radioactivity of U, Th and K in the embedding sediment or contained within the dated sample. Therefore the

geochemistry of these materials is crucial to the date.

ESR dating is based on the growth of ESR signals in the hydroxyapatite crystals of enamel of faunal teeth (Rink 1997). The radiation dose is partly derived from U and its daughter isotopes; most of the U is taken up by the tooth postmortem. Various models for the uptake history have been developed (e.g. Grün et al.1988). Eggins et al. (2003) have used laser-ablation ICP/MS to determine the distribution and isotope ratio of U and Th in teeth. A large part of the dose to enamel comes from U in the adjacent dentine, although dentine itself cannot be dated by ESR owing to organic interferences. ESR dating has been widely used to establish the chronology of early modern humans in Africa and elsewhere (Grün et al. 2003).

OSL and TL dating both measure the light output from samples which are being stimulated either by other wavelengths of light (OSL) or by heat (TL) (Wintle 2008). Quartz is the main mineral dated by OSL. Although the age limit of the method is less than 200 ka, the lower limit overlaps that of radiocarbon. The initial 'geological' signal of quartz is removed when the quartz is exposed to solar radiation ('bleached') (Roberts 1997). The dose rate is mainly derived from trace elements in the sediment, while quartz itself is essentially devoid of U and other radioisotopes. TL was once widely used for dating of ceramics in which the geological signal has been removed by firing the ceramic. K and Th are major contributors to the dose rate and in some case ^{87}Rb as well. The use of OSL dating has been important in establishing the chronology of African early modern human sites (Jacobs et al. 2003, 2006).

Cosmogenic isotopes

Beryllium-10 and ^{26}Al are produced in surficial quartz deposits by cosmic rays. Once the material is buried below a few tens of centimetres, production is stopped and we can determine the time of last exposure from present-day abundances of these isotopes determined using AMS (Granger and Muzikar 2001). Sediments transported into caves provide a possible target for dating, as was shown at the South African site of Wonderwerk Cave by Chazan et al. (2008).

PALAEOCLIMATE

Almost since its inception, stable isotope geochemistry has been used to study the climate through the Pleistocene, the period when humans evolved. Wallace Broecker was one of the early pioneers in this field, using δ^{18}O values of foraminifera to show how the climate had oscillated between glacials and interglacials (Broecker and van Donk 1970). Similar cycles found in continental ice cores confirmed that through most of the Pleistocene the Earth's surface had been much cooler than at present, a fact that archaeologists

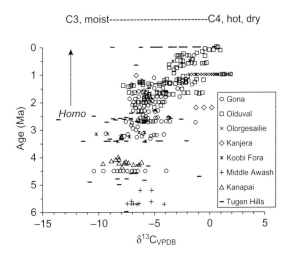

Fig. 12.3 δ^{13}C of calcitic layers in soils from East Africa. Increasing values represent increasing proportions of C_4 grasses (δ^{13}C ~ -12 ‰). From Levin et al. (2004), with permission.

were quick to recognize as having a profound influence on human evolution and migration. In addition, archaeological scientists working at specific sites have used palaeoclimatic indicators to evaluate the conditions prevailing when the sites were occupied (Hill and Vrba 1996).

The earliest members of the genus Homo evolved in Africa, a continent whose climate experienced only distant connections with the system of alternating ice ages closer to the poles. Isotopic studies of sediments from East Africa have helped to show that, through the late Pliocene and Pleistocene, the environment was gradually shifting from wetter, subtropical forests to open grasslands. This was marked by the gradual spread of C_4 grasses which left their ^{13}C -enriched signature in sequences of calcareous soil layers (Levin et al. 2004: Fig. 12.3).

Isotope palaeotemperature measurements have had other impacts on archaeology. For example, archaeologists have sought evidence for seasonality of occupation of coastal sites where humans were harvesting molluscs. Shackleton (1973) was the first to show that serial samples within single shells could reveal detailed records of seasonal temperature variation (Carré et al. 2005). The end of the record, at the exterior margin of the shell, marks the season of harvest and reveals whether a site was occupied in one season only, or throughout the year. Archaeologists working at coastal sites and riverine sites have taken up this method although seasonal fluctuations in $\delta^{18}O$ of fresh waters and varying salinity in coastal waters can also affect the isotopic record in shells (Gillikin et al. 2005).

PALAEODIET

There is considerable natural variation in the isotopic ratios of human nutrients. This allows us to use the isotopic composition of human remains to determine their past diet (DeNiro and Epstein 1978, 1981; Schwarcz and Schoeninger 1991). This information can be recovered from human bones found at archaeological sites as old as the time of the Neanderthals. Bone is a composite intergrowth of the protein collagen and the mineral hydroxyapatite. The $^{13}C/^{12}C$ and $^{15}N/^{14}N$ ratios in collagen largely reflect the corresponding ratios in dietary protein (Krueger and Sullivan 1984), while the $\delta^{18}O$ of hydroxyapatite is largely determined by the $\delta^{18}O$ of water that was consumed (Luz et al. 1984). The $\delta^{13}C$ of carbonate in the apatite reflects $\delta^{13}C$ of the total diet (Schwarcz 2000).

Figure 12.4 shows the range of variation of $\delta^{13}C$ and $\delta^{15}N$ in typical dietary items. There are two kinds of plant foods, based on the C_3 and C_4 photosynthetic mechanisms respectively, which differ greatly in $\delta^{13}C$. Marine foods are enriched in both ^{13}C and ^{15}N relative to terrestrial resources. The isotopic ratios of bone (collagen and apatite) are offset from the δ values of the food sources: $\delta^{13}C$ of collagen is approximately 5‰ higher than that of the diet while $\delta^{15}N$ of protein is enriched by 3‰ relative to dietary protein.

An important application of these methods has been the study of the role of the C_4 plant maize in the diet of Americans, starting with the first paper in 1977 by Vogel and van der Merwe. Katzenberg et al. (1995) studied humans from Ontario, Canada; starting around AD 900 maize gradually became an important source of food for the native population until more than half the C atoms in collagen were derived from this source (Fig. 12.5). The slow increase in its role in the diet was presumably a consequence of gradual modification of the genome of this subtropical plant, to generate a hardier variety able to flourish in the colder climate and shorter growing season of northeastern North America.

Ancient human populations living on seacoasts are expected to be partly dependent on marine food resources. This has been demonstrated

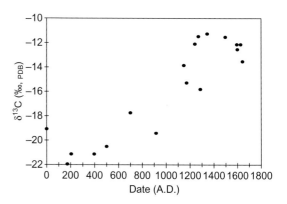

Fig. 12.5 Temporal variation in $\delta^{13}C$ of collagen of bones of native people living in Ontario, Canada prior to arrival of Europeans. The increase in $\delta^{13}C$ is due to an increase in proportion of corn (maize) in their diet. From Katzenberg et al. (1995), with permission.

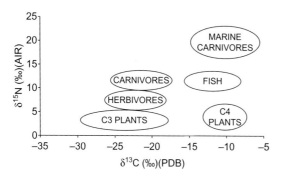

Fig. 12.4 Range of $\delta^{13}C$ and $\delta^{15}N$ in common foods.

through analysis of the $\delta^{13}C$ and $\delta^{15}N$ of bone collagen of human skeletal remains, some of whom are associated with shell middens (large mounds largely composed of the discarded shells of mollusks). Walker and DeNiro (1986) showed that humans living on the islands off Santa Barbara, California were more enriched in ^{15}N and ^{13}C than coeval people living on the mainland, as expected if they were dependent on marine foods as a protein source. Keenleyside et al. (2008) similarly showed that the diet of Roman-age residents of a coastal settlement in Tunisia was a mixture of terrestrial foods (e.g. bread, milk eggs) and marine foods, in varying proportions.

Although collagen tends to be lost from bones on a timescale of 10^4 years, it has been possible to trace protein sources in the diet of humans ranging in age up to 40,000 y BP. Richards and Trinkaus (2010) showed that the $\delta^{15}N$ values of Neanderthals were as high or higher than carnivores from the same site, suggesting that the humans were themselves highly carnivorous. They reached a similar conclusion in a study of Upper Paleolithic modern humans from Europe (Richards et al. 2001).

Non-isotopic analyses of skeletal materials have also been used to learn about palaeodiet and, specifically, the trophic level of the individual. This is based on the principle of biopurification: successive trophic levels select against elements that are biologically inactive. For example, the Sr/Ca ratio is lower in herbivores than in plants, and still lower in carnivores, because Sr is not biologically active. Therefore we can use Sr/Ca as a measure of the trophic level of ancient humans, provided that the Sr concentration has not been diagenetically altered. During their burial history, bones generally take up Sr from the soil, and thus raise the Sr/Ca ratio (Sillen et al. 1989). Nevertheless, the Sr/Ca ratio in tooth enamel appears to preserve the dietary ratio, even in 2 Ma-old hominins (Sponheimer et al. 2005)

Food residues left in ceramic vessels can also be analysed to give a more direct indication of the diet of ancient people. Using gas chromatography and non-isotopic mass spectrometry, it is possible to identify individual fatty acids and their deriva-

tives, which are indicative of their plant and animal sources (Evershed 2008). By the addition of stable isotopic analysis of the isolated lipids it is possible to further characterize their source; for example this allows the detection of the storage of milk-derived foods in ceramic vessels (Copley et al. 2005).

MIGRATION

Humans are one of the most cosmopolitan terrestrial animal species, matched only by animals that accompanied us in our migration: dogs, fleas, rats etc. We began our wandering 1.8 million years ago when *Homo erectus* left Africa, and hominins have occupied every continent, although our appearance in the Western Hemisphere appears to date from quite recent times (less than 12 ka BP).

Geochemists use the isotopic record in human teeth and bones to trace the migration of individual humans. A person acquires isotopic labels from the food and water of the regions where they have lived. From measurements of these isotopes in skeletal remains, we can tell whether a person was indigenous to the place where they were found and, if not, we can use isotopes to test hypotheses as to where they came from. People migrate for many reasons such as intermarriage, getting food and other resources, capture in warfare. Many archaeological sites contain people whose isotopes show that they are not from where they were buried.

Oxygen and strontium isotopes have proven particularly effective in tracking human movements. The $^{18}O/^{16}O$ ratio of bone mineral in humans is usually highly correlated with the oxygen-isotope ratio of local meteoric precipitation, which enters the body both in drinking water and food. Consequently, variations in $\delta^{18}O$ of bones and teeth, as well as other tissues, track corresponding inter-regional differences in $\delta^{18}O$ and δD of precipitation (Longinelli 1984, Luz et al. 1990; Cormie et al. 1994; Daux et al. 2005, Kirsanow et al. 2008). Ehleringer et al. (2008) showed that O and H isotopes in hair could also be used to trace place of residence, although hair is less commonly preserved archaeologically.

The $^{18}O/^{16}O$ and D/H ratios of precipitation ($\delta^{18}O_{ppt}$, δD_{ppt}) vary regionally, decreasing with elevation, distance from the sea, temperature of precipitation and volume of rain. In general, regions of constant $\delta^{18}O_{ppt}$, δD_{ppt} form elongated zones that extend contiguously for hundreds to thousands of kilometres. Additionally, $\delta^{18}O_{ppt}$ and δD_{ppt} are themselves highly correlated, so they do not give orthogonal data needed to pinpoint a place of origin. Additional isotopic data are needed, and the optimal choice is generally the $^{87}Sr/^{86}Sr$ ratio of tooth enamel or, less usefully, bone.

^{87}Sr is the daughter of ^{87}Rb (half-life = 5×10^{10} y). The $^{87}Sr/^{86}Sr$ ratio varies regionally and is controlled by the Rb/Sr and age of bedrock or overlying Quaternary sediment (e.g. loess, glacial drift).

Where human diet includes marine foods, these contribute a Sr component with the isotopic ratio of modern seawater (0.7092). Soil acquires a $^{87}Sr/^{86}Sr$ ratio close to that of bedrock or Quaternary cover although selective dissolution of Rb-rich minerals may cause the ratio in soil to depart from that of the substrate.

White et al. (2007) studied both oxygen and strontium isotopes in the remains of people found at the ancient Mexican city of Teotihuacan. Warriors in full military dress were buried in the Moon Pyramid. Isotopic analyses of their teeth (Fig. 12.6) show that the majority of these had been born and raised far from this site. Possible localities where they could have originated can be estimated by matching the Sr, O data to values for other areas in Mesoamerica. Because these data are effectively orthogonal, a well-resolved,

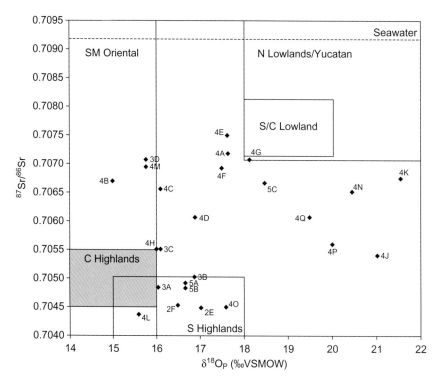

Fig. 12.6 Cross-plot of $\delta^{18}O$ and $^{87}Sr/^{86}Sr$ in teeth of humans buried at the Moon Pyramid, Teotihuacan, Mexico. Teotihuacan lies in the Central Highlands; none of the people appear to have been born and raised in that region. From White et al. (2007), with permission.

unambiguous match can be obtained. The authors conclude that these people had been recruited for military service and eventual sacrificed by the Teotihuacan rulers.

Price et al. (2001) used Sr isotopes in human remains to test for migration of Neolithic farmers in Central Europe. They analysed Sr in teeth from two sites, Flomborn and Schweitzingen. Lacking a suitable reference material, they estimated the local $^{87}Sr/^{86}Sr$ value from the mean of the population at each cemetery. It would be preferable in such studies to use the $^{87}Sr/^{86}Sr$ ratio of contemporaneous faunal material (e.g. skeletons of cows, horses; (Bentley and Knipper 2005)).

Ratios for seven of the eleven skeletons at Flomborn are greater than 2 s.d. from the mean, while 7 out of 21 at Schweitzingen are outside the local range. This supports the widely held notion that many of the people buried at these sites had migrated there, although their source area was undefined. A later study by Bentley and Knipper (2005) attempted to define possible source regions. For this purpose, they sampled archaeological pig teeth from a large region of Southern Germany, determining $^{87}Sr/^{86}Sr$ of fossil teeth as samples of biologically available Sr. They identified areas some distance to the east from the sites studied by Price et al. (2001), where identical $^{87}Sr/^{86}Sr$ ratios could be found.

The $^{87}Sr/^{86}Sr$ ratio method has also been used in the New World. At the Andean site of Tiwanaku, Peru, Knudson et al. (2004) showed that some of the individuals at the site had been born and raised away from other nearby, identifiable regions. At the Mayan site of Tikal, Guatemala, Wright (2005) analysed 83 individuals, of whom all but eight had values around 0.70812. Tikal lies geographically well within the southern lowlands, for which a somewhat lower $^{87}Sr/^{86}Sr$ ratio is expected (and supported by data from local fauna: rodents, snails: 0.7078–0.7081). Wright suggests that the higher mean value for humans arises from consumption of small amounts of sea salt containing traces of Sr with a $^{87}Sr/^{86}Sr$ ratio 0.7092. Consumption of marine foods was also invoked by Knudson et al. (2004) to account for high Sr isotope ratios in some individuals.

Price et al. (2006) studied a mass grave at the site of Talheim, Southern Germany, and found that all but four of the 22 victims analysed had $^{87}Sr/^{86}Sr$ ratios like that of the local, loess-derived soil between 0.709 and 0.710. The others have higher ratios, suggesting that they had arrived at the site from regions of more radiogenic strontium, for example sites on nearby outcrops of the Keuper sandstone.

Other radiogenic isotopes can be used but present more serious problems. Pb, like Sr, is readily taken up from the soil, although $^{206}Pb/^{207}Pb$ ratios in tooth enamel should be better preserved. Nd isotopes could be used to advantage in complementing $^{87}Sr/^{86}Sr$ data to trace human origins (Schwarcz et al. 2004).

PROVENANCE

An important aspect of the history of material culture is the search by humans for raw materials for manufacture of artifacts. Elemental geochemistry has played a large role in determining the source of these materials, including: stone used for lithic artifacts; metals; clay and temper used in ceramics, and pigments and sculptural materials used in art works. Typically, archaeochemists have built up databases containing compositions of potential source materials as obtained by one of the conventional analytical procedures, including instrumental neutron activation analysis (INAA), X-ray fluorescence (XRF) and ICP-MS. A general summary of the application of analytical chemistry in archaeology is presented by Pollard and Heron (1996).

Volcanic glass, and especially obsidian, have been prized raw materials for manufacture of sharp-edged tools from the most ancient times until the present. These materials have always been highly prized for the excellence of the sharp edges that they yield when they are fashioned into chipped stone tools. Study of the trace and major element composition of obsidian raw materials and artifacts shows that, in any relatively small region, only a few discrete sources of obsidian are detectible (Kuzmin et al. 2002; Carter et al. 2006; Negash et al. 2007; Rivero-Torres et al. 2008). Trace and major elements in these glasses have been measured by ICP-MS and neutron activation and, less accurately, by X-ray fluorescence.

The advantage of the last method is that it is non-destructive, allowing archaeologists to determine the composition of finely-shaped artifacts without damaging them. For example, de Francesco et al. (2008) studied volcanic glasses from both source areas and archaeological sites around the Mediterranean. While the compositions of these glasses could be easily discriminated using conventional XRF analysis on powders (Fig. 12.7), this method would not be appropriate for artifacts. XRF analysis of the ratios of elements can be determined with adequate precision. In another study, Acquafredda and Muntoni (2008) were able to carry out non-destructive analyses on artifacts using a scanning electron microscope equipped with non-dispersive XRF detection; the long working distance of their instrument allowed them to analyse the relatively irregular surface of the artifact.

Trace and major element provenance studies of ceramics have been widely developed around the world, with the creation of regional databases for particular regions. A typical question in the world of ceramics is: does an artifact come from a certain source region and culture, or was it made out of local raw materials according to designs copied from the other region? Instrumental neutron activation analysis (INAA) of ceramics and possible source materials have been widely used for these purposes (e.g. Gomez et al. 2002).

Cyprus was a well-known source of ceramics and ceramic designs found widely in the Middle East, and much discussion has centred around the trade in the objects as well as the replication of their designs at other manufacturing sites. For example, two-toned ceramic pottery was made on Cyprus during the late Bronze Age. Similar pottery was found at sites in Egypt dating to this period, but it was unclear whether these were from Cyprus or were Egyptian copies. Tschegg et al. (2008), using a variety of analytical and synthetic methods, concluded that the Egyptian pots were locally produced and not from Cyprus. This can be seen, for example, in the trace element distributions made using ICP-MS methods on solutions obtained by dissolving pieces of the ceramic (Fig. 12.8).

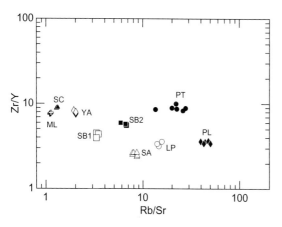

Fig. 12.7 X-ray fluorescence analyses of obsidian samples from sources around the Mediterranean, showing that measurements of ratios of trace elements can distinguish discrete sources; these measurements can be made non-destructively on artifacts. From de Francesco et al. (2008), with permission.

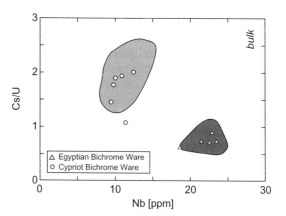

Fig. 12.8 Comparison of INAA analyses of Cypriot-style ceramic jars from Egypt with analyses of genuine Cypriot ware shows a closer connection to compositions of other Egyptian ceramics (light grey) and distinction from Cypriot ware. From Tschegg et al. (2008), with permission.

SEDIMENT ANALYSES AT SITES

Human activities also leave a geochemical record in the soils and sediments of the sites. For example, at most early prehistoric sites, hunting activities have left behind significant quantities of bone which has subsequently undergone diagenesis, generating a variety of phosphatic minerals (Goldberg and Nathan 1975). For example, at the site of Hayonim in Israel, study of the mineralogy of bulk sediment using field-based Fourier transform infrared (FTIR) spectroscopy analysis (Weiner et al. 2002) shows that where animal bones were once plentiful, the mineral dahllite (similar to born apatite) is present in the soil. The dissolution and reprecipitation of phosphate minerals would presumably also have a significant effect on the concentration of soluble, radioactive elements in the sediment (K, U), and lead to changes in the environmental dose rate. This would in turn affect age estimates based on trapped-charge dating (OSL, ESR, TL).

Agricultural activities can also leave a geochemical signal in soil. Where C_4 plants (maize, millet, sorghum) have been cultivated in soils which previously grew wild C_3 vegetation, we expect to find some enrichment of ^{13}C in the humic matter of the soil. At Mayan archaeological sites in Belize and Guatemala, terraced and valley fields have been recognized in which it is believed that large amounts of maize were raised prior to the collapse of the Maya (c. AD 900). After the collapse, most of these fields were abandoned and C_3 plants with lower $\delta^{13}C$ values took over. At the site of Caracol in Belize, Webb et al. (2004) found ^{13}C enrichment in the profiles of two of the terraces that they studied, marking the former cultivation of maize. However, the enrichment was only observed below a depth of 60 cm in the soil, owing to downward migration of the post-agricultural C_3 signal due to bioturbation. The highest terrace was developed last and did not acquire a C_4 signal.

Other geochemical markers of human activity in soils have also been demonstrated, including enrichment in ^{15}N as a result of manuring (Commisso and Nelson 2007), and phosphate build-up in soils in dwellings marking sites of human activity (Parnell et al. 2002).

CONCLUSION

In this brief review I have attempted to exemplify the wide range of methods and applications of geochemistry that have impacted on the archaeological and anthropological community. Papers in this field can be found most commonly in the *Journal of Archaeological Sciences* and *Archaeometry* but are also found widespread in the more general archaeological literature. At least a few departments of archaeology or anthropology around the world have state-of-the-art stable isotope facilities as well as other laboratories either for analysis of samples or for preparation of samples for analysis in allied geochemistry labs in the same or other institutions. This is certainly a mature field which is expected to grow in importance as it overlaps broadly with the allied field of environmental geochemistry.

ACKNOWLEDGEMENTS

I am grateful to Russell Harmon for having invited me to present the talk on which this paper was based, and to have, in fact, presented that talk for me at Oslo. My own work in this field has been the result of stimulating discussions with many archaeologists and anthropologists and archaeological scientists over the years, notably Paul Goldberg, Ofer Bar Yosef, Steve Weiner and Stanley Ambrose. In addition, my former students and postdoctoral fellows have provided me with endless food for thought: Chris White, Annie Katzenberg, Lori Wright, Elizabeth Webb, Bonnie Blackwell, Tracy Prowse, Jack Rink, Rainer Grün and Hilary Stuart-Williams.

FURTHER READING

Readers interested in further literature in the topics discussed above may consult any of the references previously cited. However, more general treatments on some of these topics can be found in survey textbooks of which the following are important examples.

In the area of chronology based on isotopic and physical methods, see Walker's 2005 book on Quaternary dating methods, which filled an important gap in this field. Uranium-series dating is reviewed in Bourdon et al. (2003). Radiocarbon as well as U-series and argon dating have been reviewed in Taylor and Aitken (1997), while argon/argon dating was reviewed by McDougall and Harrison (1999).

Stable isotopes applied to palaeoclimatic issues are extensively discussed in books on palaeoclimate, including Elias (2007) and Bradley (1999). The book titled 'Isoscapes' (West et al. 2010) is a compilation of papers on the use of isotopic measurements on regional scales. Stable isotopes in palaeodiet have been reviewed in the comprehensive treatise edited by Pollard and Heron (2008) and in the article by Katzenberg in Katzenberg and Saunders (2000).

The use of trace elements in archeological research is also covered by Pollard and Heron (2008) and Lambert (2005);. A broad overview of archaeological chemistry by Goffer (2007) may be useful.

In addition to these treatises, papers applying geochemical methods in archaeology appear regularly in the Journal of Archaeological Sciences and Archaeometry. Also, however, papers on these topics appear in more general archaeological journals including the Journal of Human Evolution, Antiquity, the American Journal of Physical Anthropology, and the Journal of Anthropological Archaeology.

Bradley, RS. (1999) Paleoclimatology: reconstructing climates of the Quaternary. San Diego, Calif.; London: Academic, 613 pp.

Elias, SA. (2007) Encyclopedia of quaternary science, v. 4, Paleoclimatology. Amsterdam: Elsevier.

Goffer, Z. (2007) Archaeological Chemistry, 2nd edn. New York, Wiley, 656 pp.

Katzenberg, Mary Anne and Saunders, Shelley Rae (2000) Biological anthropology of the human skeleton. New York: Wiley, 504 pp.

Lambert, J. (2005) Archaeological chemistry. In Maschner, HDG and Chippindale, C. (eds.), Lanham, MD: AltaMira Press, Handbook of archaeological methods, 478–500 pp.

McDougall, I and Harrison, TM. (1999) Geochronology and thermochronology by the 40Ar/39Ar method. New York: Oxford University Press, 269 pp.

Pollard, AM, Heron, Carl (eds.) (2008) Archaeological chemistry (2nd edn.) Cambridge, UK: Royal Society of Chemistry, 438 pp.

Taylor, RE and Aitken, MJ. (eds.) (1997) Chronometric dating in archaeology. New York, Plenum Press, 395 pp.

Walker, M. (2005) Quaternary Dating Methods. John Wiley, Chichester & New York.

West, JB, Bowen, GJ, Dawson, TE, Tu, KP. (eds.) (2010) Isoscapes: Understanding movement, pattern, and process on Earth through isotope mappin. Heidelberg, Springer-Verlag, 490 pp.

REFERENCES

Acquafredda, P and Muntoni, IM. (2008) Obsidian from Pulo di Molfetta (Bari, Southern Italy): provenance from Lipari and first recognition of a Neolithic sample from Monte Arci (Sardinia). *Journal of Archaeological Science*, **35**: 947–55.

Bennett, CL, Beukens, RP, Clover, MR, Elmore, D, Gove, HE, Kilius, L, Litherland, AE and Purser KH. (1978) Radiocarbon dating with electrostatic accelerators: dating of milligram samples. *Science*, **201**: 345–7.

Bentley, RA and Knipper, C. (2005) Geographical patterns in biologically available strontium, carbon and oxygen isotope signatures in prehistoric SW Germany. *Archaeometry*, **47**: 629–44.

Bischoff, J. and Fitzpatrick, J. (1991) U-series dating of impure carbonates; an isochron technique using total-sample dissolution. *Geochimica et Cosmochimica Acta*, **55**: 543–54.

Blackwell, B, Schwarcz, HP and Debenath, A. (1983) Absolute dating of hominids and Paleolithic artifacts of the cave of La Chaise de Vouthon (Charente), France. *Jour. Arch. Sci.*, **10**: 493–513.

Bourdon, B, Henderson, G, Lundstrom, C and Turner, SP. (2003) Uranium-series geochemistry. Reviews in Geochemistry, **52**, Washington, DC, Mineralogical Society of America. 656 pp.

Broecker, WS and Donk, Jan van (1970) Insolation changes, ice volumes, and the ^{18}O record in deep-sea cores. *Reviews of Geophysics and Space Physics*, **8**: 169–98.

Calvert, AT, Moore, RB and McGimsey, RG (2005) Argon geochronology of late Pleistocene to Holocene Westdahl Volcano, Unimak Island, Alaska, U. S. Geological Survey Professional Paper P 1709-D, 16 p.

Carré, M, Bentaleb, I, Blamart, D, Ogle, N, Cardenas, F, Zevallos, S, Kalin, RM, Ortlieb, L and. Fontugne, M. (2005) Stable isotopes and sclerochronology of the bivalve Mesodesma donacium: Potential application to Peruvian paleoceanographic reconstructions.

Palaeogeography, Palaeoclimatology, Palaeoecology **228**: 4–25.

Carter, T, Poupeau, G, Bressy, C and Pearce, NJG. (2006) A new programme of obsidian characterization at Çatalhöyük, Turkey. *Journal of Archaeological Science*, **33**: 893–909.

Chazan, M, Ron, H, Matmon, A, Porat, N, Goldberg, P, Yates, R, Avery, M, Sumner, A and Horwitz, L. (2008) Radiometric dating of the Earlier Stone Age sequence in Excavation I at Wonderwerk Cave, South Africa: preliminary results. *Journal of Human Evolution*, **55**: 1–11.

Commisso, RG and Nelson, DE (2007) Patterns of plant $\delta^{15}N$ values on a Greenland Norse farm. *Journal of Archaeological Science*, **34**: 440–50.

Copley MS, Berstan R, Mukherjee AJ, Dudd SN, Straker, V, Payne, S and Evershed, RP. (2005) Dairying in antiquity. III. Evidence from absorbed lipid residues dating to the British Neolithic. *Journal of Archaeological Science*, **32**: 523–46.

Cormie, A, Schwarcz, HP and Gray, J. (1994) Relation between hydrogen isotopic ratios of collagen and rain. *Geochimica et Cosmochim. Acta*, **58**: 377–92.

Dalrymple, GB and Lanphere, MA. (1969) Potassium-argon dating; principles, techniques, and applications to geochronology. San Francisco: W.H. Freeman, **258** p.

Daux, V, Lecuyer, C, Adam, F, Martineau, F and Vimeux, F. (2005) Oxygen isotope composition of human teeth and the record of climate changes in France (Lorraine) during the last 1700 years. *Climatic Change* (2005) **70**: 445–64.

De Francesco, AM, Crisci, GM and Bocci, M. (2008) Non-destructive analytic method using XRF for determination of provenance of archaeological obsidians from the Mediterranean area: a comparison with traditional XRF methods. *Archaeometry* **50**: 337–50 .

De Lumley, H, Lordkipanidze, D, Feraud, G, Garcia, T, Perrenoud, C, Falgùeres, C, Gagnepain, J, Saos, T and Voinchet, P. (2002) Datation par la méthode $^{40}Ar/^{39}Ar$ de la couche de cendres volcaniques (couche VI) de Dmanissi (Géorgie) qui a livré des restes d'hominidés fossils de 1,81Ma. *Comptes Rendus Palevol.*, **1**: 181–9.

DeNiro, MJ and Epstein, S. (1978) Influence of diet on the distribution of nitrogen isotopes in animals. *Geochimica et Cosmochimica Acta*, **42**: 495–506.

DeNiro, MJ and Epstein, S. (1981) Influence of diet on the distribution of carbon isotopes in animals. *Geochim. Cosmochim. Acta*, **45**: 341–51.

Eggins, S, Grun, R and Pike, AWG. (2003) U-238, Th-232 profiling and U-series isotope analysis of fossil teeth by laser ablation-ICP/MS. *Quaternary Science Reviews*, **22**: 1373–82.

Ehleringer, JR, Bowen, GJ, Chesson, LA,. West, AG, Podlesak, DW and Cerling,TE. (2008) Hydrogen and oxygen isotope ratios in human hair are related to geography. *Proceedings of the National Academy of Sciences (US)*, **105**: 2788–93.

Evershed RP (2008) Organic residue analysis in archaeology: The archaeological biomarker revolution. *Archaeometry* **50**: 895–924.

Falguéres, C, Yokoyama, Y, Shen, G, Bischoff, JL, Ku, TL and Lumley, H. (2004) New U-series dates at the Caune de l'Arago, France. *Journal of Archaeological Science*, **31**: 941–52.

Gillikin, DP, De Ridder, F, Ulens, H, Elskens, M, Keppens, E, Baeyens, W and Dehairs, F. (2005) Assessing the reproducibility and reliability of estuarine bivalve shells (*Saxidomus giganteus*) for sea surface temperature reconstruction: implications for paleoclimate studies. *Palaeogeography, Palaeoclimatology, Palaeoecology* **228**: 70–85.

Goldberg, P and Nathan, Y. (1975) The phosphate mineralogy of et-Tabun cave, Mount Carmel, Israel. *Mineralogical Magazine*, **40**: 253–8.

Gomez, B, Neff, H, Rautman, ML, Vaughan, SJ and Glascock, MD. (2002) The source provenance of Bronze Age and Roman pottery from Cyprus. *Archaeometry*, **44**: 23–36.

Granger, DE and Muzikar, PF. (2001) Dating sediment burial with in-situ produced cosmogenic nuclides: theory, techniques, and limitations. *Earth and Planetary Science Letters*, **188**: 269–81.

Grün, R, Beaumont, P, Tobias, PV and Eggins, S. (2003) On the age of Border Cave 5 human mandible. *Journal of Human Evolution*, **45**: 155–67.

Grün, R, Schwarcz, HP and Chadam, J. (1988) ESR dating of tooth enamel: Coupled correction for U-uptake and U-series disequilibrium. *Nuclear Tracks and Radiation Measurement*, **14**: 237–41.

Hill, A and Vrba, ES. (1996) Faunal and environmental change in the Neogene of East Africa; evidence from the Tugen Hills Sequence, Baringo District, Kenya. In ES Vrba, GH Denton, TC Partridge and LH Burckle (eds.) Paleoclimate and evolution, with emphasis on human origins, 1996. New Haven, Yale University Press, pp. 178–93.

Howell, FC.(1962) Potassium-Argon Dating at Olduvai Gorge. *Current Anthropology*, **3**: 306–8.

Jacobs, Z, Wintle, A and Duller, G. (2003) Optical dating of dune sand from Blombos Cave, South Africa: I – multiple grain data. *Journal of Human Evolution*, **44**: 599–612.

Jacobs, Z, Duller, GAT, Wintle, AG and Henshilwood, CS. (2006) Extending the chronology of deposits at Blombos Cave, South Africa, back to 140 ka using optical dating of single and multiple grains of quartz. *Journal of Human Evolution*, **51**: 255–73.

Katzenberg, A, Schwarcz, HP, Knyf, M and Melbye, FJ. (1995) Stable isotope evidence for maize horticulture and paleodiet in southern Ontario, Canada. *American Antiquity*, **60**: 335–50.

Keenleyside, A, Schwarcz, HP, Stirling, L and Lazreg, Nejib Ben (2008) Stable isotopic evidence for diet in a Roman and late Roman population from Leptiminus, Tunisia. *Journal of Archaeological Science*, **36**: 51–63.

Kirsanow, K, Makarewicz, C and Tuross, N. (2008) Stable oxygen ($\delta^{18}O$) and hydrogen (δD) isotopes in ovicaprid dentinal collagen record seasonal variation. *Journal of Archaeological Science*, **35**: 3159–67.

Knudson, K, Price, TD, Buikstra, J and Blom, D. (2004) The use of strontium isotope analysis to investigate Tiwanaku migration and mortuary ritual in Bolivia and Peru. *Archaeometry*, **46**: 5–18.

Krueger, HW and Sullivan, CH. (1984) Models for carbon isotope fractionation between diet and bone. In: JE Turnlund, PE Johnson (eds.), Stable Isotopes in Nutrition. American Chemical Society Symposium Series **258**: 205–22.

Kuzmin, YV, Popov, VK, Glascock, MD and Shackley, MS. (2002) Sources of archaeological volcanic glass in the Primorye (Maritime) Province, Russian Far East. *Archaeometry*, **44**: 505–15.

Levin, NE, Quade, J, Simpson, SW, Semaw, S and Rogers, M. (2004) Isotopic evidence for Plio-Pleistocene environmental change at Gona, Ethiopia. *Earth and Planetary Science Letters*, **219**: 93–110.

Li, W-X, Lundberg, J, Dickin, AP, Ford, DC, Schwarcz, HP and Williams, D. (1989) High_precision mass-spectrometric U-series dating of cave deposits and implications for paleoclimate studies. *Nature*, **339**: 534–6.

Libby WF. (1961) Radiocarbon dating. *Science*, **133**: 621–9.

Longinelli A. (1984) Oxygen isotopes in mammal bone phosphate. A new tool for paleohydrological and paleoclimatological research. *Geochimica et Cosmochimica Acta*, **48**: 385–90.

Luz, B, Cormie, A and Schwarcz, HP. (1990) Oxygen isotope variations in phosphate of deer bones. *Geochimica et Cosmochimica Acta*, **54**: 1723–8.

Luz, B, Kolodny, Y and Horowitz, M. (1984) Fractionation of oxygen isotopes between mammalian bone-phosphate and environmental drinking water. *Geochimica et Cosmochimica Acta*, **48**: 1689–93.

McDougall, I and Brown, FH. (2008) Geochronology of the pre-KBS Tuff sequence, Omo Group, Turkana Basin. *Journal of the Geological Society*, **165**: 549–62.

Merrihue, C and Turner, G. (1966). Potassium-argon dating by activation with fast neutrons. *Journal of Geophysical Research*, **71**: 2852–7.

Negash, A, Alene, M, Brown, FH, Nash, BP and Shackley, MS. (2007) Geochemical sources for the terminal Pleistocene/early Holocene obsidian artifacts of the site of Beseka, central Ethiopia . *Journal of Archaeological Science*, **34**: 1205–10.

Nelson, DE, Korteling, RG and Stott, WR. (1977) Carbon-14: Direct detection at natural concentrations. *Science*, **198**: 507–8.

Parnell, JJ, Terry, RE and Nelson, Z. (2002) Soil chemical analysis applied as an interpretive tool for ancient human activities in Piedras Negras, Guatemala. *Journal of Archaeological Science*, **29**: 379–404.

Pike, AWG and Pettitt, PB. (2003) U-series dating and human evolution. In B Bourdon, S Turner, GM. Henderson, and CC. Lundstrom (Eds) *Uranium Series Geochemstry*. Reviews in Mineralogy and Geochemistry v. 52. Washington DC; Mineralogical Society of America, pp. 607–630, 2003.

Pollard, AM and Heron, C. (eds.) (1996) Archaeological Chemistry. Cambridge, The Royal Society of Chemistry, 1996, 438 pp.

Price, TD, Bentley, RA, Gronenborn, D, Lüning, J and Wahl, J. (2001) Human migration in the Linear-bandkeramik of Central Europe, *Antiquity*, **75**: 593–603.

Price, TD, Wahl, J and Bentley RA. (2006) Isotopic evidence for mobility and group organization among Neolithic farmers at Talheim, Germany, 5000 BC. *European Journal of Archaeology*, **9**: 259–84.

Richards, MP, Pettitt, PB, Stiner, MC and Trinkaus, E. (2001) Stable isotope evidence for increasing dietary breadth in the European mid-Upper Paleolithic, *The Proceedings of the National Academy of Sciences, USA* **98**: 6528–6532.

Richards, MP and Trinkaus, E. (2010) Isotopic evidence for the diets of European Neanderthals and early modern humans. The Proceedings of the National Academy of Sciences, **106**: 16034–9.

Rink, WJ. (1997) Electron spin resonance (ESR) dating and ESR applications in Quaternary science and archaeometry. *Radiation Measurements*, **27**: 975–1025.

Rivero-Torres, S, Calligaro, T,, Tenorio, D and Jiménez-Reyes, M. (2008) Characterization of archaeological obsidians from Lagartero, Chiapas Mexico by PIXE. *Journal of Archaeological Science*, **35**: 3168–71.

Roberts, RG. (1997) Luminescence dating in archaeology: from origins to optical. *Radiation Measurements*, **27**: 819–92.

Schwarcz, HP. (1997) Uranium series dating. In Taylor, RE and MJ Aitken (eds.), Chronometric Dating in Archaeology, Plenum, New York, pp. 159–82.

Schwarcz, HP. (2000) Some biochemical aspects of carbon isotopic paleodiet studies. In Ambrose, S and Katzenberg, MA (eds.), Biogeochemical Approaches to Paleodietary Analysis, New York, Kluwer Academic, pp. 189–210.

Schwarcz, HP and Schoeninger, M. (1991) Stable isotope analyses in human nutritional ecology. *Yearbook of Physical Anthropology* **34**: 283–321.

Schwarcz, H, Dickin, A, Holck, P and Walker, P. (2004) Isotopic evidence for the birthplaces of early Icelanders. *Soc. for American Anthropology*, Annual Meeting.

Schwarcz, HP and Latham, AG. (1989) Dirty Calcites, 1. Uranium series dating of contaminated calcites using leachates alone. *Isotope Geoscience*, **80**: 35–43.

Shackleton, NJ. (1973) Oxygen isotope analysis as a means of determing season of occupation of prehistoric midden sites. *Archaeometry*, **15**: 133–41.

Sillen, A, Sealy, JC and Van der Merwe, NJ. (1989) Chemistry and paleodietary research – no more easy answers. *American Antiquity*, **54**: 504–12.

Smith, JR, Giegengack, R, Schwarcz, HP, McDonald, MA, Kleindienst, MR, Hawkins, AL and Churcher, CS. (2004) Reconstructing Quaternary pluvial environments and occupation through the stratigraphy and geochronology of fossil-spring tufas, Kharga Oasis, Egypt. *Geoarchaeology*, **19**: 407–39.

Sponheimer, M, de Ruiter, D, Lee-Thorp, J and Späth A. (2005) Sr/Ca and early hominin diets revisited: new data from modern and fossil tooth enamel. *Journal of Human Evolution*, **48**: 147–56.

Stafford Jr, TW, Brendel, K and Duhamel, RC. (1988) Radiocarbon, ^{13}C and ^{15}N analysis of fossil bone: Removal of humates with XAD-2 resin. *Geochimica et Cosmochimica Acta*, **52**: 2257–67.

Swisher, CC, Curtis, GH, Jacob, T, Getty, AG, Suprijo, A and Widiasmoro, S. (1994) Age of the earliest known hominids in Java, Indonesia. *Science*, **263**: 1118–21.

Taylor, RE. (1987) Radiocarbon dating: an archaeological perspective. Academic Press, 1987, 212 pp.

Tschegg, C, Hein, I and Ntaflos, T. (2008) State of the art multi-analytical geoscientific approach to identify Cypriot bichrome wheelmade ware reproduction in the Eastern Nile delta (Egypt). *Journal of Archaeological Science*, **35**: 1134–47.

van Klinken, GJ, Bowles, AD and Hedges, REM. (1994) Radiocarbon dating of peptides isolated from contaminated fossil bone collagen by collagenase digestion and reversed-phase chromatography. *Geochimica et Cosmochimica Acta*, **58**: 2543–51.

Vogel, JC and van der Merwe, N. (1977) Isotopic evidence for early maize cultivation in New York State. *Am. Antiquity*, **42**: 238–42.

Walker, PL and DeNiro, M. (1986) Stable nitrogen and carbon isotope ratios in bone collagen as indices of prehistoric dietary dependence on marine and terrestrial resources in Southern California. *American Journal of Physical Anthropology*, **71**: 51–61.

Walter, R. (1997) Potassium-argon/argon-argon dating methods. In RE Taylor. and MJ Aitken (eds.) Chronometric Dating in Archaeology. New York: Plenum, pp. 97–26.

Webb, EA, Schwarcz, HP and Healy, PF. (2004) Detection of ancient maize agriculture in the Maya lowlands using the stable carbon isotope compositions of soil organic matter: Evidence from Caracol, Belize. *Journal of Archaeological Science*, **31**: 1039–52.

Weiner, S, Goldberg, P and Bar-Yosef, O. (2002) Three-dimensional distribution of minerals in the sediments of Hayonim Cave, Israel: Diagenetic processes and archaeological implications. *Journal of Archaeological Science*, **29**: 1289–308.

White, CD, Price, TD and Longstaffe, FJ. (2007) Residential histories of the human sacrifices at the Moon Pyramid, Teotihuacan. *Ancient Mesoamerica*, **18**: 159–172.

Wintle, AG. (2008) Fifty years of luminescence dating. *Archaeometry*, **50**: 276–312.

Wright, LE. (2005) Identifying immigrants to Tikal, Guatemala: Defining local variability in strontium isotope ratios of human tooth enamel. *Journal of Archaeological Science*, **32**(4): 555–66.

Index

Frontiers in Geochemistry: Contribution of Geochemistry to the Study of the Earth, First edition. Edited by Russell S. Harmon and Andrew Parker. © 2011 Blackwell Publishing Ltd. Published 2011 by Blackwell Publishing Ltd.